Arduino and Kinect Projects

Design, Build, Blow Their Minds

Enrique Ramos Melgar
Ciriaco Castro Díez
with Przemek Jaworski

Apress®

Arduino and Kinect Projects

ISBN-13 (pbk): 978-1-4302-4167-6

ISBN-13 (electronic): 978-1-4302-4168-3

President and Publisher: Paul Manning
Lead Editor: Gwenan Spearing
Technical Reviewer: Cliff Wootton

Coordinating Editor: Corbin Collins
Copy Editor: Mary Behr
Compositor: Mary Sudul
Indexer: SPi Global
Cover Designer: Anna Ishchenko

Distributed to the book trade worldwide by Springer Science+Business Media New York, 233 Spring Street, 6th Floor, New York, NY 10013. Phone 1-800-SPRINGER, fax (201) 348-4505, e-mail `orders-ny@springer-sbm.com`, or visit `www.springeronline.com`.

For information on translations, please e-mail `rights@apress.com`, or visit `www.apress.com`.

Apress and friends of ED books may be purchased in bulk for academic, corporate, or promotional use. eBook versions and licenses are also available for most titles. For more information, reference our Special Bulk Sales–eBook Licensing web page at `www.apress.com/bulk-sales`.

Any source code or other supplementary materials referenced by the author in this text is available to readers at `www.apress.com`. For detailed information about how to locate your book's source code, go to `http://www.apress.com/source-code/`.

To the memory of Alasdair Turner

Contents at a Glance

Contents

About the Authors

Enrique Ramos Melgar is an architect specializing in computational design. His research interests include natural interaction, physics-based generative design, and their impact on architecture. He is co-founder of esc-studio, an architecture and computational design practice based in London and an Adjunct Professor at McGill School of Architecture in Montreal.

Enrique previously worked at leading international architecture firms. He was an associate at Foster+Partners in London and an on-site architect at Ateliers Jean Nouvel in Paris. His interest in computing started while working on complex architecture projects and hasn't withered since.

Enrique studied architecture at Seville University and the Ecole d'Architecture de Strasbourg. He holds a Master of Science in Adaptive Architecture and Computation from the Bartlett, UCL.

Ciriaco Castro Díez is an architect with a special interest in new technologies and programming. His research focuses on human-robotic interaction and architecture design, examining methods of interaction and the use of robotics in architecture. He is co-founder of esc-studio together with Enrique Ramos and Silvia Polito. He teaches at MIATD Master program at Seville University and has been collaborating in physical computing workshops at the Bartlett, UCL.

He has worked for Foster+Partners and Amanda Levete Architects, and he has been involved in high-profile international projects and research groups on parametric design, programming, and sustainability issues.

Ciriaco studied architecture at Seville University and ETSA Valles (Barcelona) and holds a Master of Science in Adaptive Architecture and Computation from the Bartlett, UCL.

Przemek Jaworski is an architect and computational designer. Since 2011, he is also the owner of Jawor Design Studio, a small consultancy dealing with computational techniques in architecture. He teaches parametric design and computation at Wrocław University of Technology, Poland. He worked as a member of the Specialist Modeling Group at Foster+Partners from 2006 to 2010. He graduated from Bartlett, UCL (Adaptive Architecture and Computation Msc) and Wroclaw University of Technology (MSc in Architecture and Urban Planning). His interests include semi-intelligent and self-organizing systems, real-time interactive simulations, and physical computing. Currently he is heavily involved in research on digital fabrication techniques.

About the Technical Reviewer

Cliff Wootton is a former Interactive TV systems architect at BBC News. The News Loops service developed there was nominated for a BAFTA and won a Royal Television Society Award for Technical Innovation. He has been an invited speaker on pre-processing for video compression at the Apple WWDC Conference. Cliff has taught postgraduate MA students about real-world computing, multimedia, video compression, metadata, and researching the deployment of next-generation interactive TV systems based on open standards. He is currently working on R&D projects, investigating new Interactive TV technologies, involved with MPEG standards working groups, writing more books, and speaking at conferences—when he's not lecturing on multimedia at the University of the Arts in London.

Acknowledgments

We would like to thank Alasdair Turner, who passed away last year, for creating the Msc AAC at the Bartlett. We had the great chance of being taught by him, Sean Hanna, and Ruairi Glynn, and so this work is partly theirs. Thanks also to Miriam Dall'Igna, without whom the Kinect-controlled delta robot project wouldn't exist.

The three authors met while working at Foster+Partners in London. We would like to thank everybody there for maintaining such a creative environment.

We would also like to thank Michelle Lowman at Apress for giving us this great opportunity and our editors, Corbin Collins, Gwenan Spearing, and Cliff Wootton, for all of their help and encouragement through the writing process.

Finally, we would like to thank the whole Arduino, Processing, and Kinect communities for their endless sharing. Creativity is a collective process and this book, although written by three people, is the result of a common, worldwide effort. Thank you.

—The Authors

My friend Carlos Miras gave me my first Arduino board for my 29th birthday in an Italian restaurant in London. The couple next to us looked pretty puzzled but that Arduino would be the first of many to come. Years earlier, Rafa Vázquez taught me how to program cellular automata in a Parisian cafe. I didn't know either that I would go on to program many others. I would like to thank all my friends for teaching me so many things throughout the years.

I would also like to thank my family for always pushing me to do whatever I wanted at any given moment, which has often changed directions almost randomly. And I would like to thank Mariève who, for the last five months, lived surrounded by wires and strange contraptions and still found it "l'fun."

—Enrique Ramos

I would like to thank Enrique for all the knowledge and creativity shared in the process of building this book. I would like to thank my family for their support during these months. I would like to thank my friends, especially Ollie Palmer, Stefano, Filippo, Max, Paula, Fer, Edu, and Maria for their critical viewpoint.

I would also like to thank Silvia, who has been part of this book by giving advice and support and by asking questions about this little green circuit that can do so many things.

—Ciriaco Castro

Introduction

If you've done some Arduino tinkering and wondered how you could incorporate the Kinect in your projects — or the other way around — this book is for you.

If you haven't actually done any tinkering but you are looking for a good way to get started, this book might also be for you. Even though this is not intended as an introductory book, we have tried to assume as little knowledge from the reader as possible, starting from the installation of every tool you will be using and thoroughly explaining every new concept appearing in the book.

The Structure of This Book

This is a practical book. As such, you are going to be driven through a series of projects evolving from simple to complex, learning all you need to know along the way. The book starts with three introductory chapters oriented to getting you acquainted with Arduino, Kinect, and Processing in the least amount of time possible so you can go straight into building cool projects. From Chapter 4, you will be led through a series of 10 fun projects increasing in complexity, starting with the Arduino and Kinect equivalent of "Hello World" and finishing with the construction and programming of a Kinect-driven delta robot.

The Content of the Chapters

Each chapter will lead you step-by-step through the construction of the physical project, the building of the necessary circuits, the programming of the Arduino board, and the implementation of the Processing programs that connect the Kinect data to your Arduino board. Most projects will involve the implementation of two separate programs, the Arduino program and the Processing program. Arduino code will be displayed in bold monospace typeface, like this:

```
digitalRead(A0); // This is Arduino Code
```

Processing programs will be written in normal monospace font style, like this:

```
fill(255,0,0);  // This is Processing Code
```

In each chapter, you will be introduced to the specific concepts and techniques that you will need to build that particular project—and probably some of the following ones. If you are an experienced programmer, you might want to read this book non-sequentially, starting by the project that interests you the most. If you are a programming beginner, you will find it easier to start with the first project and build up your knowledge as you progress through the book.

This is a list of topics that will be introduced in each chapter, so if you are interested in a specific concept you can jump straight to the right project.

- **Chapter 1:** You will learn everything you need to know about the Arduino platform, you will install the necessary drivers and software, and you will write your first Arduino program.
- **Chapter 2:** This chapter will help you discover what's inside that amazing new device that has changed human-computer interfaces: the Kinect.

- **Chapter 3:** You will discover the Processing programming language and IDE. You will install Processing, build your first Processing programs, and learn a great deal about working in 3D.
- **Chapter 4:** You will learn about communicating with Arduino and Kinect through serial, you will develop your own communication protocol, and you will use hand tracking for the first time. You will also learn how to use pulse width modulation and how to work with LEDs and light sensors.
- **Chapter 5:** This chapter will teach you to hack a remote control and use body gestures to control your TV set. You will learn how to use relays and how to build a circuit on a prototyping shield. You will even develop your own gesture recognition routine.
- **Chapter 6:** You will learn to work with servos and how to communicate through networks and over the Internet. You will also use Kinect's skeleton tracking capabilities to drive a puppet with your body gestures.
- **Chapter 7:** The Arduino Nano and the XBee wireless module will be introduced in this chapter. You will also learn all about resistors and color LEDs. You will take the skeleton tracking to three dimensions and use it to control a series of lamps responding to your presence.
- **Chapter 8:** Przemek Jaworski has contributed this amazing drawing arm project. In this chapter, you will work with Firmata, a library that allows you to control the Arduino from Processing without the need to write any Arduino code. You will also learn how to build a tangible table to control your robotic arm.
- **Chapter 9:** You will be introduced to DC motors and how to control them using H-bridges. You will also learn how to use proximity sensors. You will use all these techniques to control a RC car with hand gestures.
- **Chapter 10:** This project will teach you how to hack a bathroom scale to provide user weight data wirelessly. You will learn how to acquire data from a seven-segment LCD display, and you will then combine the data with the Kinect skeleton tracking to implement user recognition and body mass index calculation.
- **Chapter 11:** You will build a wireless, wearable circuit on a glove using the Arduino LilyPad, flex sensors, and an XBee module. You will then implement your own simple computer-assisted design (CAD) software, and you will use your wireless interface to draw 3D geometries by moving your hand in space.
- **Chapter 12:** This chapter will teach you to parse, transform, and recompose raw point clouds in order to perform 360-degreee scans of objects using just one Kinect and a turntable that you will build. Then you will learn how to write your own `.ply` file export routine and how to import the point data into Meshlab to prepare it to be 3D printed or rendered.
- **Chapter 13:** This final project will teach you the basics of inverse kinematics and how to use all the techniques from throughout the book to build a Kinect-controlled delta robot.

This book has been intentionally built upon multi-platform, open source initiatives. All the tools utilized in the book are free and available for Mac OSX, Windows, and Linux on commercial licenses.

Because of the three-dimensional nature of the data that you can acquire with the Kinect, some of the more advanced projects rely on the use of trigonometry and vector math. We have tried to cover the necessary principles and definitions, but if your mathematical skills are somewhat rusty, you might want to consider having a reference book at hand. John Vince's *Mathematics for Computer Graphics* (Springer, 2006) is an amazing resource. Web sites like Wolfram Alpha (`www.wolframalpha.com`) or mathWorld (`http://mathworld.wolfram.com`) can be helpful as well.

Every chapter will include the necessary code to make the project work. You can copy this code from the book or you can download the necessary files from Apress (`www.apress.com`) or the book's web site (`www.arduinoandkinectprojects.com`).

If you need to contact the authors, you can find us via the following addresses:

- Enrique Ramos Melgar: `enrique@esc-studio.com`
- Ciriaco Castro Díez: `ciriaco@esc-studio.com`
- Przemek Jaworski: `studio@jawordesign.com`

CHAPTER 1

Arduino Basics

by Enrique Ramos

This first chapter is dedicated to Arduino, one of the two cornerstones of this book. The following pages cover the basics of the Arduino hardware and software. By the end of the chapter, you will have installed the platform on your computer and you will have programmed two examples using the Arduino integrated development environment (IDE).

The structure of an Arduino sketch will be analyzed and each function explained. You will learn about the Arduino input and output pins. You will also learn the basic concepts of electricity, circuits, and prototyping techniques that you will be using throughout the book.

It is highly likely that this is not the first book on Arduino that you have held in your hands. If you are an experienced Arduino user, you should consider reading these pages as a refresher. If you are a complete beginner, read carefully!

Figure 1-1. Arduino Uno

What is Arduino?

Arduino is an open source electronics prototyping platform composed of a microcontroller, a programming language, and an IDE. Arduino is a tool for making interactive applications, designed to simplify this task for beginners but still flexible enough for experts to develop complex projects.

Since its inception in 2005, more than 200,000 Arduino boards (see Figure 1-1) have been sold, and there is a rapidly increasing number of projects using Arduino as their computing core. The Arduino community is vast, and so is the number of tutorials, reference books, and libraries available. Being open source means that there are Arduino clones that you can choose for your projects. Advanced Arduino users are constantly developing shields for the available Arduinos as well as completely new Arduino-compatible boards specializing in different tasks.

The intentional simplicity of approach to the Arduino platform has permitted an access to physical computing for people who would have never thought of using or programming a microcontroller before the Arduino/Wiring era.

A Brief History of Arduino

Arduino was born in 2005 at the *Interaction Design Institute Ivrea*, Italy, as a fork of the open source Wiring Platform. The founders of the project, Massimo Banzi and David Cuartielles, named the project after Arduin of Ivrea, the main historical character of the town.

Hernando Barragán, a student at the same Institute, along with Diego Gonzalez Joven, had developed Wiring in 2003 as his master's thesis, which was supervised by Massimo Banzi and Casey Reas (one of the initiators of Processing). The idea behind Wiring was to allow an easy introduction to programming and sketching with electronics for artists and designers—in a similar mindset in which Casey Reas and Ben Fry had developed Processing some years earlier (you will learn more on the history of Processing in Chapter 3).

Arduino was built around Wiring but developed independently from 2005. The two projects are still very much alive, so everything you will be doing in this book with Arduino could also be done with Wiring boards and IDE.

Installing Arduino

The first thing you need to do if you want to work with Arduino is to buy an Arduino board and a standard USB cable (A-to-B plug if you are using an Arduino Uno). Well, of course you will need more than this if you want to build any reasonable useful application, but for the moment you'll just work with the bare essentials.

Arduino runs on Windows, Mac OS X, and Linux, so there's a version of Arduino for you whatever your OS. Go to the Arduino software web site at `http://arduino.cc/en/Main/Software` and download the version of the software compatible with your system. If after reading the following sections you are still having trouble with the installation, you can have more detailed information at `http://arduino.cc/en/Guide/HomePage`.

Installation on Mac OS X

After you have downloaded the .zip file, uncompress it. You can drag the Arduino application to your Applications folder, and then run from there. Pretty easy, right?

If you are using a board older than Arduino Uno (Duemilanove, Diecimila, or older), you need to install the drivers for the FTDI chip on the board. You can find the drivers and the instructions at `http://www.arduino.cc/en/Guide/MacOSX`.

Installation on Windows

First, download the Windows .zip file and uncompress it. You will see a folder called arduino-1.0. This is your Arduino folder and you need to store it somewhere in the computer from where you can later run the program. The Program Files folder would seem like a reasonable place for the Arduino software, but feel free to choose an alternative location.

You need to install the Arduino drivers before you can start working with your board. Assuming you are installing an Arduino Uno, follow these steps:

- Plug your board to your computer, and wait for Windows to start the driver installation process. After a while, it will fail. No problem.
- Click Start menu and open the Control Panel.
- Go to System and Security, then System and then Device Manager.
- Find the Arduino Uno port listed under Ports (COM & LPT).
- Right-click on it, and choose "Update Driver Software," selecting "Browse my Computer for Driver Software."
- Finally, navigate and select the Arduino Uno's driver file named `ArduinoUNO.inf` located in the Drivers folder of the Arduino software folder you have just downloaded. Windows will successfully install the board now.

Installation on Linux

If you are working on Linux, the installation process is slightly different depending on your Linux distribution, and unfortunately we don't have the space to detail all of them in this book. You can find all the information you need at `http://www.arduino.cc/playground/Learning/Linux`.

Testing the Arduino

Once you have installed the Arduino software and drivers, you need to perform a little test to make sure that everything is working properly.

Arduino has a built-in LED connected to pin 13, so you can actually test the board without having to connect any extra hardware to it. Among the numerous Arduino examples included with the Arduino IDE, there is an example called Blink that makes this built-in LED to blink every second. You will learn more about pins, LEDs, and Arduino code in the following sections, but for the moment, you are going to use this example as a way to find out if your Arduino is communicating properly with your computer and if the sketches can be uploaded to the board.

Connect your Arduino board to your computer and run the Arduino software. On the Tools menu, select Board, and make sure that the right kind of board is selected. In our case, we want to check that the Arduino Uno board is ticked (see Figure 1-2).

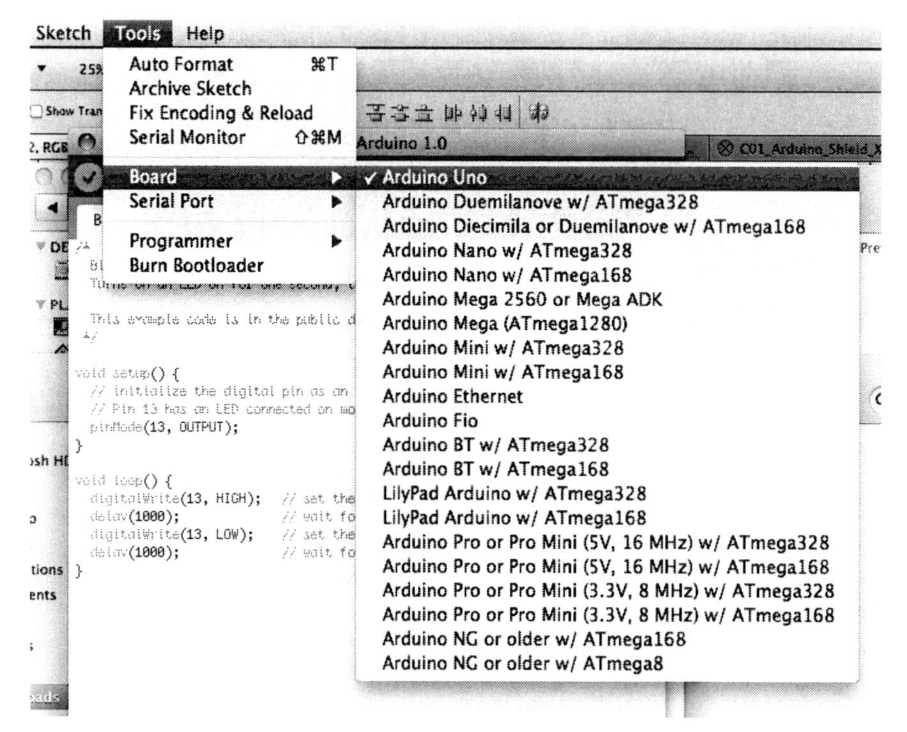

Figure 1-2. Board selection

After this, select the serial device under the Serial Port option in your Tools menu. If you are working on Mac OS X, the port should be something with `/dev/tty.usbmodem` (for the Uno or Mega 2560) or `/dev/tty.usbserial` (for older boards) in it (see Figure 1-3). If you are using Windows, the port is likely to be COM 3 or higher.

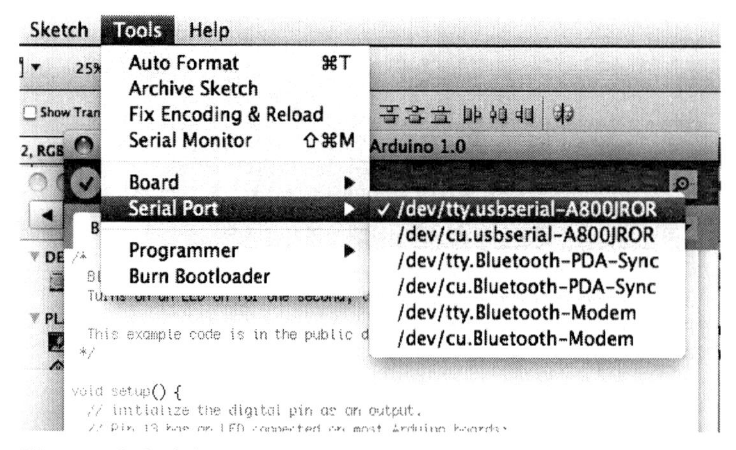

Figure 1-3. Serial port

Now everything should be ready for you to upload your first program to your Arduino board. You will be using one of the classic Arduino examples for this test. Click the Open button on your toolbar and navigate to 1.Basics ➤ Blink, as shown in Figure 1-4. Then upload the sketch by clicking the Upload button (the one with the arrow pointing right on your toolbar). After some seconds, you will get a message saying "Done Uploading" on the status bar, and the LED marked with a capital "L" on your board will start blinking every second. If you succeeded in this, your Arduino is correctly set up, and you can move on to learning about the Arduino hardware and software.

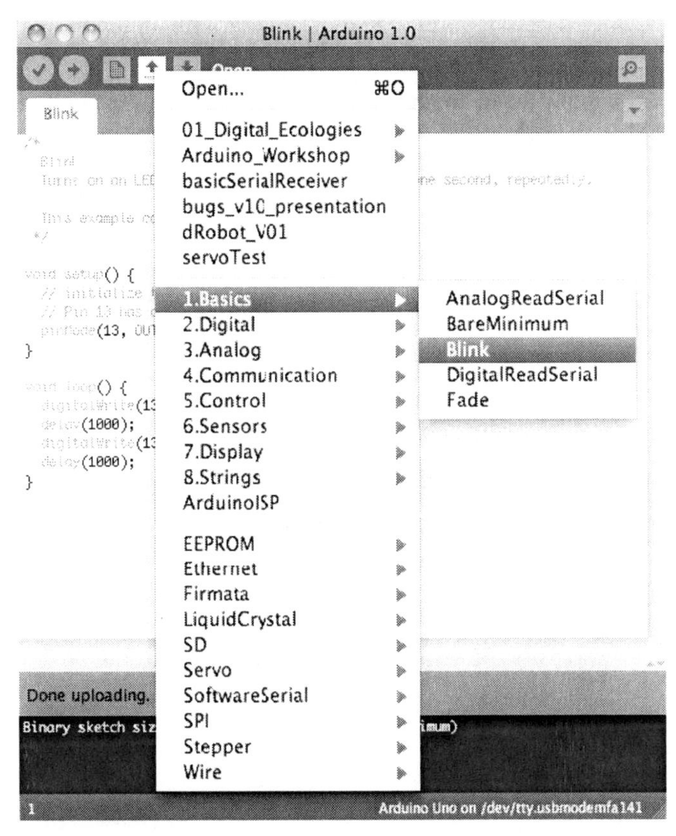

***Figure 1-4.** Opening the Blink example*

Arduino Hardware

You can think of the Arduino board as a little brain that allows you to connect together a vast range of sensors and actuators. The Arduino board is built around an 8-bit Atmel AVR microcontroller. Depending on the board, you will find different chips including ATmega8, ATmega168, ATmega328, ATmega1280, and ATmega2560. The Arduino board exposes most of the microcontroller's input and output pins so they can be used as inputs and outputs for other circuits built around the Arduino.

■ **Note** A microcontroller is just a small computer designed for embedded applications, like controlling devices, machines, appliances, and toys. Atmel AVR is an 8-bit single chip microcontroller developed by Atmel in 1996.

If you go to `http://arduino.cc/en/Main/Hardware`, you will find a surprisingly vast range of Arduino boards. Each one of them has been designed with a different purpose in mind and thus has different amount of pins and electronics on board (see Figure 1-5).

The Arduino Uno is, at the time of writing, the simplest Arduino board to use. For the rest of the book, whenever we refer to Arduino board, we actually mean an Arduino Uno unless explicitly stated (we will also work with Arduino Nano and Arduino Lilypad in further chapters).

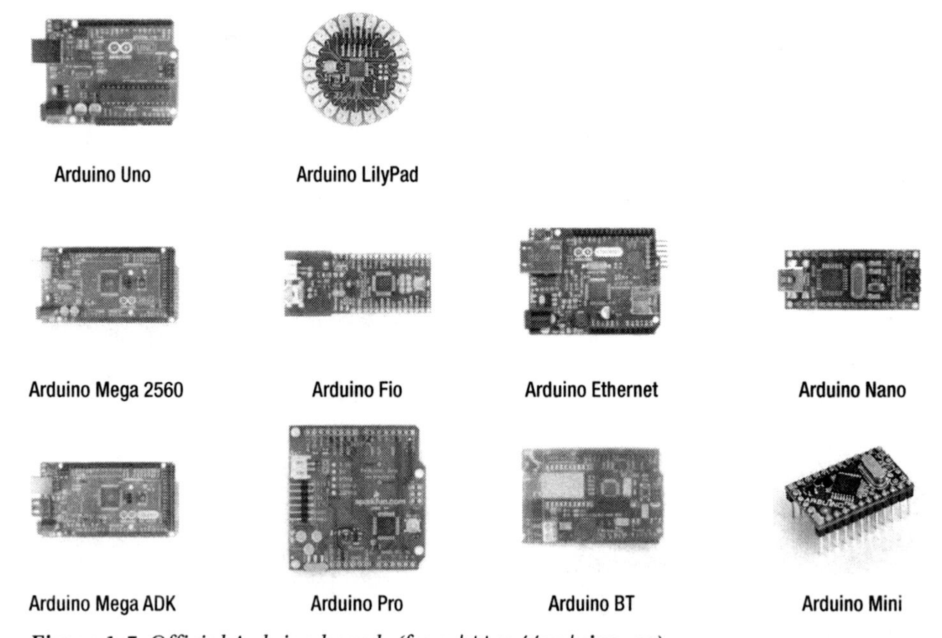

Figure 1-5. *Official Arduino boards (from* `http://arduino.cc`*)*

Arduino is open source hardware and as such, every official board's schematics are available to the public as EAGLE files (EAGLE is a PCB layout, autorouter, and CAM software by Cadsoft). In this way, if none of the board designs seem to satisfy your needs, or if you want to have access to cheaper boards, you can actually make your own Arduino-compatible board. If you want to know how to do this, get *Arduino Robotics* (Apress, 2011) and you will learn all you need to know to make your own Arduino clone. If you are thinking of creating your own Arduino clones for commercial purposes, make sure you read the Arduino policy at `http://arduino.cc/en/Main/Policy`.

Arduino Input and Output Pins

The Arduino Uno has 14 digital input/output pins and six analog input pins, as shown in Figure 1-6. As mentioned, these pins correspond to the input/output pins on your ATmega microcontroller. The architecture of the Arduino board exposes these pins so they can be easily connected to external circuits.

Arduino pins can be set up in two modes: input and output. The ATmega pins are set to input by default, so you don't need to explicitly set any pins as inputs in Arduino, although you will often do this in your code for clarity.

Figure 1-6. Arduino digital pins (top) and analog pins (bottom)

Digital Pins

Arduino has 14 digital pins, numbered from 0 to 13. The digital pins can be configured as `INPUT` or `OUTPUT`, using the `pinMode()` function. On both modes, the digital pins can only send or receive digital signals, which consist of two different states: `ON` (`HIGH`, or 5V) and `OFF` (`LOW`, or 0V).

Pins set as `OUTPUT` can provide current to external devices or circuits on demand. Pins set as `INPUT` are ready to read currents from the devices connected to them. In further chapters, you will learn how to set these pins to `INPUT` so you can read the state of a switch, electric pulses, and other digital signals. You will also learn how to work with digital pins set as `OUTPUT` to turn LED lights on and off, drive motors, control systems, and other devices.

Six of these pins can also be used as pulse width modulation pins (PWM). Chapter 4 covers what this means and how to use PWM pins.

Analog Input Pins

The Atmega microcontrollers used in the Arduino boards contain a six-channel analog-to-digital converter (ADC). The function of this device is to convert an analog input voltage (also known as potential) into a digital number proportional to the magnitude of the input voltage in relation to the reference voltage (5V).

The ATmega converter on your Arduino board has 10-bit resolution, which means it will return integers from 0 to 1023 proportionally based on the potential you apply compared to the reference 5V. An input potential of 0V will produce a 0, an input potential of 5V will return 1023, and an input potential of 2.5V will return 512.

These pins can actually be set as input or output pins exactly as your digital pins, by calling them A0, A1, etc. If you need more than six digital input or output pins for a project, you can use your analog pins, reading or writing to them as if they were digital pins. You will be using this trick in Chapter 10 when you need 16 digital input pins.

Pull-Up Resistors

Whenever you are reading a digital input from one of your Arduino pins, you are actually checking the incoming voltage to that pin. Arduino is ready to return a 0 if no voltage is detected and a 1 if the voltage is close to the supply voltage (5V). You need to ensure that the incoming voltages are as close as possible to 0 or 5V if you want to get consistent readings in your circuit.

However, when the pins are set to input, they are said to be in a *high-impedance* state, which means it takes very little current to move the pin from one state to another. This can be used to make applications that take advantage of this phenomenon, such capacitive touch sensors or reading LEDs as photodiodes, but for general use this means that the state of a disconnected input pin will oscillate randomly between 1 and 0 due to electrical noise.

To avoid this, you need to somehow bias the input to a known state if no input is present. This is the role of the ATmega built-in pull-up resistors.

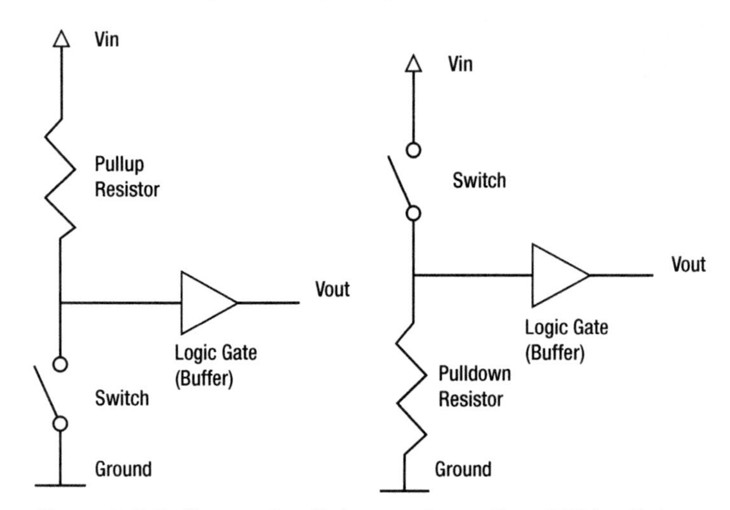

Figure 1-7. *Pull-up and pull-down resistors (from Wikipedia)*

Looking at Figure 1-7, you can see that whenever the switch is open, the voltage Vout, which is the voltage you will read at the input pin, is very near Vin, or 5V. When the switch is closed, all the current flows through the resistor to ground, making Vout very close to ground, or 0V. In this way, your input pin is always driven to a consistent state very close to 0 or 5V.

The pull-down resistor works the same way, but the signal will be biased or pulled down to ground when the switch is open and to 5V when the switch is closed.

The ATmega chip has 20K pull-up resistors built in to it. These resistors can be accessed from software in the following manner:

```
pinMode(pin, INPUT);          // set pin to input
digitalWrite(pin, HIGH);      // turn on pullup resistors
```

Arduino Shields

Arduino shields are boards that have been designed to perform a specific task and can be plugged on top of a compatible Arduino board directly, extending the Arduino pins to the top so they are still accessible.

There are a few official Arduino shields and a vast number of unofficial ones that you can find on the Internet (see Figures 1-8 and 1-9). And the best thing, you can create your own Arduino shields by using an Arduino-compatible Protoshield or by printing your own PCB compatible with the Arduino design.

Figure 1-8. *Arduino XBee Shield (left) and Arduino Prototyping Shield (right)*

Arduino Ethernet Shield

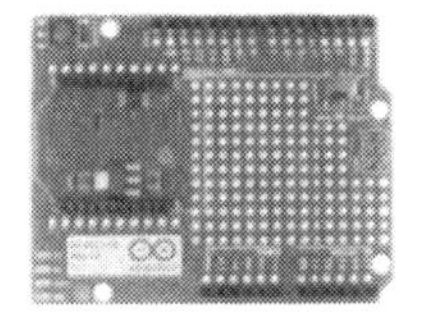

Arduino Wireless Proto Shield

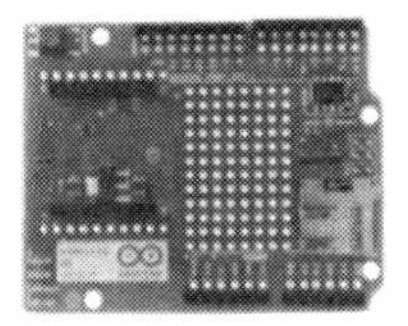

Arduino Wireless SD Shield

Arduino Motor Shield

Arduino Proto Shield

Figure 1-9. *Official Arduino shields*

Arduino IDE

The Arduino IDE is a cross-platform application written in Java and derived from the IDE for Processing and Wiring. An IDE is basically a program designed to facilitate software development. The Arduino IDE includes a source code editor (a text editor with some useful features for programming, such as syntax highlighting and automatic indentation) and a compiler that turns the code you have written into machine-readable instructions.

But all you really need to know about the Arduino IDE is that it is the piece of software that will allow you to easily write your own Arduino sketches and upload them to the board with a single click. Figure 1-10 shows what the IDE looks like.

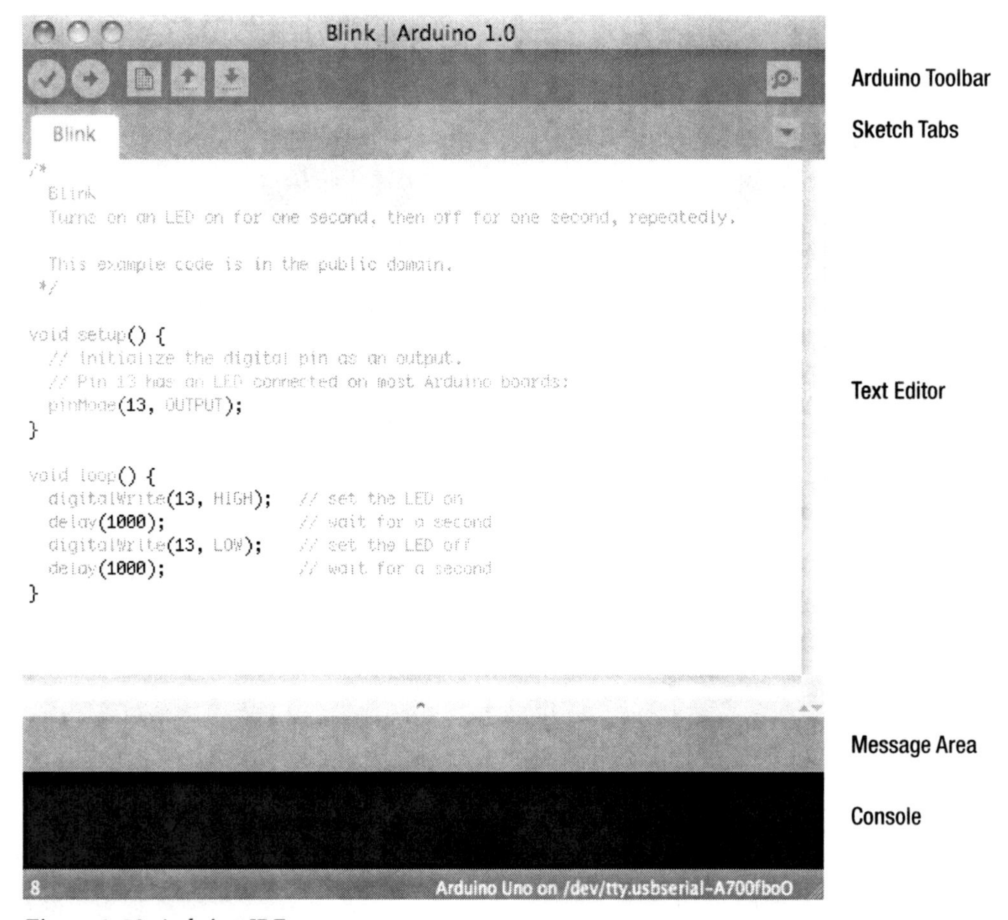

Figure 1-10. *Arduino IDE*

Take a moment to familiarize yourself with the IDE. Table 1-1 offers a description of all the buttons you will find in the toolbar.

Note that the line of text that you see at the bottom-right of the IDE is actually telling you the Arduino board and the serial port you have selected. If they don't correspond to what you expect, you can change them following the steps from the section "Testing the Arduino."

Table 1-1. *Arduino Development Environment (from `http://arduino.cc`)*

Symbol	Description
	Verify Checks your code for errors.
	Upload Compiles your code and uploads it to the Arduino I/O board.
	New Creates a new sketch.
	Open Presents a menu of all the sketches in your sketchbook. Clicking one will open it within the current window.
	Save Saves your sketch.
	Serial Monitor Opens the serial monitor.

If these symbols look somewhat different from those in your Arduino IDE, you probably have an old version of the Arduino tools. Go to `http://arduino.cc/en/Main/Software` and download the latest version. At the time of writing, Arduino 1.0 is the current software version.

Note In Arduino version 1.0, the sketches are saved with the extension `.ino` instead of the old extension `.pde` (the same as Processing sketches). You can still open `.pde` sketches created with previous versions.

Serial Monitor

The Serial Monitor (at the top right of your Arduino IDE) is a tool to display the serial data being sent from the Arduino board to the computer (see Figure 1-11). You need to match the baud rate on your serial monitor to the baud rate you specify in the Arduino sketch sending data. The baud rate is the number of symbol changes transmitted per second.

The serial monitor is very useful in the debugging of Arduino programs (the process of finding and fixing the defects in the program so it behaves as expected). Whenever you're not sure what is wrong with the code, ask Arduino to print messages to the serial monitor so you can peek into the state of the program at that precise moment.

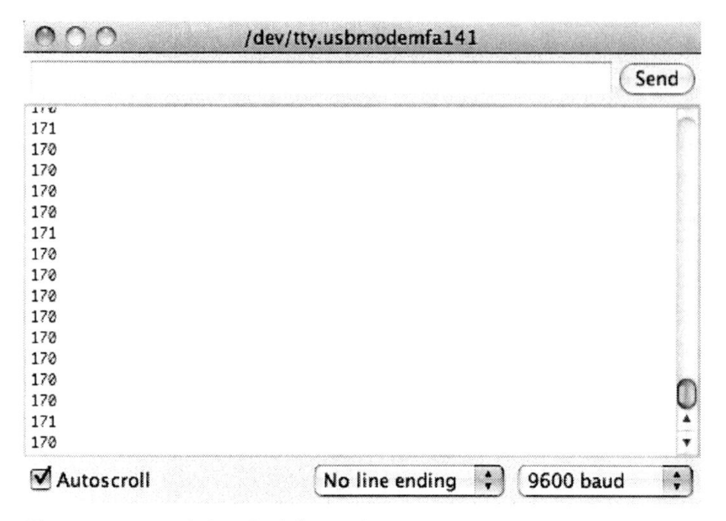

Figure 1-11. Arduino Serial Monitor

You will make extensive use of serial communication throughout the book, as it is the main way to get Arduino to communicate with Processing and thus with the Kinect data.

Arduino Language

The Arduino language is implemented in C/C++ and based in Wiring. When you write an Arduino sketch, you are implicitly making use of the Wiring library, which is included with the Arduino IDE. This allows you to make runnable programs by using only two functions: `setup()` and `loop()`. As mentioned, the Wiring language is inspired by Processing, and the Arduino language structure is inherited from the Processing language, where the equivalent functions are called `setup()` and `draw()`, as you will see in Chapter 3.

You need to include both functions in every Arduino program, even if you don't need one of them. Let's analyze the structure of a simple Arduino sketch using again the Blink example. Open your Arduino IDE, and select Open ➤ 1.Basics ➤ Blink.

The setup() Function

The `setup()` function is run only once, right after you upload a new sketch to the board, and then each time you power up your Arduino. You use this function to initialize variables, pin modes, start libraries, etc. In this example, the `setup()` function is used to define pin 13 as output:

```
void setup() {
    pinMode(13, OUTPUT);
}
```

The loop() Function

The loop() function is run continuously in an endless loop while the Arduino board is powered. This function is the core of most programs.

In this example, you turn pin 13's built-in LED on using the function digitalWrite() to send a HIGH (5V) signal to the pin. Then you pause the execution of the program for one second with the function delay(). During this time, the LED will remain on. After this pause, you turn the LED off, again using the function digitalWrite() but sending a LOW (0V) signal instead, and then pause the program for another second. This code will loop continuously, making the LED blink.

```
void loop() {
  digitalWrite(13, HIGH);
  delay(1000);
  digitalWrite(13, LOW);
  delay(1000);
}
```

This code would not actually be seen as a valid C++ program by a C++ compiler, so when you click Compile or Run, Arduino adds a header at the top of the program and a simple main() function so the code can be read by the compiler.

Variables

In the previous example, you didn't declare any variables because the sketch was rather simple. In more complex programs, you will declare, initialize, and use variables.

Variables are symbolic names given to a certain quantity of information. This means you will associate a certain piece of information with a symbolic name or identifier by which you will refer to it.

Variable Declaration and Initialization

In order to use a variable, you first need to declare it, specifying at that moment the variable's data type. The data type specifies the kind of data that can be stored in the variable and has an influence in the amount of memory that is reserved for that variable.

As Arduino language is based on C and C++, the data types you can use in Arduino are the same allowed in C++. Table 1-2 lists some of the data types you will use throughout this book. For a complete listing of C++ data types, you can refer to C++ reference books and Internet sites.

Table 1-2. *Arduino Fundamental Data Types*

Name	Description	Size	Range
char	Character or small integer	1byte	Signed: -128 to 127 Unsigned: 0 to 255
boolean	Boolean value (true or false)	1byte	True or false
int	Integer	4bytes	Signed: -2147483648 to 2147483647 Unsigned: 0 to 4294967295

Name	Description	Size	Range
long	Long integer	4bytes	Signed: -2147483648 to 2147483647 Unsigned: 0 to 4294967295
float	Floating point number	4bytes	+/- 3.4e +/- 38 (~7 digits)
double	Double precision floating point number	8bytes	+/- 1.7e +/- 308 (~15 digits)

When you declare the variable, you need to write the type specifier of the desired data type followed by a valid variable identifier and a semicolon. In the following lines, you declare an integer, a character, and a Boolean variable:

```
int myInteger;
char myLetter;
boolean test;
```

This code has declared the variables, but their value is undetermined. In some cases, you want to declare the variable and set it to a default value at the same time. This is called *initializing* the variables. If you want to do this, you can do it in the following way:

```
int myInteger = 656;
char myLetter = 'A';
boolean test = true;
```

By doing this, your three variables are set to specific values and you can start using them from this moment on.

In other cases, you want to declare the variables first and initialize them later. You will see this happening frequently, as it's necessary when you want to declare the variable as global but its value depends on other variables or processes happening within the `setup()`, `draw()`, or other functions.

```
int myInteger;
char myLetter;
boolean test;

setup(){
 myInteger = 656;
 myLetter = 'A';
 test = true;
}
```

Variable Scope

When you declare your variables, the place you do it within the code will have an impact on the scope in which the variable will be visible. If you declare your variable outside of any function, it will be considered a *global* variable and will be accessible from anywhere in the code. If you declare your variable within a function, the variable will be a *local* variable and will only be accessible from within the same function. Let's do an example from the Blink sketch.

```
int ledPin = 13;
void setup() {
```

```
  pinMode(ledPin, OUTPUT);
  int myOtherPin = 6;
  pinMode(myOtherPin, OUTPUT);
}

void loop() {
  digitalWrite(ledPin, HIGH);   // set the LED on
  delay(1000);                  // wait for a second
  digitalWrite(ledPin, LOW);    // set the LED off
  delay(1000);                  // wait for a second
}
```

This code will upload and run smoothly, behaving exactly as the original Blink example. You have declared and initialized the integer `ledPin` at the beginning of your program, outside of any function, so it is considered a global variable visible from within the other functions in the code.

The variable `myOtherPin`, on the other hand, has been defined within the `setup()` function. You have used the variable to set the pin defined by the variable to `OUTPUT`. If you tried to use the variable from the `loop()` function, Arduino would throw an error at you, saying "'myOtherPin' was not declared in this scope". And that's right; you have declared and initialized the variable within the `setup()` function, making it a local variable and only visible from within the function in which it was declared.

Your First Arduino Project

You are going to write an example Arduino sketch involving all the elements you just learned in the previous sections. This is only a very first step in the development of an Arduino project, but it will be enough to summarize this chapter on Arduino basics.

You are going to build a very simple circuit first, and then you will program it using your Arduino IDE. You will need your Arduino board, a breadboard, an LED, a 220-ohm resistor, a potentiometer, and some wires to build this example.

Breadboard

When working with electronics, it is often useful to "sketch" the circuit quickly so you can test it before building it in a more permanent form. Using a solderless breadboard, you can easily prototype electronics by plugging elements together. As the elements are not soldered to the board, the board can actually be reused as many times as you want (see Figures 1-12 and 1-13).

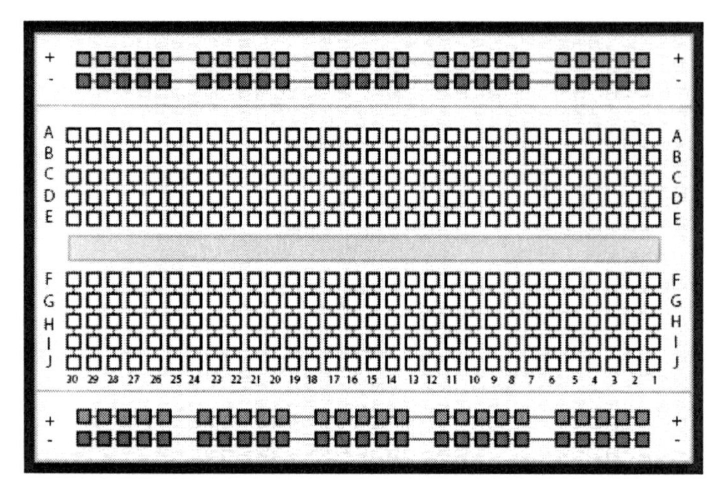

Figure 1-12. Bredboard connection diagram

Figure 1-13. Project sketched on a breadboard

Note that the pins flanked by blue and red lines on your breadboard (the top two rows and the bottom two rows in Figure 1-13) are connected together horizontally, so you can use them to provide 5V and ground to all the elements on the project. The other pins are connected in vertical strips broken in the middle. You use these linked strips to connect the resistor to the LED and the potentiometer pins to 5V, ground, and the Arduino Analog Input Pin 0.

Building the Circuit

Try to replicate this circuit on your breadboard and connect it to the Arduino as you see happening in Figure 1-14. The LED is connected to analog pin 11, and the middle pin of the potentiometer is connected to analog pin 0. The ground and 5V cables are connected to the Arduino GND and 5V pins. Note that the LED has a shorter leg (cathode), which needs to be connected to ground. The longer one is connected to the resistor. If you are curious about the functioning of LEDs and why you are using a 220-ohm resistor, there is a whole section on resistors in Chapter 7.

The goal of this project is to control the brightness of the LED by turning the potentiometer. A potentiometer is basically a variable resistor that you control by turning the knob. Note that the potentiometer is not connected to the LED or the resistor at all; it's just connected to the Arduino board. You acquire the values coming from the potentiometer from Arduino and use these values to drive the brightness of the LED. All of this is done in the Arduino code.

Programming the Arduino

Now you are going to write the code that will permit you to drive the LED by turning your potentiometer. You will use serial communication to check the values that you are getting from the potentiometer in runtime using the Serial Monitor.

First, you declare a global integer variable called `ledPin` that defines the LED pin number during the execution of the program.

```
int ledPin = 11;
```

The setup() Function

After this, you write the `setup()` function in which you start serial communication at a baud rate of 9600 and define the `pinMode` of your LED pin to `OUTPUT` so you can write to it.

```
void setup() {
  Serial.begin(9600);
  pinMode(ledPin, OUTPUT);
}
```

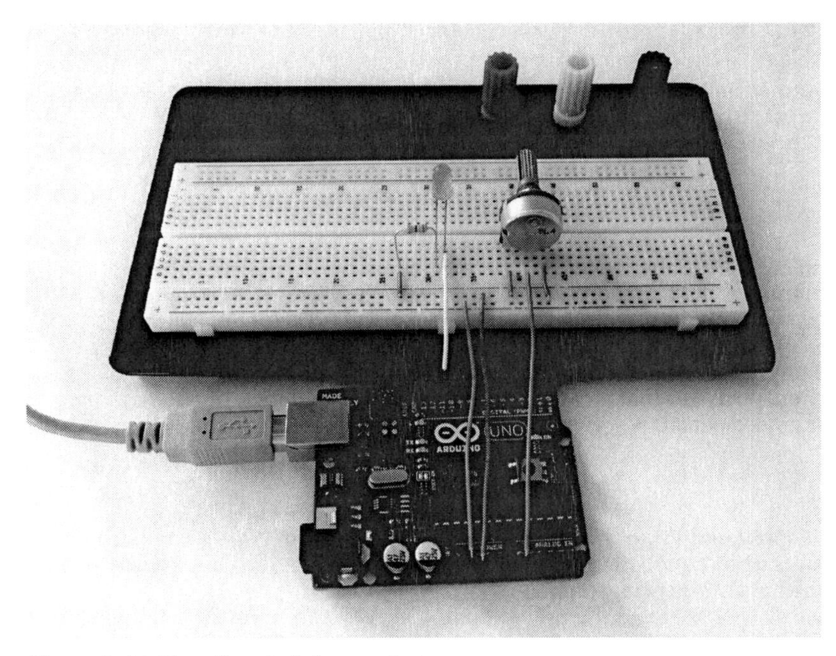

***Figure 1-14.** Your first Arduino project*

The loop() Function

In the `loop()` function, the first thing you do is declare a new local integer variable called `sensorValue` to store the value of the potentiometer connected to analog pin 0. You initialize the variable to this value by using the `analogRead()` function and specifying A0 (Analog 0) as your pin. If you were to read analog pin 1, you would write A1, and so on.

Then you print the value from the potentiometer to the serial buffer using the function `Serial.println()`. If you open the Serial Monitor while the program is running, you will see that the values coming from the potentiometer range from 0 to 1023.

You want to use this value to set the brightness of your LED, but you have stated that the scope of the brightness needs to be in a range of 0 to 255. The solution to this issue will be *mapping* the values of the potentiometer from its range to the range of the potentiometer. You will use this technique very often in your projects; luckily, Arduino has a function to perform this operation called `map()`.

The `map()` function takes five arguments: `map(value, fromLow, fromHigh, toLow, toHigh)`.

The argument `value` is the current value to be mapped. `fromLow` and `fromHigh` define the lower and higher values of the source range, and `toLow` and `toHigh`, the lower and higher values of the target range.

In your case, your LED brightness needs to be defined as `map(sensorValue,0,1023,0,255)`, so if the sensor value is 1023, the LED brightness is 255. The `map()` function will always return an integer, even in the case that the mapping would result in a floating point number.

Once you have determined the brightness of the LED, you write this value to the LED pin using the function `analogWrite(pin, value)`.

```
void loop() {
  int sensorValue = analogRead(A0);
  Serial.println(sensorValue);

  int ledBrightness = map(sensorValue,0,1023,0,255);
  analogWrite(ledPin, ledBrightness);
}
```

Upload this program to your Arduino board, and you will be able to change the brightness of your LED by turning the potentiometer. You will develop more advanced control techniques in Chapter 4, but before you move on, let's go through some basic concepts on electronics.

Circuit Diagrams

We just described the circuit using photos and text. But when we get to more complicated circuits, this will not be enough. The way circuits should be represented is with circuit diagrams, or schematics.

Circuit diagrams show the components of a circuit as simplified standard symbols and the connections between the devices. Note that the arrangement of the elements on a diagram doesn't necessarily correspond to the physical arrangement of the components.

There are numerous software packages that facilitate the drawing of schematics, but there is one designed to be used by non-engineers, and it works nicely with Arduino projects: Fritzing.

Fritzing

Fritzing is an open source tool developed at the University of Applied Sciences of Potsdam in Germany; it's aimed at artists and designers who want to develop their prototypes into real projects. You can (and should) download Fritzing from `http://fritzing.org`.

Fritzing has a graphical interface that allows the user to build the circuit as in the physical world, connecting together breadboards, Arduinos, LEDs, resistors, etc. Then you turn this visual representation into a circuit diagram just by changing the view, as shown in Figures 1-15 and 1-16.

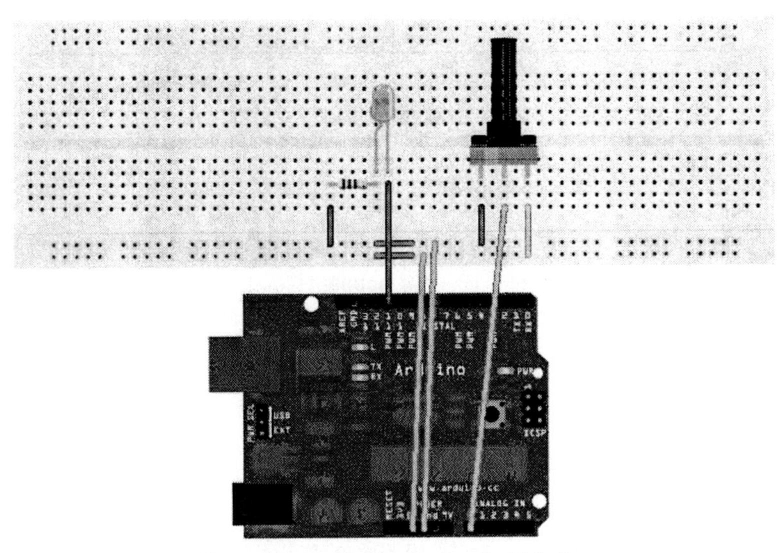

Figure 1-15. Breadboard view of your project in Fritzing

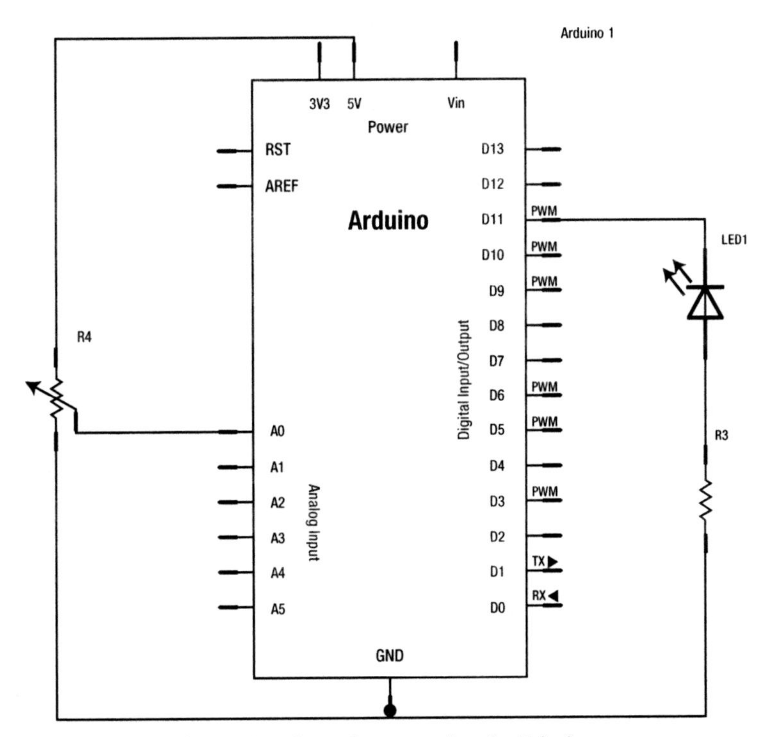

Figure 1-16. Schematics view of your project in Fritzing

Electronic Symbols

Electronic symbols, or circuit symbols, are the pictograms used to represent electrical components and devices in the schematics. There are several international standards for these graphical symbols. In this book, we will use ANSI in all the circuit diagrams and symbols. Table 1-3 is a list of some electronics symbols that you will see used in this book and some of the IEC equivalents for reference.

Table 1-3. *Electronic Symbols and IEC Equivalents*

Circuit Symbol	Component	Circuit Symbol	Component
	Wire		Resistor (ANSI)
	Wires Joined		Resistor (IEC)
	Wires Not Joined		Rheostat, or Variable Resistor (ANSI)
	Cell		Rheostat, or Variable Resistor (IEC)
	Battery		Potentiometer (ANSI)
	Earth (Ground)		Potentiometer (IEC)
	On-Off Switch	M	Motor
NO COM NC	Relay		Light Emitting Diode (LED)

Electricity

Throughout the rest of the book, you will be building circuits, using motors, and talking a lot about electronics in general. One of the advantages of Arduino is that it makes for an easy introduction to electronics for beginners, but if you are completely new to electronics, here are some basic concepts.

AC/DC

Mains electricity is alternating current (AC), but you will be using direct current (DC) for all your projects. AC is actually cheaper to produce and transmit from the power plants to your home. That is the reason why all the plugs in our households give us alternating current. But AC is also more difficult to control, so most electronic devices work with a power supply that converts AC into DC.

The current provided by your USB is DC, as is the current you get from a battery or a computer power supply, so you won't work with AC at all in this book. Every time we speak about electricity or current in the book, we are referring to DC.

Ohms Law

Electricity, or electrical *current* is, very generally, the movement of electrons through a conductor. The *intensity* (I) of this current is measured in amperes (A).

The force pushing the electrons to flow is called *voltage*, which is the measure of the potential energy difference relative between the two ends of the circuit. Voltage is measured in Volts, or V.

The conductor through which the electrons are flowing will oppose some resistance to their passage. This force is called *resistance*, and it is measured in ohms (Ω).

Georg Ohm published a treatise in 1827 describing measurements of voltage and current flowing through circuits. Today, one of these equations has become known as Ohm's law, and it ties together voltage, current, and resistance.

$$I = \frac{V}{R}$$

Putting this into words, the equation states that the current flowing through a conductor (I) is directly proportional to the potential difference (V) across the two ends. The constant of proportionality is called resistance (R).

Resistance is then the measure of the opposition to the passage of an electric current through an element. The opposite of resistance is conductance (G).

$$R = \frac{V}{I} \qquad G = \frac{I}{V} \qquad G = \frac{1}{R}$$

Ohm's law can be rewritten in three different ways depending on which elements of the equation you already know and which one you are trying to calculate.

$$I = \frac{V}{R} \qquad R = \frac{V}{I} \qquad V = R \cdot I$$

Joule's Law

Joule's first law, also known as Joule Effect, was declared by James Prescott Joule in the 1840s. This law states that the power dissipated in a resistor, in terms of the current through it and its resistance, is

$$P = I^2 \cdot R$$

If you apply the previous equations to Joule's law, you will be finally able to calculate the power in watts as a value derived by multiplying current by voltage.

$$P = I \cdot V$$

This helps you to choose the correct power rating for elements such as resistors and to calculate the power consumption from battery-driven supplies. This, in turns, lets you estimate how long your project will run on a set of batteries.

Summary

This first chapter has taken you through an overview of the Arduino platform in which both the Arduino hardware and software have been thoroughly analyzed.

Your Arduino board should be properly configured now, the Arduino IDE should be installed on your computer, and you should be communicating seamlessly with your board via USB. You have had a brief refresher course on basic electrical concepts, and you have warmed up with a couple of examples.

This introduction doesn't intend to be a substitute for more introductory works. If this is your first contact with Arduino, you might want to consider having a look at some other titles like *Beginning Arduino* (Apress, 2010) before you go deeper into this book, or at least have one at hand just in case.

After finishing this chapter, hopefully you are craving more exciting projects, but don't rush into it! You still have to go through some fundamental concepts about the other cornerstone of this book: the Kinect.

CHAPTER 2

Kinect Basics

by Enrique Ramos

The Kinect, shown in Figure 2-1, was launched on November 4, 2010 and sold an impressive 8 million units in the first 60 days, entering the Guinness World Records as the "fastest selling consumer electronics device in history." The Kinect was the first commercial sensor device to allow the user to interact with a console through a natural user interface (using gestures and spoken commands instead of a game controller). The Kinect is the second pillar of this book, so we will spend this chapter getting you acquainted with it.

In the following pages, you will learn the story of this groundbreaking device, its hardware, software, and data streams. You will learn what a structured-light 3D scanner is and how it is implemented in the Kinect sensor. You will learn about the different data you can acquire from your Kinect (such as RGB, infrared, and depth images), and how they are combined to perform motion tracking and gesture recognition. Welcome to the world of Kinect!

Figure 2-1. *The Kinect*

BUYING A KINECT

There are currently three ways to buy a Kinect sensor.

- Kinect for Windows ($249.99)
- Kinect for Xbox 360 as a standalone peripheral to be used with your previously purchased Xbox 360 ($129.99)
- Kinect for Xbox 360 bundled with the Xbox 360 (from $299.99, depending on the bundle)

If you purchase the new Kinect for Windows or the Kinect for Xbox 360 as a standalone peripheral, both are ready to connect to your computer.

But maybe you already have a Kinect device that you are using with your Xbox 360, or you are planning to buy the Kinect together with an Xbox 360. If you purchase the Kinect bundled with the Xbox 360, you will also need to buy a separate power supply ($34.99 from Microsoft).

When used with the latest Xbox, the Kinect device doesn't need the AC adapter, because the console has a special USB port that delivers enough power for the Kinect to function. Your computer USB can't deliver as much current, though, and the special USB from Kinect doesn't fit into your standard USB port. You will need the AC power supply to connect your Kinect to an electrical outlet, and to convert the special USB plug to a standard USB that you will be able to plug to your computer (Figure 2-2).

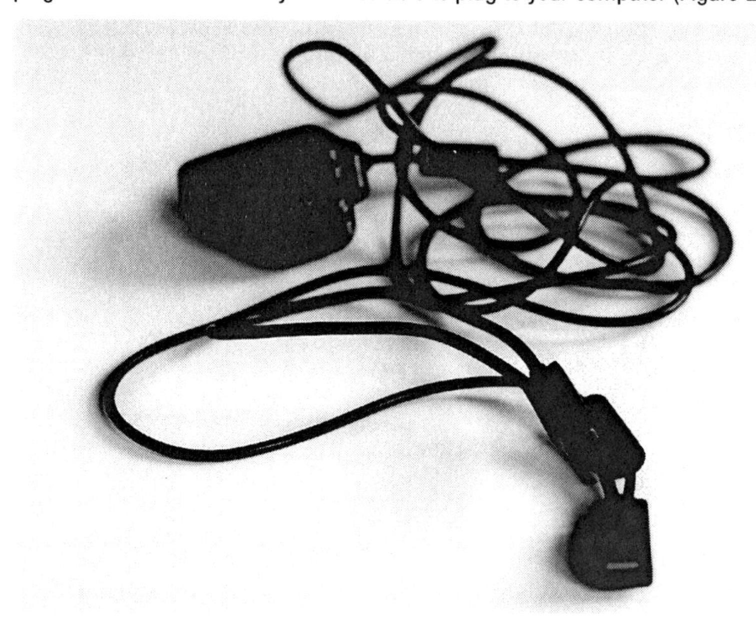

Figure 2-2. Kinect AC adapter with standard USB connection

A Brief History of the Kinect

Microsoft announced the Kinect project on June 1, 2009 under the code name Project Natal. The name was changed to Kinect on June 13, 2010; it is derived from the words "kinetic" and "connect" to express the ideas behind the device. The motto of the marketing campaign for the launch was, very appropriately, *"You are the controller."*

This first launch caused a worldwide frenzy, and hackers soon found the one "open door" in this amazing device. Unlike other game controllers, the Kinect connection was an open USB port, so it could potentially be connected to a PC. Unfortunately, Microsoft had not released any PC drivers for the device and didn't seem to be willing to do so in the near future.

Note *Drivers* are computer programs that allow other higher-level computer programs to interact with hardware devices. They convert a well-known and predictable API (application programming interface) to the native API built into the hardware, making different devices look and behave similarly. Think of drivers as the translators between the hardware and the application or the operating system using it. Without the proper drivers, the computer wouldn't know how to communicate with any of the hardware plugged into it!

Hacking the Kinect

Right after the Kinect was released in November 2010, and taking advantage of the Kinect's open USB connection, Adafruit Industries offered a bounty of $1,000 to anybody who would provide "open source drivers for this cool USB device. The drivers and/or application can run on any operating system—but completely documented and under an open source license." After some initial negative comments from Microsoft, Adafruit added another $2,000 to the bounty to spice it up.

The winner of the $3,000 bounty was Héctor Martín, who produced Linux drivers that allowed the use of both the RGB camera and the depth image from the Kinect.

The release of the open source drivers stirred a frenzy of Kinect application development that caught the attention of the media. Web sites dedicated to the new world of Kinect applications, such as `http://www.kinecthacks.com`, mushroomed on the Internet. The amazed public could see a growing number of new applications appear from every corner of the world.

Official Frameworks

But the Kinect hackers were not the only ones to realize the incredible possibilities that the new technology was about to unleash. The companies involved in the design of the Kinect soon understood that the Kinect for Xbox 360 was only a timid first step toward a new technological revolution—and they had no intention of being left behind.

In 2010, PrimeSense (the company behind Kinect's 3D imaging) released its own drivers and programming framework for the Kinect, called OpenNI. Soon after that, it announced a partnership with ASUS in producing a new Kinect-like device, the Xtion.

In 2011, Microsoft released the non-commercial Kinect SDK (software development kit). In February 2012, it released a commercial version, accompanied by the Kinect for Windows device.

And this is where we stand now. Laptops with integrated Kinect-like cameras are most probably on their way. This is only the beginning of a whole range of hardware and applications using the technology behind Kinect to make computers better understand the world around them.

The Kinect Sensor

The Kinect sensor features an RGB camera, a depth sensor consisting of an infrared laser projector and an infrared CMOS sensor, and a multi-array microphone enabling acoustic source localization and ambient noise suppression. It also contains an LED light, a three-axis accelerometer, and a small servo controlling the tilt of the device.

■ **Note** The infrared CMOS (complementary metal–oxide semiconductor) sensor is an integrated circuit that contains an array of photodetectors that act as an infrared image sensor. This device is also referred to as IR camera, IR sensor, depth image CMOS, or CMOS sensor, depending on the source.

Throughout this book we will focus on the 3D scanning capabilities of the Kinect device accessed through OpenNI/NITE. We won't be talking about the microphones, the in-built accelerometer, or the servo, which are not accessible from OpenNI because they are not part of PrimeSense's reference design.

The RGB camera is an 8-bit VGA resolution (640 x 480 pixels) camera. This might not sound very impressive, but you need to remember that the magic happens in the depth sensor, which is completely independent of the RGB camera.

The two depth sensor elements, the IR projector and IR camera, work together with the internal chip from PrimeSense to reconstitute a 3D motion capture of the scene in front of the Kinect (Figures 2-3 and 2-4). They do this by using a technique called *structured-light 3D scanning*, which we will discuss in depth at the end of this chapter. The IR camera also uses a VGA resolution (640 x 480 pixels) with 11-bit depth, providing 2,048 levels of sensitivity.

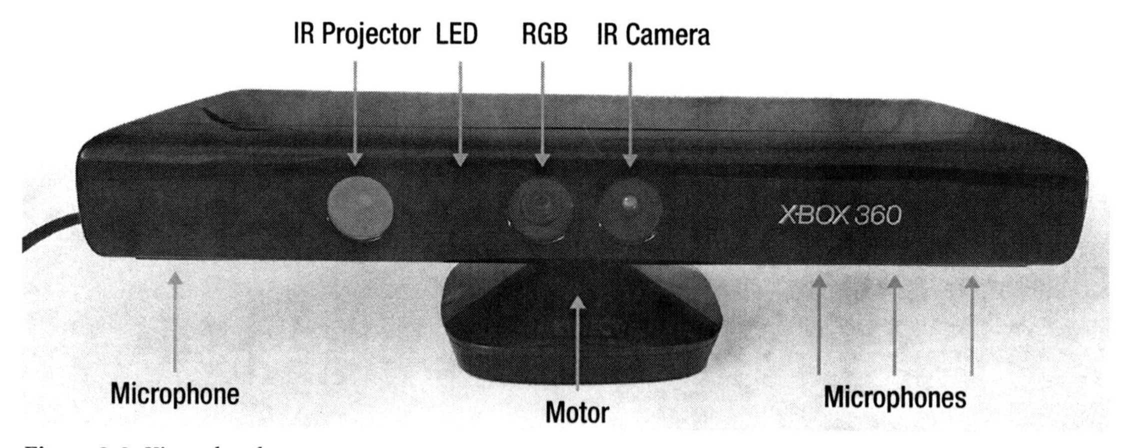

Figure 2-3. Kinect hardware

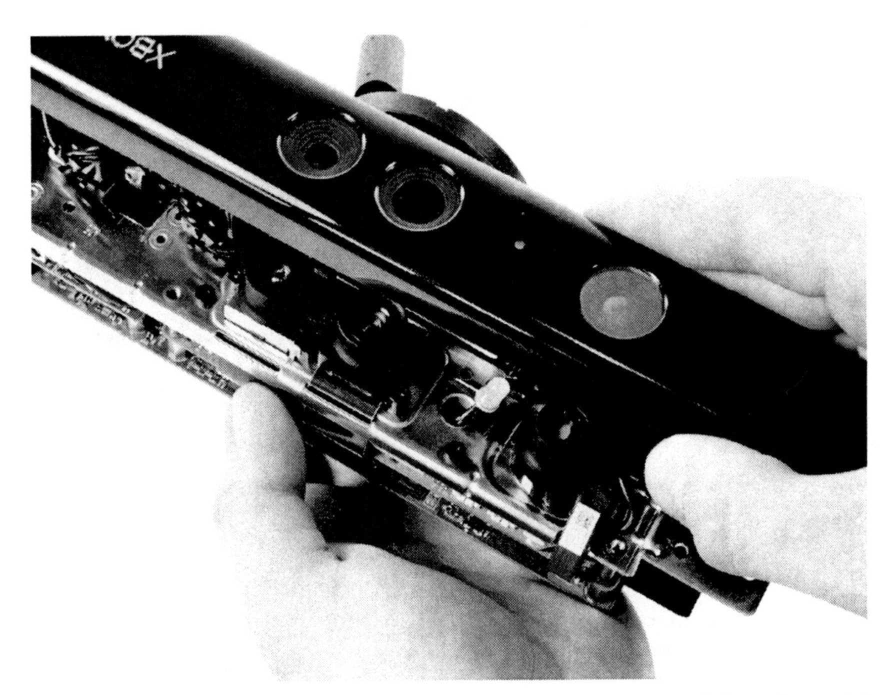

Figure 2-4. *Left to right: Kinect IR camera, RGB camera, LED, and IR projector (photo courtesy of ifixit)*

Of course, there is much more going on within the sensor. If you are interested in the guts of the Kinect device, the web site ifixit featured a teardown of the Kinect in November 2010 (`http://www.ifixit.com/Teardown/Microsoft-Kinect-Teardown/4066`).

Positioning your Kinect

The Kinect's practical range goes from 1.2m to 3.5m. If the objects stand too close to the sensor, they will not be scanned and will just appear as black spots; if they are too far, the scanning precision will be too low, making them appear as flat objects. If you are using the Kinect for Windows device, the range of the camera is shorter, from 40cm to 3m.

Kinect Capabilities

So what can you do with your Kinect device and all of the hi-tech stuff hidden inside? Once the Kinect is properly installed and communicating with your computer, you will be able to access a series of raw data streams and other capabilities provided by specific middleware. In this book, we will use the OpenNI drivers and NITE middleware.

RGB Image

Yes, you can use the Kinect as a 640 x 480 pixel webcam. You will learn how to access Kinect's RGB image using OpenNI in Chapter 3.

IR Image

As Kinect has an infrared CMOS, you can also access the 640 x 480 IR image using OpenNI. This is also covered in Chapter 3.

Depth Map

The depth map is the result of the operations performed by PrimeSense's PS1080 chip on the IR image captured by Kinect's IR CMOS. This VGA image has a precision of 11 bits, or 2048 different values, represented graphically by levels of grey from white (2048) to black (0).

In Chapter 3, you will learn how to access and display the depth image from Processing using the OpenNI framework and how to translate the gray scale image into real space dimensions. There is one detail that you should take into consideration: the sensor's distance measurement doesn't follow a linear scale but a *logarithmic scale*. Without getting into what a logarithmic scale is, you should know that the precision of the depth sensing is lower on objects further from the Kinect sensor.

Hand and Skeleton Tracking

After the depth map has been generated, you can use it directly for your applications or run it through a specific middleware to extract more complex information from the raw depth map.

In the following chapters, you will be using NITE middleware to add hand/skeleton tracking and gesture recognition to your applications. Chapter 4 will teach you how to use hand tracking to control LED lights through an Arduino board. Chapter 5 will introduce NITE's gesture recognition, and you will even program your own simple gesture recognition routine. Chapter 6 will teach you how to work with skeleton tracking.

The algorithms that NITE or other middleware use to extract this information from the depth map fall way beyond the scope of this book, but if you are curious, you can read *Hacking the Kinect* by Jeff Kramer et al (Apress, 2012), in which you will find a whole chapter on gesture recognition.

Kinect Drivers and Frameworks

In order to access the Kinect data streams, you will need to install the necessary drivers on your computer. Because of the rather complicate story of this device, there are a series of choices available that we will detail next.

OpenKinect: Libfreenect Drivers

Soon after creating the Kinect open source drivers for Adafruit, Héctor Martín joined the OpenKinect community (`http://openkinect.org`), which was created by Josh Blake with the intention of bringing together programmers interested in natural user interfaces (NUIs). OpenKinect develops and maintains the libfreenect core library for accessing the Kinect USB camera. It currently supports access to the RGB and depth images, the Kinect motor, the accelerometer, and the LED. Access to the microphones is being worked on.

PrimeSense: OpenNI and NITE

Throughout this book, we will be using OpenNI and NITE to access the Kinect data streams and the skeleton/hand tracking capabilities, so here is a little more detail on this framework. The Israeli company PrimeSense developed the technology behind Kinect's 3D imaging and worked with Microsoft in the development of the Kinect device. In December 2010, PrimeSense created an industry-led, not-for-profit organization called OpenNI, which stands for open natural interaction (`http://www.openni.org`).

This organization was formed to "certify and promote the compatibility and interoperability of natural interaction (NI) devices, applications, and middleware." The founding members of the OpenNI organization are PrimeSense, Willow Garage, Side-Kick, ASUS, and AppSide.

OpenNI

In order to fulfill its goal, OpenNI released an open source framework called OpenNI Framework. It provides an API and high-level middleware called NITE for implementing hand/skeleton tracking and gesture recognition.

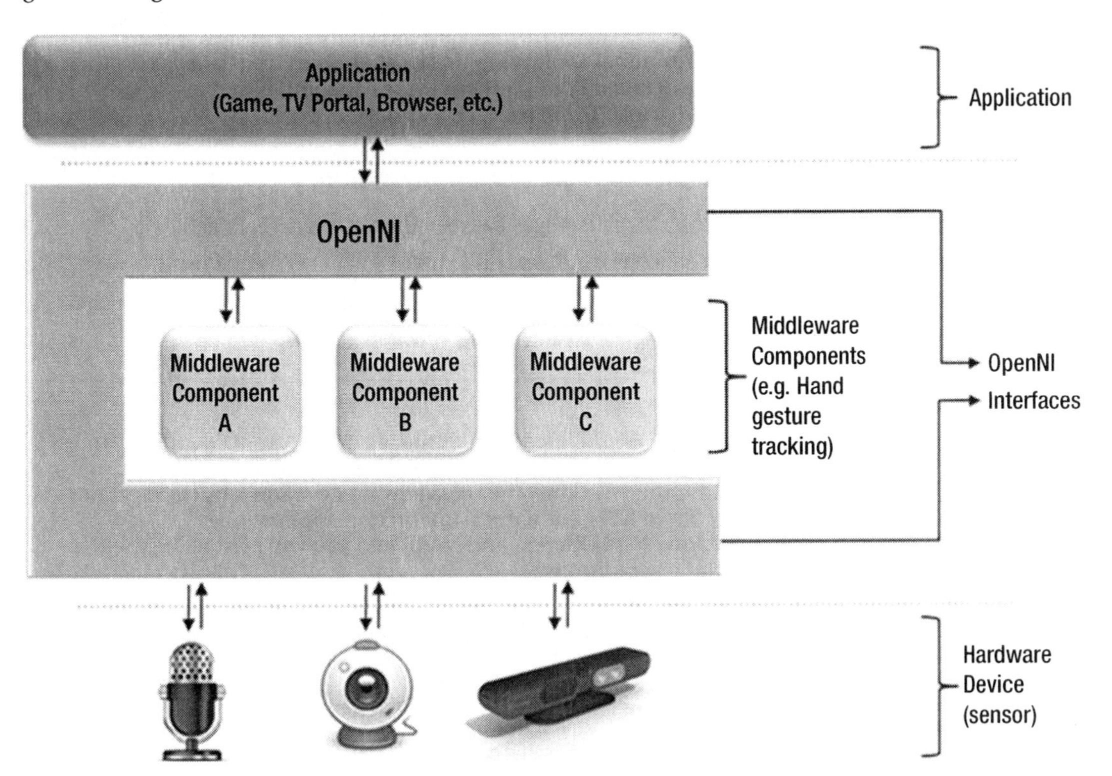

Figure 2-5. OpenNI abstract layered view (courtesy of PrimeSense)

Because OpenNI breaks the dependency between the sensor and the middleware (see Figure 2-5), the API enables middleware developers to develop algorithms on top of raw data formats, independent of the sensor device that is producing the data. In the same way, sensor manufacturers can build sensors that will work with any OpenNI-compliant application.

Kinect was the first implementation of the PrimeSense reference design; its optics and microchip were developed entirely by PrimeSense. Then Microsoft added a motor and a three-axis accelerometer to the design. This is why OpenNI doesn't provide access to the motor or the accelerometer; they are specific to the Kinect implementation.

PrimeSense is a fabless (fabrication-less) semiconductor company. It makes its revenue by selling hardware and semiconductor chips while outsourcing fabrication. It's mainly a B2B (business to business): it sells solutions to manufacturers who insert these solutions into consumer products. This is exactly the way in which it was involved in the development of the Kinect with Microsoft.

PrimeSense then sells its technology to manufacturers such as ASUS and other computer or television manufacturers. But for this market to be developed, there needs to exist an ecosystem of people creating natural interaction-based content and applications. PrimeSense created OpenNI as a way to empower developers to add natural interaction to their software and applications so this ecosystem would flourish.

NITE

For natural interaction to be implemented, the developer needs more than the 3D point cloud from Kinect. The most useful features come from the skeleton and hand tracking capabilities. Not all developers have the knowledge, time, or resources to develop these capabilities from scratch, as they involve advanced algorithms. PrimeSense decided to implement these capabilities and distribute them for commercial purposes but keep the code closed, and so NITE was developed.

■ **Note** There has been much confusion about the differences between OpenNI and NITE. OpenNI is PrimeSense's framework; it allows you to acquire the depth and RGB images from the Kinect. OpenNI is open source and for commercial use. NITE is the middleware that allows you to perform hand/skeleton tracking and gesture recognition. NITE is not open source, but it is also distributed for commercial use.

This means that without NITE you can't use skeleton/hand tracking or gesture recognition, unless you develop your own middleware that processes the OpenNI point cloud data and extracts the joint and gesture information. Without a doubt, there will be other middleware developed by third parties in the future that will compete with OpenNI and NITE for natural interaction applications.

In Chapter 3, you will learn how to download OpenNI and NITE, and you will start using them to develop amazing projects throughout the rest of the book.

Microsoft Kinect for Windows

On June 16, 2011, six months after PrimeSense released its drivers and middleware, Microsoft announced the release of the official Microsoft Kinect SDK for non-commercial use. This SDK offered the programmer access to all the Kinect sensor capabilities plus hand/skeleton tracking. At the time of writing, Kinect for Windows SDK includes the following:

- Drivers for using Kinect sensor devices on a computer running Windows 7 or Windows 8 developer preview (desktop apps only)
- APIs and device interfaces, along with technical documentation
- Source code samples

Unfortunately, the non-commercial license limited the applications to testing or personal use. Also, the SDK only installs on Windows 7, leaving out the Linux and Mac OSX programmer communities. Moreover, the development of applications is limited to C++, C#, or Visual Basic using Microsoft Visual Studio 2010.

These limitations discouraged many developers, who chose to continue to develop applications with OpenNI/NITE plus their OS and developing platform of choice, with an eye towards the commercialization of their applications.

From February 2012, the Kinect for Windows includes a new sensor device specifically designed for use with a Windows-based PC and a new version of the SDK for commercial use. The official Microsoft SDK will continue to support the Kinect for Xbox 360 as a development device.

Kinect Theory

The technique used by PrimeSense's 3D imaging system in the Kinect is called structured-light 3D scanning. This technique is used in many industrial applications, such as production control and volume measurement, and involves highly accurate and expensive scanners. Kinect is the first device to implement this technique in a consumer product.

Structured-Light 3D Scanning

Most structured-light scanners are based on the projection of a narrow stripe of light onto a 3D object, using the deformation of the stripe when seen from a point of view different from the source to measure the distance from each point to the camera and thus reconstitute the 3D volume. This method can be extended to the projection of many stripes of light at the same time, which provides a high number of samples simultaneously (Figure 2-6).

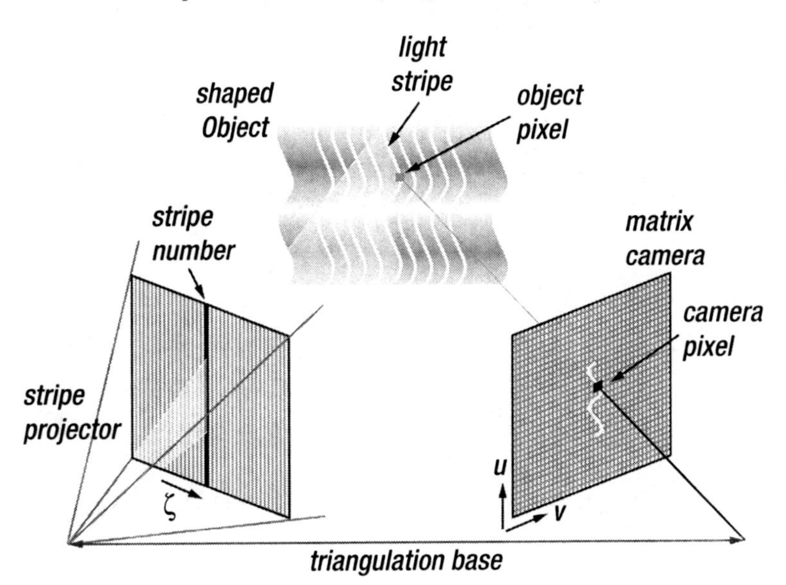

Figure 2-6. Triangulation principles for structured-light 3D scanning (from Wikipedia, `http://en.wikipedia.org/wiki/Structured_Light_3D_Scanner`, licensed under the Creative Commons Attribution-Share Alike 3.0 Unported license)

The Kinect system is somewhat different. Instead of projecting stripes of visible light, the Kinect's IR projector sends out a pattern of infrared light beams (called an *IR coding image* by PrimeSense), which bounces on the objects and is captured by the standard CMOS image sensor (see Figure 2-7). This captured image is passed on to the onboard PrimeSense chip to be translated into the depth image (Figure 2-8).

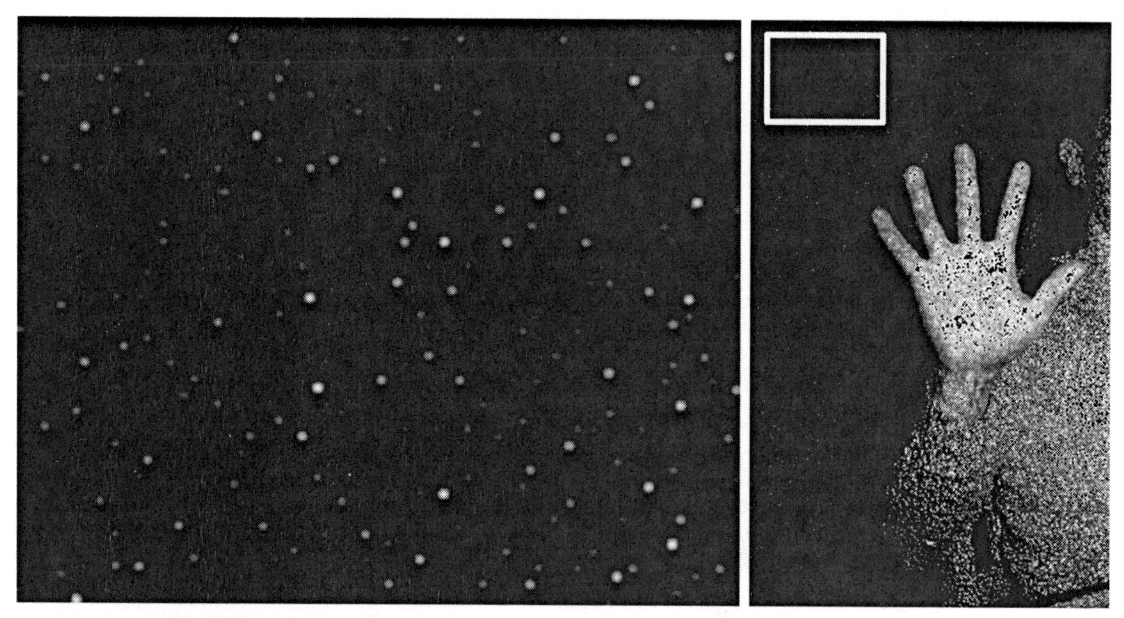

Figure 2-7. *Kinnect IR coding image (detail)*

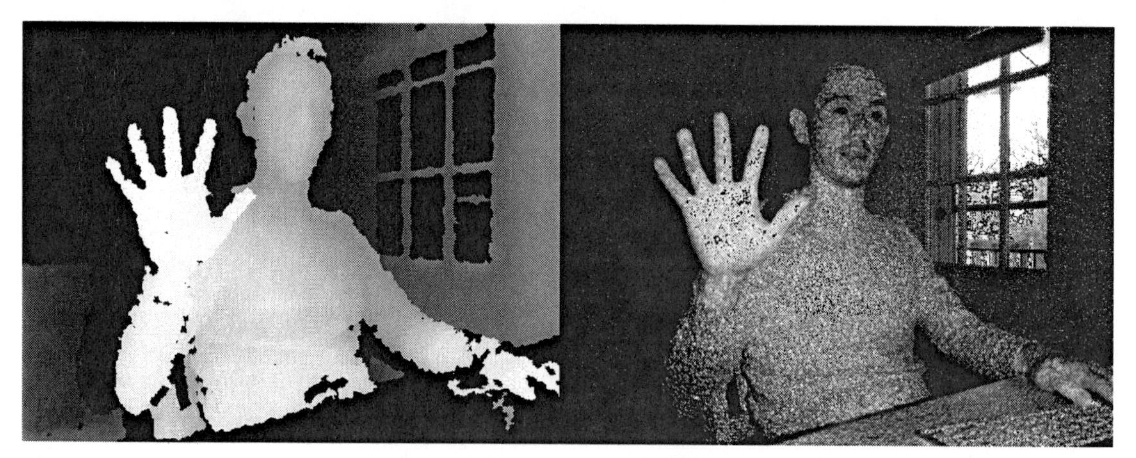

Figure 2-8. *Depth map (left) reconstituted from the light coding infrared pattern (right)*

Converting the Light Coding Image to a Depth Map

Once the light coding infrared pattern is received, PrimeSense's PS1080 chip (Figure 2-9) compares that image to a reference image stored in the chip's memory as the result of a calibration routine performed on each device during the production process. The comparison of the "flat" reference image and the incoming infrared pattern is translated by the chip into a VGA-sized depth image of the scene that you can access through the OpenNI API.

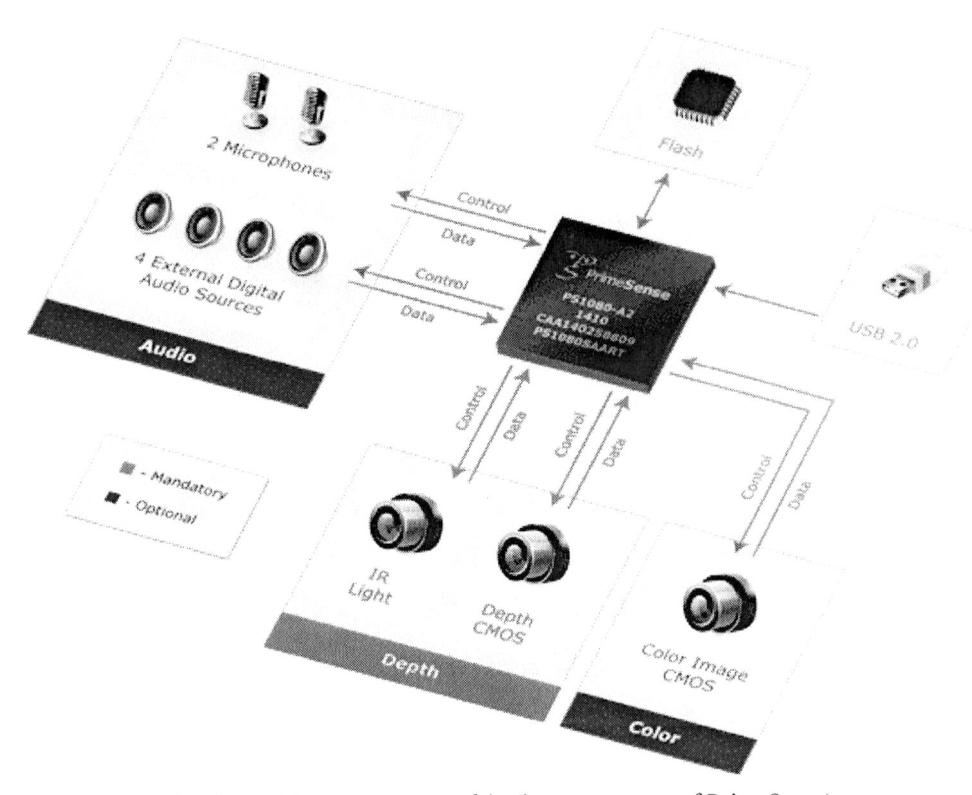

Figure 2-9. PrimeSense PS1080 system on chip (image courtesy of PrimeSense)

Kinect Alternative: ASUS Xtion PRO

After PrimeSense released the OpenNI framework, ASUS and PrimeSense announced their intention to release a PC-compatible device similar to Kinect. In 2012, ASUS revealed the Xtion PRO, an exact implementation of PrimeSense's reference design that only features a depth camera. It was followed by the Xtion PRO LIVE that, like Kinect, includes an RGB camera well as the infrared camera (Figure 2-10). ASUS claims its device is "the world's first and exclusive professional PC motion-sensing software development solution" because the Xtion is designed to be used with a PC (unlike the Kinect, which was initially designed for the Xbox 360). ASUS is also creating an online store for Xtion applications where developers will be able to sell their software to users.

■ **Note** ASUS Xtion is OpenNI- and NITE-compatible, which means that all the projects in this book can also be implemented using an Xtion PRO LIVE camera from ASUS!

Figure 2-10. *ASUS Xtion PRO LIVE (image courtesy of ASUS)*

Summary

This chapter provided an overview of the Kinect device: its history and capabilities, as well as its hardware, software, and the technical details behind its 3D scanning. It was not the intention of the authors to give you a detailed introduction to all of the technical aspects of the Kinect device; we just wanted to get you acquainted with the amazing sensor that will allow you to build all the projects in this book—and the ones that you will imagine afterwards.

In the next chapter, you will learn how to install all the software you need to use your Kinect. Then you will implement your first Kinect program. Now *you are the controller!*

CHAPTER 3

Processing

by Enrique Ramos

This book is about making projects with Kinect and Arduino, and one of the most direct ways of interfacing them is through Processing, a Java-based, open source programming language and IDE. You should be familiar with Arduino by now, so you will be glad to know that the Processing IDE much resembles Arduino's. In fact, the Arduino IDE was based on the Processing IDE (see Figure 3-1).

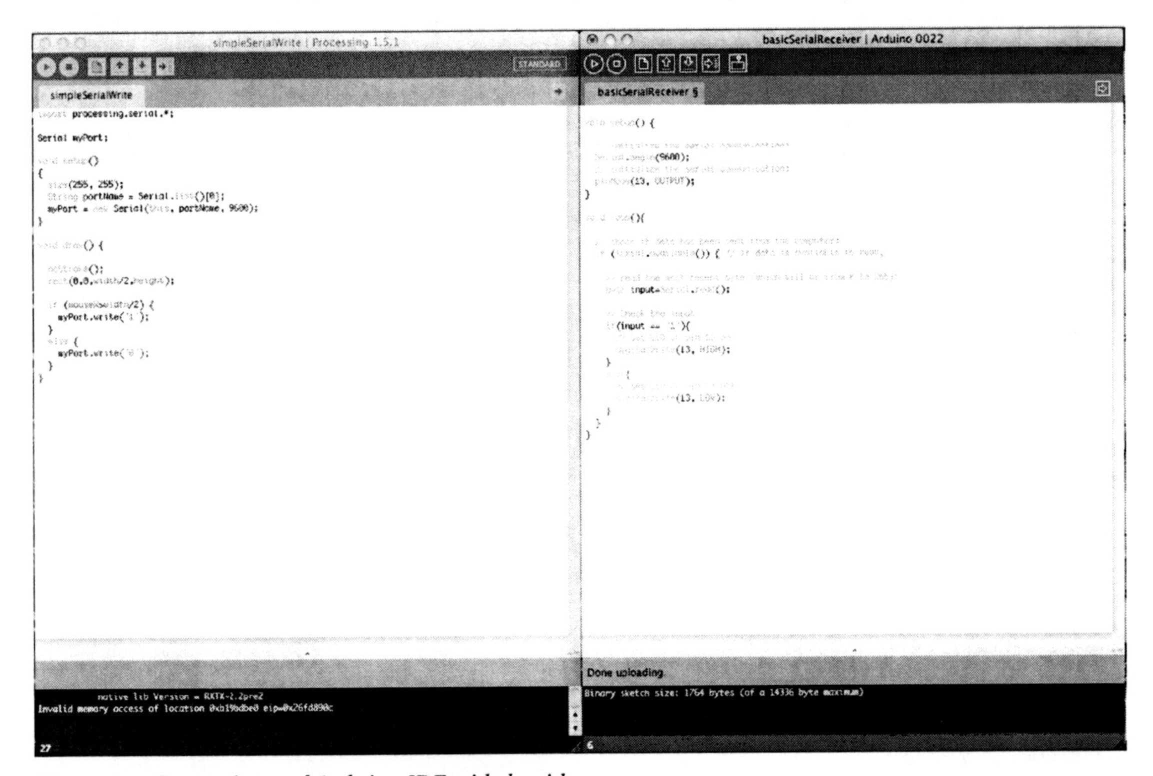

Figure 3-1. Processing and Arduino IDEs side by side

One of the best features of Processing is that, being built on Java, there is a vast community of people creating Processing libraries that you can use for your projects without having to worry (too much) about the scary, dark places hidden behind every class and method. Kinect is no exception to this, and thus Processing is able to talk to Kinect devices through several libraries available from its web site. Processing is also capable of talking to Arduino using serial communication.

This chapter will get you acquainted with the basics of the Processing language and Simple-OpenNI, one of Processing's Kinect libraries. Throughout the rest of the book, you will be creating Kinect and Arduino projects that talk through Processing and Simple-OpenNI, so the understanding of these techniques is essential to the development of any of the upcoming projects.

Processing Language

Casey Reas and Ben Fry, both formerly of the Aesthetics and Computation Group at the MIT Media Lab, created Processing in 2001 while studying under John Maeda. One of the stated aims of Processing is to act as a tool to get non-programmers started with programming. On the Processing web site (`http://www.processing.org`), you can read this introduction:

Processing is an open source programming language and environment for people who want to create images, animations, and interactions. Initially developed to serve as a software sketchbook and to teach fundamentals of computer programming within a visual context, Processing also has evolved into a tool for generating finished professional work. Today, there are tens of thousands of students, artists, designers, researchers, and hobbyists who use Processing for learning, prototyping, and production.

Processing is based in Java, one of the most widespread programming languages available today. Java is an object-oriented, multi-platform programming language. The code you write in Java is compiled into something called bytecode that is then executed by the Java Virtual Machine living in your computer. This allows the programmer (you) to write software without having to worry about the operating system (OS) it will be run on. As you can guess, this is a huge advantage when you are trying to write software to be used in different machines.

Processing programs are called *sketches* because Processing was first designed as a tool for designers to quickly code ideas that would later be developed in other programming languages. As the project developed, though, Processing expanded its capabilities, and it is now used for many complex projects as well as a sketching tool.

We won't have space in this book to discuss the ins and outs of the Processing language, but if you are familiar with Java, C++, or C#, you will have no problem following the code in this book. For a thorough introduction to Processing, I strongly recommend Ira Greenberg's monumental *Processing: Creative Coding and Computational Art.*

Installing Processing

Installing Processing is definitely not much of a hassle. Go to `http://www.processing.org` and download the appropriate version for your OS of choice. If you are a Windows user and you aren't sure if you should download the Windows or Windows without Java version, go for the Windows version. The other version will require you to install the Java Development Kit (JDK) separately. Once you have downloaded the .zip file, uncompress it; you should find a runnable file. If you run it, you should get a window like the one depicted in Figure 3-2.

Processing will work on any OS from wherever you store it. If you are a Mac OS X user, you might want to put it in your Applications folder. On Windows, Program Files would seem like a good place for a new program to live. On Linux, users tend to store their programs in `/usr/bin` or `/usr/local`, but you can create your own folder.

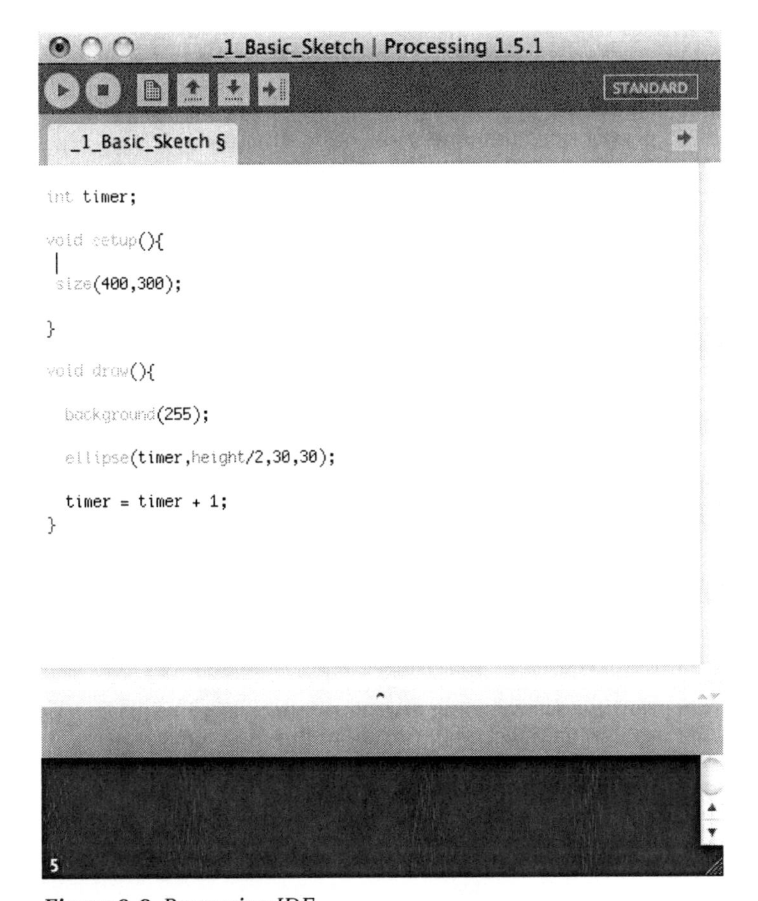

Figure 3-2. *Processing IDE*

Processing IDE

As mentioned, the Processing IDE is very similar to the Arduino IDE described in Chapter 2. There is a console at the bottom of the Processing IDE; a message area that will warn you about errors in the code and at runtime; a text editor where you will write your code; plus a tab manager and a toolbar at the top of the IDE.

The Processing toolbar has a distribution of buttons similar to Arduino, yet with some differences (see Figure 3-2). The New, Open, and Save buttons play the same roles as in Arduino. But instead of having Verify and Upload buttons at the left, there are Run and Stop buttons. You might have guessed what they are for, but just in case, the Run button will run your application and the Stop button will halt it. There is another button called Export PApplet; pressing it will convert the Processing code into Java code and then compile it as a Java Applet, which can be embedded in a web browser. This is the way you'll find Processing Applets, or PApplets, displayed on the Internet.

The Standard button on the right side of the toolbar is used to change the mode from Standard to Android (this is for Android Application development, and we are not going to get anywhere near Android Apps in this book, so you can forget about it.)

A First Processing Sketch

Processing was designed for simplicity, and it allows you to create runnable programs with just a few lines of code. Your first Processing sketch will get you acquainted with the basic structure on a Processing program. But first, let's get into some general programming concepts.

Processing Variables

Variables are symbolic names used to store information in a program. There are eight primitive data types in Java, and they are all supported in Processing: `byte`, `short`, `int`, `long`, `float`, `double`, `boolean`, and `char`. A detailed explanation of each one of these data types would be beyond the scope of this book, but we will state some general rules.

Integers (`int`) will be used to store positive and negative whole numbers (this means numbers without a decimal point, like 42, 0, and -56. An integer variable can store values ranging from -2,147,483,648 to 2,147,483,647 (inclusive). If you need to store larger numbers, you need to define your variable as a `long`.

Whenever you need more precision, you will be using floating-point numbers, or `floats`, which are numbers with a decimal place (23.12, 0.567, -234.63). The data type `double` works the same way as `float`, but it is more precise. In general, Processing encourages the use of `floats` over `doubles` because of the savings in memory and computational time.

The data type `boolean` stores two values: true and false. It is common to use `booleans` in control statements to determine the flow of the program.

The character data type, or `char`, stores typographic symbols (a, U, $). You will use this data type when you want to display or work with text.

Even though `String` is not a proper Java primitive, we are going to discuss it here, as you will be using it to store sequences of characters. `String` is actually a class (more about this in Chapter 4), with methods for examining individual characters, comparing and searching strings, and a series of other tools that you will use to work with text and symbols in your programs.

Variable Scope

Whenever you declare a variable, you are implicitly setting a scope, or realm of validity for the variable. If you declare the variable within a function, you will be able to access it only from within that specific function. If you define the variable outside of any function, the variable will be set as a global variable and every function in the program will be allowed to "see" and interact with the new variable.

Structure of a Processing Sketch

Open your Processing IDE and declare a `timer` integer variable outside of any function. As mentioned, variables defined like this are treated as global variables in Processing, so you can call them from any method and subclass.

```
int timer;
```

setup() Function

Next, include a `setup()` function. This function is called once in the lifetime of a program (unless you call it explicitly from another function). For this project, you are only going to specify the size of your sketch within this function.

```
void setup(){
  size(800,600);
}
```

You can include comments in your code as a means to remind yourself—or inform anybody else reading your code—about specific details of the code. There are two main types of comments: single line comments, which start with two forward slash characters, and multi-line comments, used for large descriptions of the code. You can include some comments in your `setup()` function in the following way:

```
void setup(){
  // The next function sets the size of the sketch
  size(800,600);
  /*
    This is a multiline comment, anything written between the two multiline commment
    delimiters will be ignored by the compiler.
  */
}
```

There is a third kind of comment, the doc comment (/** comment */), which can be used to automatically generate Javadocs for your code. You won't be using this sort of comment in this book, but they can be extremely useful when writing programs intended to be used by other people.

draw() Function

Next, you need to include a `draw()` function that will run as a loop until the program is terminated by the user. You are going to draw a circle on screen that will move to the right of the screen as time passes. Remember to include a `background()` function to clear the frame with a color at every iteration. If you comment it out (temporarily remove it by adding // to the beginning of the line), you will see that the circles are drawn on top of the previous ones.

```
void draw() {
  background(255);
  ellipse(timer, height/2, 30, 30);
  timer = timer + 1;
}
```

Now you are ready to run your sketch. Press the Run button at the top-left of your Processing IDE (marked with a red arrow in Figure 3-3) and you will see a new window open with a white circle running to the right side.

Well, that was your first Processing sketch! I take it you are much impressed. You're going to rush into a much deeper discussion on vector classes and 3D processing in the next few pages, so if you are completely new to Processing, you should have one of the introductory books to Processing at hand.

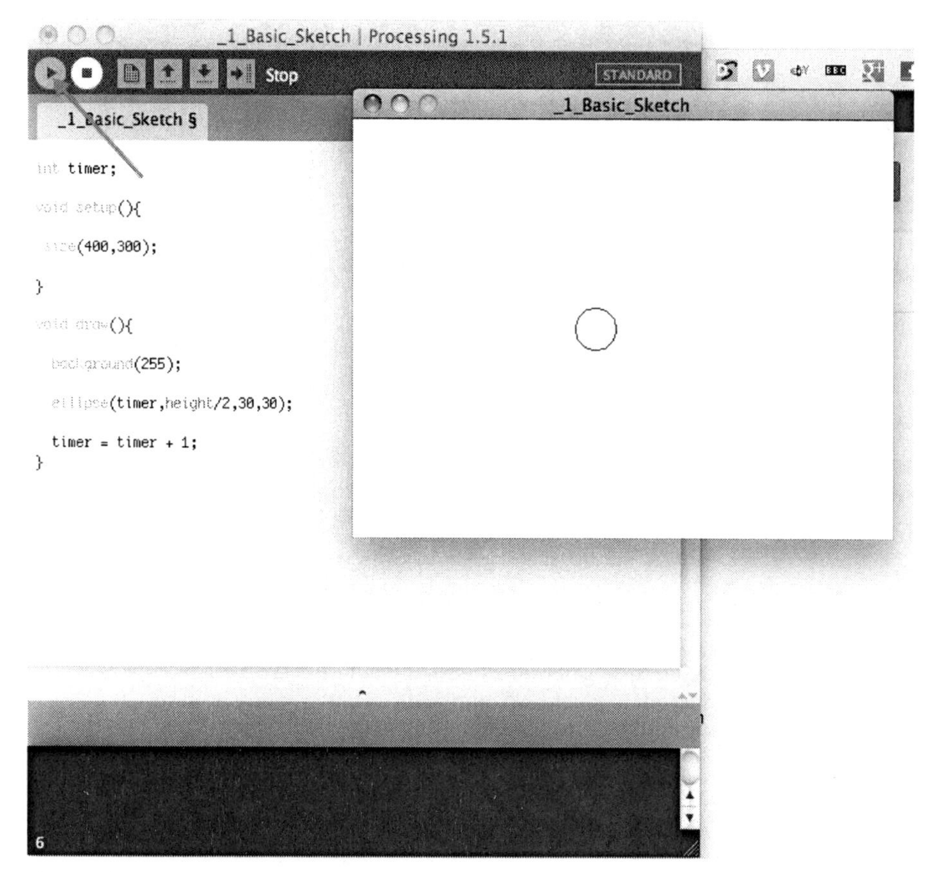

Figure 3-3. Your first Processing sketch

Processing Libraries

Processing libraries are lines of Java code that have been packaged into a `.jar` file and placed in the libraries folder in your system so you can call its functions from any Processing sketch. There are two sorts of libraries: core libraries, like OpenGL and Serial, which are included with Processing when you download it, and contributed libraries, which are created and maintained by members of the Processing community. Contributed libraries need to be downloaded and stored in the libraries folder in your system. If you are a Mac OS X user and you have saved the Processing application in the Applications folder, the path will be the following:

```
/Applications/Processing/Contents/Resources/Java/modes/java/libraries
```

To open your Processing application on Mac OS X (it is essentially a folder, so to find the libraries folder), right-click the icon and select "Show Package Contents" from the pop-up menu.

If you are working on a Windows machine and you have put the Processing folder in the Program Files folder, this will be your path:

```
C:\Program Files\processing-1.5.1\modes\java\libraries
```

If you are a Linux user, wherever you save your Processing folder, the path to the libraries folder will be the following:

```
processing-1.5.1/modes/java/libraries
```

At the time of writing, there are three Kinect libraries available for Processing (shown in Table 3-1). Two of them are OS-specific, and one works on Mac OS X, Linux, and Windows. They also differ in the functions they implement.

Table 1-1. *Kinect Libraries*

Library	Author	Based on	OS
Openkinect	Daniel Shiffman	OpenKinect/Libfreenect	Mac OS X
dLibs	Thomas Diewald	OpenKinect/Libfreenect	Windows
simple-openni	Max Rheiner	OpenNI/NITE	Mac OS X, Windows, Linux

The previous chapter covered how Kinect was developed and why there are several ways to access the Kinect data stream depending on the drivers and middleware you choose. If you look at the list of Kinect libraries for Processing in Table 3-1, you will realize that the two of them are based on OpenKinect and available only for one operating system. In this book, you are going to focus on Max Rheiner's Simple-OpenNI because it is multi-platform and built upon Primesense's OpenNI/NITE middleware, which means you can perform skeletonizing and hand recognition off the shelf. If you are doing projects where you only need the point cloud, you can use one of the other two libraries, which, as an advantage, provide access to the motor built inside your Kinect device.

Simple-OpenNI

Simple-OpenNI is a NITE and OpenNI wrapper for Processing implemented by Max Rheiner, a media artist and software developer currently lecturing at Zurich University.

What does this mean? Well, NITE and OpenNI are implemented in C++, and Rheiner has wrapped all the functions and methods in such a way that they are accessible from Java; then he wrote a library to make the Java code available from Processing. This allows us, Processing lovers, to have all the functionality of OpenNI and NITE available from our favorite programming language.

Installing Simple-OpenNI

You learned how to install a Processing library in the previous section, but there are some specific prerequisites for this library to work on your computer. After you have installed the prerequisites described next, you can go to the download web site at `http://code.google.com/p/simple-openni/downloads/list` to download the library and install it in your Processing libraries folder.

Installing Simple-OpenNI requires the previous installation of OpenNI and NITE on your machine. There is a very detailed description on how to do this on the Simple-OpenNI web site at `http://code.google.com/p/simple-openni/`, which we have reproduced here for you to follow.

Installation on Windows

Follow these steps if you are running Simple-OpenNI on a Windows machine:

- Download and install OpenNI.

http://www.openni.org/downloadfiles/opennimodules/openni-binaries/20-latest-unstable

- Download and install NITE.

http://www.openni.org/downloadfiles/opennimodules/openni-compliant-middleware-binaries/33-latest-unstable

The installer will ask you for the key; use this one:
0KOIk2JeIBYClPWVnMoRKn5cdY4=

Install the Primesense driver (Asus Xtion Pro, etc.).

- Download and install Primesensor.

http://www.openni.org/downloadfiles/opennimodules/openni-compliant-hardware-binaries/31-latest-unstable

Install the Kinect driver.

- Download SensorKinect, unzip it, and open the file at /avin2/SensorKit/SensorKinect-Win-OpenSource32-?.?.?.msi

https://github.com/avin2/SensorKinect

If everything worked out, you should see the camera in your Device Manager (under Primesensor: Kinect Camera, Kinect Motor). You can now download and install the Simple-OpenNI Processing library.

Installation in Mac OS X

If you are working on Mac OSX, this is your lucky day! Max Rheiner made an installer that goes through the whole installation process automatically. This should at least work with Mac OS X 10.6.

Install OpenNI the Short Way (OpenNI v1.1.0.41 and NITE v1.3.1.5)

- Download the Installer.

http://code.google.com/p/simple-openni/downloads/detail?name=OpenNI_NITE_Installer-OSX-0.20.zip&can=2&q=#makechanges

- Unzip the file and open a terminal window.
- Go to the unzipped folder, OpenNI_NITE_Installer-OSX.

```
> cd ./OpenNI_NITE_Installer-OSX
```

- Start the installer.

```
> sudo ./install.sh
```

If this installation didn't work for you, try the following, longer method.

Install OpenNI the Official Way

Those programs have to be on your system in order to install OpenNI:

- Xcode

http://developer.apple.com/xcode/

- MacPorts

http://distfiles.macports.org/MacPorts/MacPorts-1.9.2-10.6-SnowLeopard.dmg

- Java JDK (This will be already installed after you installed Processing)

When you have finished installing these programs, you can install OpenNI. Open the terminal and type in these commands:

```
> sudo port install git-core
> sudo port install libtool
> sudo port install libusb-devel +universal
> mkdir ~/Development/Kinect
> cd Kinect
> tar xvf OpenNI-Bin-MacOSX-v1.0.0.25.tar.bz2
> sudo ./install.sh
> mkdir ~/Development/Kinect
> cd ~/Development/Kinect
> mkdir OpenNI
> mkdir Nite
> cd OpenNI
> tar xvf OpenNI-Bin-MacOSX-v1.0.0.25.tar.bz2
> sudo ./install.sh
```

You can now download and install the Simple-OpenNI Processing library.

Installation on Linux

If you are a Linux user, these are the steps to install Simple-OpenNI. First, download the install files (stable or unstable, 32-bit or 64-bit):

- OpenNI

http://openni.org/downloadfiles/2-openni-binaries

- Nite

http://openni.org/downloadfiles/12-openni-compliant-middleware-binaries

- Kinect drivers

https://github.com/avin2/SensorKinect

- Primesense drivers (Asus Xtion Pro, etc.)

http://openni.org/downloadfiles/30-openni-compliant-hardware-binaries

- Uncompress all files to a folder.
- Install OpenNI.

```
> cd ./yourFolder/OpenNI-Bin-Linux64-v1.1.0.41
> sudo install
```

- Install NITE (the installer asks you for the key; enter the following: **0KOIk2JeIBYClPWVnMoRKn5cdY4=**)

```
> cd ./yourFolder/Nite-1.3.1.5
> sudo install
```

- Install the Kinect driver.

  Decompress ./yourFolder/avin2-SensorKinect-28738dc/SensorKinect-Bin-Linux64-v5.0.1.32.tar.bz2

```
> cd ./yourFolder/SensorKinect-Bin-Linux64-v5.0.1.32
> sudo install
```

- Install the Primesense driver (only if you use an Asus Xtion or the Primesense Developer Kit).

```
> cd ./yourFolder/Sensor-Bin-Linux64-v5.0.1.32
> sudo install
```

You can now download and install the Simple-OpenNI Processing library.

Accessing the Depth Map and RGB Image

Now that you have installed the Simple-OpenNI Processing library, let's try a very simple example that will access the data streams from Kinect from Processing.

First, import the Simple-OpenNI library, and declare the variable kinect that will contain the Simple-OpenNI object, like so:

```
import SimpleOpenNI.*;
SimpleOpenNI kinect;
```

Within setup(), initialize the kinect object, passing this (your PApplet) as an argument. As you saw in Chapter 2, the Kinect has a standard RGB camera and an infrared camera on board, used in combination with an infrared projector to generate the depth image for 3D scanning. Enable the depth map and RGB images in your kinect object and the mirroring capability that will mirror the Kinect data so it's easier for you to relate to your image on screen. (Try the same sketch without enabling mirroring, and you will see the difference.) Then, set your sketch size to the total size of the RGB and depth images placed side by side, so you can fit both in your frame.

```
void setup() {
  kinect = new SimpleOpenNI(this);
  // enable depthMap and RGB image
  kinect.enableDepth();
  kinect.enableRGB();
  // enable mirror
  kinect.setMirror(true);

  size(kinect.depthWidth()+kinect.rgbWidth(), kinect.depthHeight());
}
```

The first thing you need to do in the draw loop is to update the kinect object so you will get the latest data from the Kinect device. Next, display the depth image and the RGB images on screen, as in Figure 3-4.

```
void draw() {
  kinect.update();
  // draw depthImageMap and RGB images
  image(kinect.depthImage(), 0, 0);
  image(kinect.rgbImage(),kinect.depthWidth(),0);

}
```

Figure 3-4. Depth map and RGB image

Alright, so in 14 lines of code you are accessing the RGB and depth images of the Kinect. If you want to access the infrared image, you can use `enableIR` instead of `enableRGB`, and then `kinect.irWidth()` instead of `kinect.rgbWidth()`, and `kinect.irImage()` for `kinect.rgbImage()`.

■ **Note** You can't capture an RGB image and an IR image at the same time with the current firmware. You can, however, capture the depth map and RGB or the depth map and IR.

Both images are 640x480 pixels. Well, that's not very impressive for the RGB camera, just an average low-quality webcam. The magic comes from the depth map. You are actually getting a 640x480 pixel three-dimensional scan in real time (30fps). The levels of grey you see in Figure 3-4 represent the distances to the Kinect camera. So from this image, you are getting a two-dimensional image in which every pixel has embedded information of its distance to the camera. Could you reconstruct the spatial volume in the image? Yes, you can, and you're going to do it. But let's first make clear what we're talking about when we talk about 3D.

The Third Dimension

This might sound less intriguing than talking about the fourth or the fifth dimension, but it is still a subject that provokes many a headache to anybody getting into the world of computer graphics, so let's take a close look at what we mean when we talk about 3D.

We all live in a three-dimensional world, though our screens and human interfaces have traditionally been two-dimensional. With the advent of real-time 3D scanners like the Kinect, this limitation is somewhat left behind. We can interact with the computer using the three dimensions of physical space (or at least the three we are aware of, if you ask some physicists!).

Our screens are still 2D, though, so representing the three dimensions of space on the display requires some nasty math dealing with perspective and projections—which luckily we don't have to work out by ourselves because somebody did for us already. Processing has full 3D support and two different 3D renderers, P3D and OpenGL. You're going to focus on OpenGL because it is the more powerful of the two.

Processing in 3D

Let's do a quick example of a Processing 3D sketch. There are two 3D primitives already implemented in Processing, `box()` and `sphere()`. You're going to draw a three-dimensional box on screen.

For this, you need to import the OpenGL Processing core library into your sketch so you can set your renderer to `OPENGL` in the `size()` function within `setup()`. Then, in the `draw()` loop, set the background to white, move to the center of the screen, and draw a cube of 200 units each side (see Figure 3-5).

```
import processing.opengl.*;
void setup()
{
  size(800, 600, OPENGL);
}

void draw()
{
  background(255);
  noFill();
  translate(width/2,height/2);
  box(200);
}
```

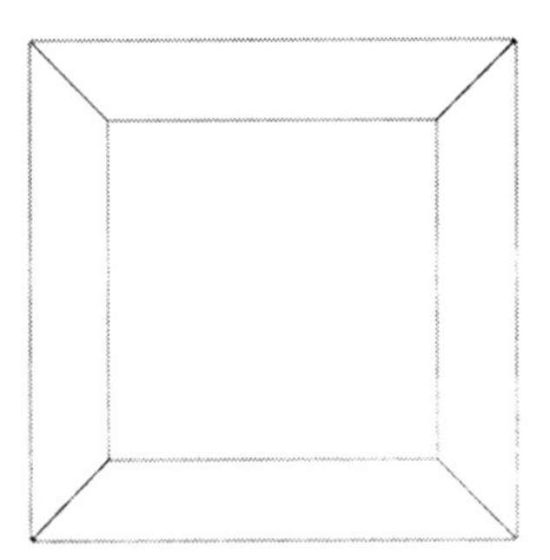

Figure 3-5. A cube in Processing

The function `translate()` that you have used in the previous example makes part of the Processing transform functions. These functions have in common the fact that they affect the current transformation matrix.

Matrix Transforms

The transformation matrix is the frame of reference that Processing uses to print objects on screen. If you don't alter the transformation matrix, the origin of coordinates (the zero point to which all the coordinates refer) will be placed at the top-left corner of your screen, the x-axis pointing right, and the y-axis pointing down.

If you draw a rectangle on screen, you can specify the x and y coordinates of its center in the parameters because the function is `rect(x, y, width, height)`. But if you want to draw a box, it has no position parameters, being just `box(width, height, depth)` or even just `box(size)`.

So if you want to draw a box at a different position than the global origin of coordinates, you need to change the frame of reference to the place and orientation that you want the box to be in, so that when you draw the box, it will appear at the desired position. This will be performed with matrix transforms.

Look at the following code and see what it does on screen (shown in Figure 3-6):

```
import processing.opengl.*;
void setup()
{
  size(800, 600, OPENGL);
```

```
}

void draw()
{
  background(255);
  noFill();
  translate(mouseX,mouseY);
  rotate(45);
  box(200);
}
```

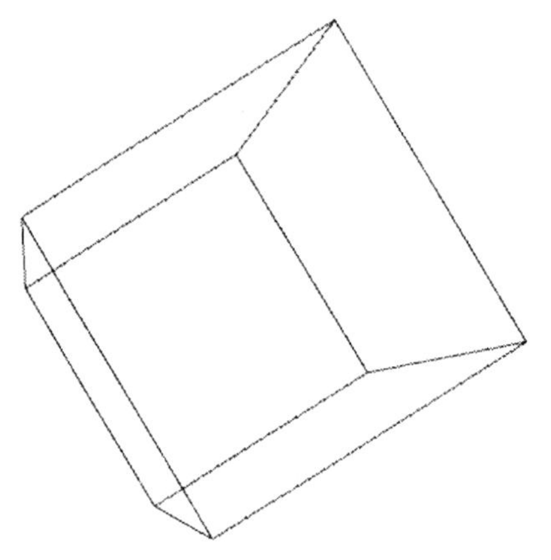

Figure 3-6. *Controlling a box with the mouse position*

If you want to modify the transformation matrix for some operations and then return it back to where it was, you can use the methods `pushMatrix()` and `popMatrix()`. The first method will push the current matrix onto something called the *matrix stack*, from where you can recover it using `popMatrix()`. You don't need to worry too much about the matrix stack at the moment; you just need to understand that you can save the current matrix with `pushMatrix()`, do some transformations and draw some objects, and then restore the previous matrix with `popMatrix()`, so you can continue drawing objects within the previous frame of reference. This is illustrated by the following code and shown in Figure 3-7:

```
import processing.opengl.*;
void setup()
{
  size(800, 600, OPENGL);
}

void draw()
{
  background(255);
  noFill();

  pushMatrix();
  translate(mouseX,mouseY);
  rotate(45);
```

```
  box(200);
  translate(300,0);
  box(50);
  popMatrix();

  translate(200,200);
  box(100);
}
```

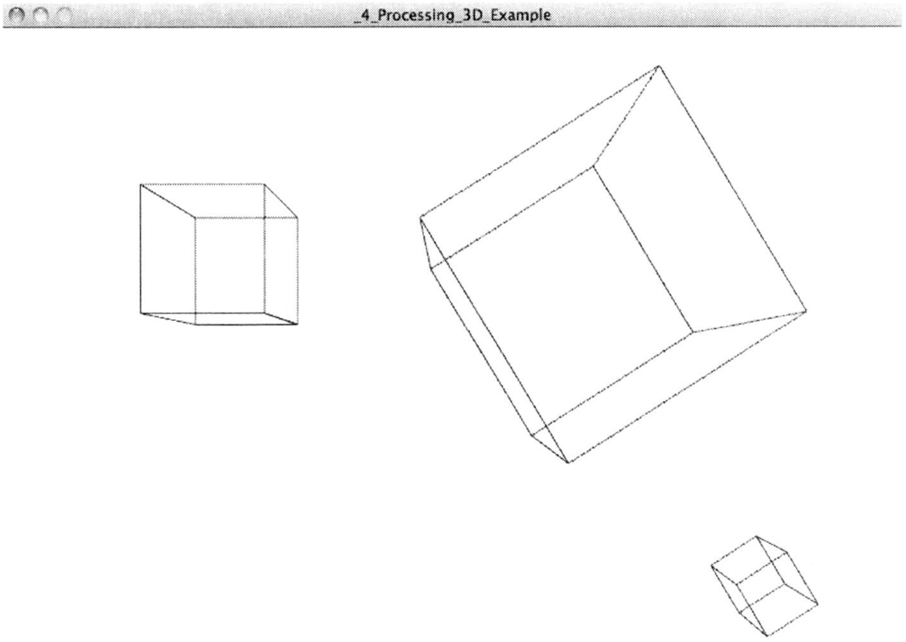

Figure 3-7. Three boxes drawn on different reference systems

If you play with the previous applet for a while, you'll realize that the two larger cubes follow the mouse and are rotated 45 degrees, while the intermediate cube remains static in its place. Analyzing the code, you will understand why. First, you pushed the matrix, and then you translated it to the mouse coordinates, rotated it by 45 degrees, and drew the largest box. Then you translated the matrix again on the x-axis, but because the matrix was already translated and rotated, it resulted in a diagonal movement, so the second cube is drawn rotated at the bottom right of the first.

Then, you popped the matrix, so you came back to the original reference system; you then translated the restored matrix by a certain distance in x and y, and drew the third box, which will not move at all with the mouse. The last transform doesn't accumulate because every time you start a new `draw()` loop, the matrix gets reset.

For now, you will be drawing three-dimensional geometry on screen and looking at it from one point of view. But the best part of 3D geometry is that it can be seen from any point of view in space. This is what you are going to explore in the next section.

Camera Control Libraries

In order to visualize 3D space, all CAD packages make use of mouse-driven camera control capabilities, so you can orbit around your objects and get a better understanding of the geometry. As you're dealing with 3D data acquisition with the Kinect, you are going to include a camera control library into your 3D sketches. At the time of writing, there are several camera control libraries on the Processing web site.

- Peasycam by Jonathan Feinberg
- Obsessive Camera Detection (OCD) by Kristian Linn Damkjer
- Proscene by Jean Pierre Charalambos

In this book, you are going to use the KinectOrbit library for your examples. This library is quite simple and somewhat rudimentary, but it is pretty straightforward to use. Also, it incorporates the possibility of saving viewpoints, so you don't need to be orbiting around every time you open a sketch. Furthermore, it keeps the camera roll to zero to avoid the Kinect point cloud appearing upside-down or strangely rotated on screen. The camera in the library is especially set up to work with the Kinect point cloud, so you won't need to worry about clipping planes, unit scales, direction of the axes, and other issues that can be off-putting when beginning to work in 3D.

You can download the KinectOrbit library from the book's web site at `http://www.arduinoandkinectprojects.com/kinectorbit`. If you have a preference for another camera control library, or want to implement your own, feel free to substitute it for KinectOrbit in every project using it.

KinectOrbit Example

Let's do a quick test with this library. After installing the KinectOrbit library following the directions in the "Processing Libraries" section, open a new sketch and import the KinectOrbit and OpenGL libraries. Then declare a `KinectOrbit` object and call it `myOrbit`.

```
import processing.opengl.*;
import kinectOrbit.*;

KinectOrbit myOrbit;
```

Within `setup()`, specify the size of the sketch and the 3D renderer, and initialize the `myOrbit` object.

```
void setup()
{
  size(800, 600, OPENGL);
  myOrbit = new KinectOrbit(this);
}
```

Within the `draw()` method, set a background color, push the `myOrbit` object, draw a box at the origin of the coordinate system with a side length of 500 units, and pop the orbit. After this, draw a square on screen using the Processing function `rect()` for comparison.

```
void draw()
{
  background(255);
  myOrbit.pushOrbit(this);
  box(500);
  myOrbit.popOrbit(this);
  rect(10,10,50,50);
}
```

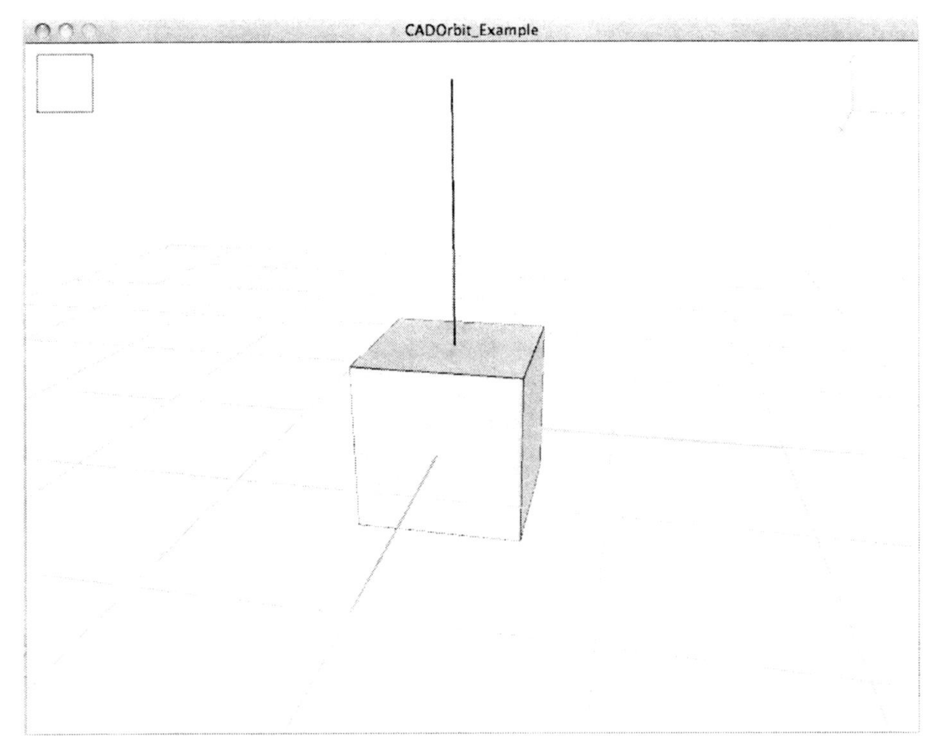

***Figure 3-8.** 3D View Controlled by KinectOrbit*

Run the sketch and observe that the box appears to be in perspective from a defined viewpoint, while the rectangle has been drawn in screen coordinates (see Figure 3-8). The library is actually designed to work in a very similar way as the `pushMatrix()` and `popMatrix()` functions introduced earlier in this chapter. This means that your `KinectOrbit` object will allow you to control the perspective from which you are seeing everything drawn inside the `pushOrbit/popOrbit` loop, while anything drawn outside that loop will be drawn from the default coordinate system. This is why the rectangle is drawn on screen as if no orbit library were used.

You can control the point of view by using the mouse. The controls are the following:

- **Right button drag**: Pan the camera.
- **Left Button drag:** Orbit around the objects.
- **Wheel:** Zoom in and out.

If you like a view in particular, you can save it by pressing the P key, and then retrieve it by pressing the O key. When you start the sketch again, it will start in the latest viewpoint that you saved. The viewpoint parameters will be stored in a file named `orbitSet_0.csv` in your data folder. If you delete this file at some point, it will be created again when you run the sketch and set to the default values.

You can also control the gizmos and visualization style of your orbit object. For more information, refer to the Javadoc of KinectOrbit on the web site. Now let's do a test with the Kinect in 3D.

Kinect Space

You have already accessed Kinect's depth map in the "Accessing the Depth Map and RGB Image" section and wondered if you can get that map translated back into the real-scale 3D coordinates of the objects. The following sketch does that for you. It starts pretty much like the depth map example but includes a function that shows the real coordinates of the points. First, import the appropriate libraries, declare Simple-OpenNI and KinectOrbit objects, and initialize them in the setup() function, enabling the depth map only.

```
import processing.opengl.*;
import SimpleOpenNI.*;
import kinectOrbit.*;

KinectOrbit myOrbit;
SimpleOpenNI kinect;

void setup()
{
  size(800, 600, OPENGL);
  myOrbit = new KinectOrbit(this, 0);
  kinect = new SimpleOpenNI(this);
  // enable depthMap generation
  kinect.enableDepth();
}
```

In the draw() function, include a pushOrbit/popOrbit loop; inside it, add the drawPointCloud() function and draw the Kinect frustum with the method drawCamFrustum(), already included in SimpleOpenNI.

■ **Note** In computer graphics, the term "frustum" describes the three-dimensional region visible on the screen. The Kinect frustum is actually the region of space that the Kinect can see. The Simple-OpenNI method drawCamFrustum() draws this region on screen, so you know exactly what portion of the space is being tracked. This method also draws a three-dimensional representation of the Kinect device for reference.

```
void draw()
{

  kinect.update();
  background(0);

  myOrbit.pushOrbit(this);
  drawPointCloud();
  // draw the kinect cam and frustum
  kinect.drawCamFrustum();

  myOrbit.popOrbit(this);
}
```

Now, you need to implement the function drawPointCloud(), which draws each pixel from the depth map in its real-space coordinates. You first store the depth map into an array of integers. You realize that it is not a bi-dimensional array, but a linear array; Processing images are linear arrays of pixels. If you want to access a pixel with coordinates x and y, use the following code:

```
index = x + y * imageWidth;
```

You are going to represent only a number of pixels in the array because the full 640x480 image gets pretty heavy for your processor when represented as 3D points. The integer steps is going to be used as a step size within the for() loops in order to skip some of the pixels.

```
void drawPointCloud() {
  // draw the 3d point depth map
  int[]   depthMap = kinect.depthMap();
  int     steps   = 3;  // to speed up the drawing, draw every third point
  int     index;
```

Next, initialize a PVector that will contain the x, y, and z coordinates of the point you want to draw on screen and then run through the depth image pixels using two nested for loops.

Note A PVector is a Processing implementation of a vector class. A vector is a set of three values representing the x, y, and z coordinates of a point in space. Vectors are used extensively in mathematics and physics to represent positions, velocities, and other more complicated concepts. The Processing PVector class implements several methods that you can use to get the vector's magnitude and coordinates, and to perform vector math operations, like adding and subtracting vectors or dot and cross products.

```
  PVector realWorldPoint;
  stroke(255);
  for (int y=0;y < kinect.depthHeight();y+=steps)
  {
    for (int x=0;x < kinect.depthWidth();x+=steps)
    {
```

If you want to see each point with the depth map color, include the following line:

```
      stroke(kinect.depthImage().get(x,y));
```

Within the loop, you first find out the index associated with the x and y coordinates in the depth map. Then, if the depthMap point with the current index is not 0, call the function depthMapRealWorld()[index] to return the 3D coordinates of the current point, which you display with Processing's point() function. (Note that all dimensions returned are in millimeters.)

```
      index = x + y * kinect.depthWidth();
      if (depthMap[index] > 0)
      {
        realWorldPoint = kinect.depthMapRealWorld()[index];
        point(realWorldPoint.x, realWorldPoint.y, realWorldPoint.z);
      }
    }
  }
}
```

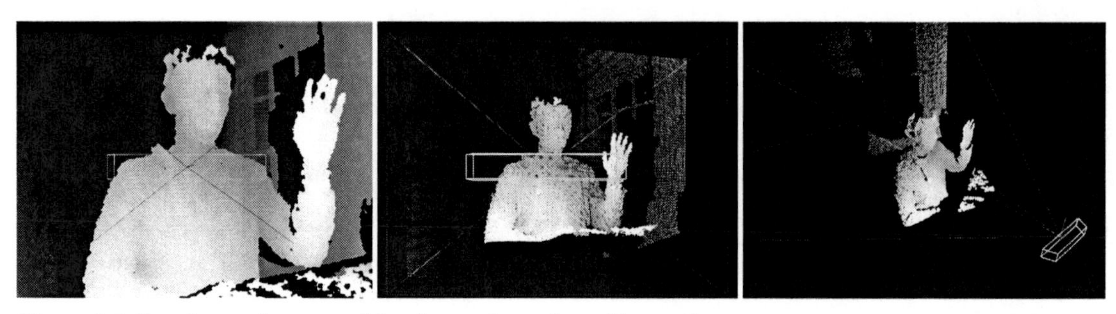

Figure 3-9. *Depth map image and depth map in real-world coordinates*

Run the example and spend some time spinning around the model (see Figure 3-9). Have a look at the Kinect frustum in relation to the Orbit gizmo. The horizontal axis is X (red), the vertical axis is Y (green), and the depth axis is Z (blue), as you can see in Figure 3-10. The Kinect considers itself the origin of coordinates (a little pretentious, I would say), so all coordinates are a function of the Kinect's position in space.

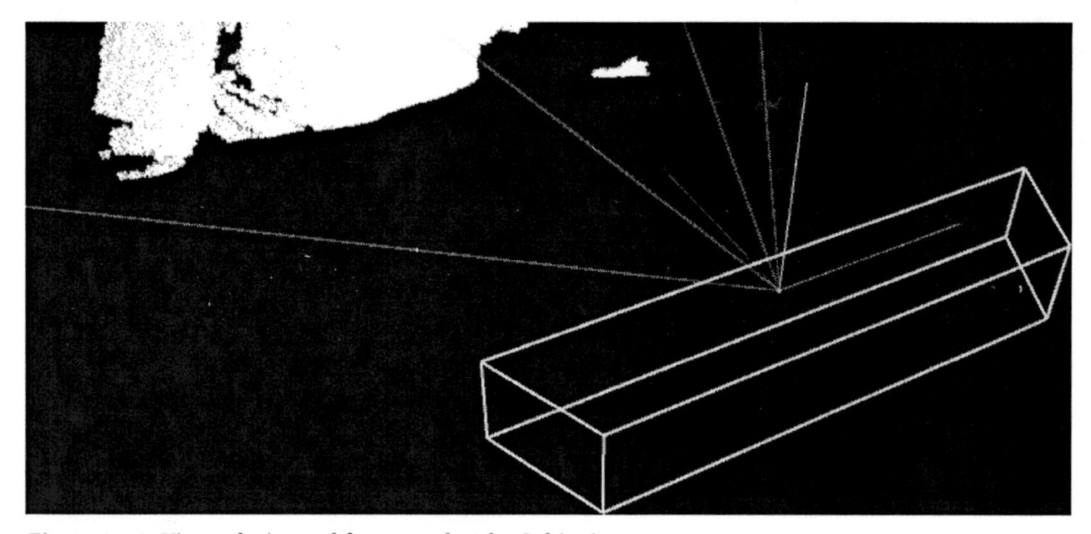

Figure 3-10. *Kinect device and frustum plus the Orbit gizmo*

Linear and Two-Dimensional Arrays

If you're now wondering why you have been going from linear arrays to two-dimensional arrays and back, note that you could have done the deal with the following code:

```
void drawPointCloud () {
  for (int i=0; i<kinect.depthMapRealWorld().length;i+=3) {
    stroke(255);
    point(kinect.depthMapRealWorld()[i].x, kinect.depthMapRealWorld()[i].y,
kinect.depthMapRealWorld()[i].z);
  }
}
```

But if you get closer to your point cloud, you can see that you have lost control of your "pixel dropping." In the previous code, you were drawing one every steps rows of pixels and one every steps columns of pixels, so you were still drawing a consistent grid of points. With the linear array approach, you are drawing one every steps pixels on the linear array representing the image, whichever column and row that pixel is placed in. In this way, you aren't producing a consistent grid. Working with two-dimensional arrays gives you more control over the way you parse and represent your Kinect point cloud, so you'll continue to do it this way.

So you now have a 3D point cloud that represents the real world position of every point scanned by the Kinect. You will be doing many things with this information later in the book. For starters, as you also have an RGB camera installed on your Kinect device, could you associate the color image from the RGB camera to the point cloud to get a 3D picture of what the Kinect sees?

Coloring the Point Cloud

Coloring the point cloud should be as straightforward as setting the color of each point as the color of the equivalent point in the RGB Image. This could be done by enabling the RGB image in the setup() and then changing the stroke color to the color of the RGB image at the same index. You should comment out the following line:

```
//stroke(kinect.depthImage().get(x,y)); This is commented out
```

And you should add this line to the code after setting the value of index:

```
stroke(kinect.rgbImage().pixels[index]);
```

Your points are now colored according to the RGB camera, but to your surprise, the image and the 3D point cloud don't match! This is easy to explain. The RGB camera and the infrared cameras can't occupy the same spot in the Kinect, so they are close to each other but slightly separated. The images they are capturing are thus slightly off. In order to make them match, you need to call a Simple-OpenNI method to align the point cloud with the RGB image. The setup() function looks this:

```
void setup()
{
  size(800, 600, OPENGL);
  myOrbit = new KinectOrbit(this, 0);
  myOrbit.setPanScale(3);
  kinect = new SimpleOpenNI(this);
  // enable depthMap and RGB generation
  kinect.enableDepth();
  kinect.enableRGB();
  // align depth data to image data
  kinect.alternativeViewPointDepthToImage();
}
```

Now you should get a pretty neat point cloud that displays the real colors of every pixel.

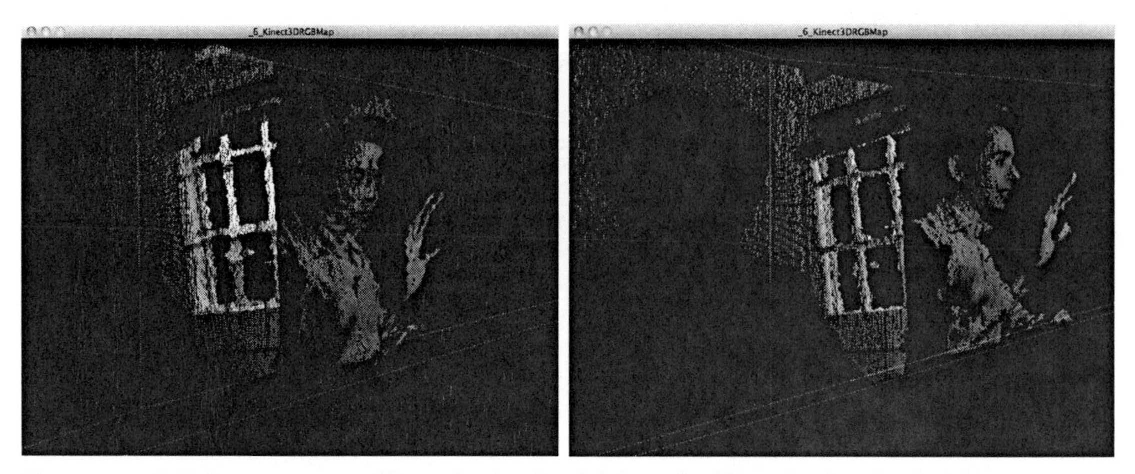

Figure 3-11. RGB image on non-calibrated point cloud (left) and calibrated point cloud (right)

If you do the same thing and render the results on screen, you will realize what's happening. If you don't add this method, the depth map image is different from your RGB image. It appears slightly bigger and displaced (have a look at Figure 3-11). If you add the method, you can observe in Figure 3-12 that the depth map has been modified as to be more similar to the RGB image.

Figure 3-12. Depth map calibrated to fit RGB image

NITE Functions

Up until now, everything you have been doing could have been done with any of the other two libraries based in OpenKinect. But you have seen a lot of cool applications on the Internet in which the programmers make use of hand and skeleton tracking to allow user interaction. Well, these functions are not included with OpenNI itself; they require an external module called NITE, which is not open source but is still free and publicly available, and which includes all the algorithms that Primesense has developed for user interaction.

Luckily, Max Rheiner has also wrapped nearly all of NITE functions within Simple-OpenNI, so you can access them from Processing. Let's see how.

Hand Tracking

Because of our skillful use of our upper limbs, hand tracking is a very useful feature for natural interaction. We can very precisely position our hands in space, and we can perform a vast amount of gestures with them. These abilities can be used to drive parameters or trigger different behaviors in your natural interaction applications. OpenNI/NITE includes a framework for hand tracking and hand gesture recognition that you can use off the shelf. This is a great feature that doesn't exist in the official Microsoft SDK, at least in the beta version.

To take advantage of this, you need to include a series of functions or methods in your main Processing sketch that will be called (the proper term is "invoked," which sounds a little spooky) from the Simple-OpennNI library when certain events are triggered.

■ **Note** The way you interact with OpenNI/NITE events is by making use of reflection, an advanced programming technique that allows you to invoke methods from a superclass. This is similar to the way you deal with key and mouse events in Processing. You include a `void keyPressed()` function that you don't explicitly call from your `draw()` function, but you know it will be called whenever you press a key on the keyboard. The hand tracking methods work the same way, triggered by hands or gestures recognized by NITE middleware.

Let's have a look at these methods and their application. The `onCreateHands()` method is called when a hand is detected and a hand object is initialized. You get an `integer` containing the hand ID, a `PVector` defining its position on recognition, and a `float` with the timestamp value. You can add more code here that will run when a hand is recognized.

```
void onCreateHands(int handId, PVector pos, float time){

}
```

The function `onUpdateHands()` is called every time Kinect updates the hand data. You get a new position vector and time for the hand. Add here the code you want to run when a hand is updated.

```
void onUpdateHands(int handId, PVector pos, float time){

}
```

When a hand is destroyed (when it disappears from screen or is not recognizable any more), the function `onDestroyHands()` will be invoked.

```
void onDestroyHands(int handId, float time){

}
```

NITE also calls two methods related to gesture recognition. The `onRecognizeGesture()` function is called when NITE recognizes a gesture. The string `strGesture` contains the name of the gesture recognized (Wave, Click, or RaiseHand). You also get the start and end positions of the gesture in two three-dimensional `PVectors`.

```
void onRecognizeGesture(String strGesture, PVector idPosition, PVector endPosition){

}
```

Finally, the function onProgressGesture() is triggered while the gesture is in progress but before it has been recognized. It gives you the percentage of completion of the gesture, along with the gesture type and the current position of the hand. This is normally a very short period of time, but it can be useful for some applications.

```
void onProgressGesture(String strGesture, PVector position,float progress){

}
```

You will learn how to work practically with these techniques later in this book. In the next chapter, you will be using hand tracking to drive the brightness of a couple of LED lights; in Chapter 5, you will use it to create a gesture-based remote control. You will even control a robot using hand tracking in the last chapter of the book.

Skeleton Tracking

Skeleton tracking is probably the most impressive of NITE's, and therefore of Kinect's, capabilities. This framework allows the computer to understand a person's body position in 3D and to have a quite accurate idea of where the person's joints stand in space at every point in time. This is quite an overwhelming feature if you think that not so long ago you needed to buy a daunting amount of material including several cameras, special suits, and so on in order to acquire the same data.

The only downside of skeleton tracking (for certain projects) is the need for the user to stand in a certain "start pose" to allow the middleware to detect the user's limbs and start the tracking (see Figure 3-13). This inconvenience seems to have been fixed in the current version of the official Microsoft SDK and will likely disappear from NITE, but for the time being, you'll have to make do with it.

Figure 3-13. *User in start pose, calibrated, and tracked*

Let's have a look at the callback functions that you need for skeleton tracking. You can add any code you want to all of these functions and it will run when the functions are called by Simple-OpenNI. In each of the functions, add a println() function to print out the user ID and some other messages to the console, so you know what methods are being triggered in runtime.

When a new user is recognized, the function onNewUser() is called. The user hasn't been "skeletonized" yet, just detected, so you need to start the pose detection routine if you want to track the user's limbs.

```
void onNewUser(int userId){
        println("onNewUser - userId: " + userId);
        kinect.startPoseDetection("Psi",userId);
}
```

The function onLostUser() is invoked when Simple-OpenNI isn't able to find the current user. This usually happens if the user goes off-screen for a certain amount of time.

```
void onLostUser(int userId){
        println("onLostUser - userId: " + userId);
}
```

When you recognize the user, you ask Simple-OpenNI to start looking for a starting pose. When this pose is detected, the method onStartPose() is called. At this moment, you can stop the pose detection routine and start the skeleton calibration.

```
void onStartPose(String pose,int userId){
        println("onStartPose - userId: " + userId + ", pose: " + pose);
        kinect.stopPoseDetection(userId);
        kinect.requestCalibrationSkeleton(userId, true);
}
```

If the user abandons the start pose before the calibration has finished, Simple-OpenNI calls the method onEndPose().

```
void onEndPose(String pose,int userId){
        println("onEndPose - userId: " + userId + ", pose: " + pose);
}
```

The method onStartCalibration() marks the beginning of the calibration process, which can take some seconds to perform.

```
void onStartCalibration(int userId){
        println("onStartCalibration - userId: " + userId);
}
```

Finally, if the user has stood long enough for the calibration to take place, the method onEndCalibration() is called, indicating the end of the calibration process. If the process was indeed successful, Simple-OpenNI starts tracking the skeleton; from that point on, you can abandon your pose and start moving. The newly created skeleton closely follows each one of your movements.

If the calibration process was not successful for some reason, Simple-OpenNI returns to the pose detection mode so you can restart the whole process.

```
void onEndCalibration(int userId, boolean successfull){
        println("onEndCalibration - userId: " + userId + ", successfull: " + successfull);
  if (successfull) {
    println("  User calibrated !!!");
    kinect.startTrackingSkeleton(userId);
  }
  else {
    println("  Failed to calibrate user !!!");
    println("  Start pose detection");
    kinect.startPoseDetection("Psi", userId);
  }
}
```

As with hand tracking, you will be learning how to effectively work with skeleton tracking in this book using practical examples. In Chapter 6, you will be controlling a puppet with skeleton tracking; in

Chapter 7, you will do three-dimensional skeleton tracking to develop a user control system for a lighting installation.

Summary

This chapter has packed many important concepts that will be further explored in practical projects throughout the rest of the book. You learned how to install Processing on your computer and how to install the Simple-OpenNI library that you will be using to acquire the data from Kinect.

You went through a couple of simple examples to illustrate the structure of Processing sketches and the use of the Simple-OpenNI library. You will be using Processing and Simple-OpenNI to interface with Kinect and Arduino in every single project, so this is a very important tool.

You also learned about the concept of three-dimensional space, which you will be using in some of the more advanced projects in the second half of the book, and you implemented a couple of examples of 3D applications based on the Kinect point clouds.

As you move forward in this book, you may want to re-read some of the sections of this chapter to better understand the basic concepts.

CHAPTER 4

Arduino and Kinect: "Hello World"

by Enrique Ramos

In previous chapters, you learned to work with Arduino, Processing, and Kinect independently. Now you are going to build a simple project to illustrate the power of the combination of the three. This is the Arduino and Kinect equivalent of the typical "Hello World" programming exercise.

Imagine you have several LED lights and you want to turn them on and off—or even dim them at will from your favorite sofa—just by waving a hand at them. You need sensors that can recognize the waving gesture and the brightness control gestures. Here is where the Kinect hand-tracking and gesture recognition capabilities come in handy! You also need to be able to send the order to the lights to turn on or off and to set their brightness to a certain level; this is done through the Arduino board. In the course of this project, you will learn several techniques such as serial communication and pulse width modulation (PWM) that you will be using in many other projects throughout this book.

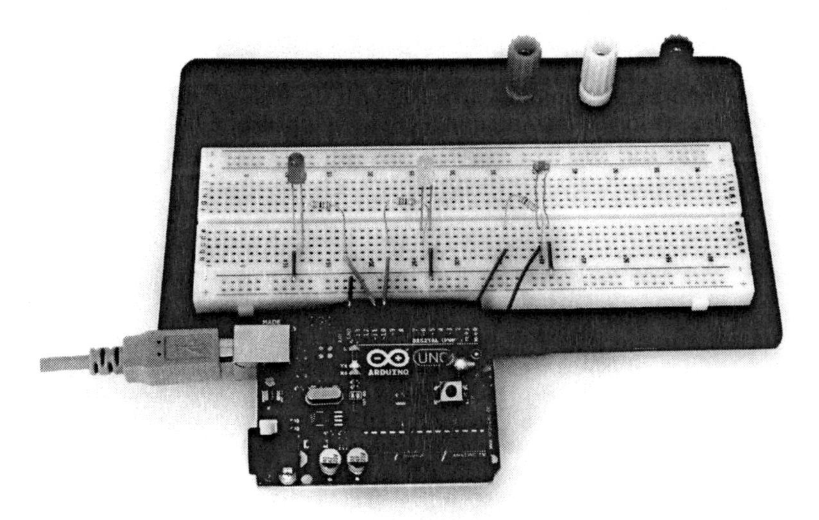

Figure 4-1. "Hello World" built on a breadboard

Parts List for Hello World

This is a very simple project so you are going to build it on a breadboard (see Figure 4-1). All you need is a couple of LEDs, your Kinect, an Arduino board, and a light-sensing resistor or photocell. Table 4-1 lists the required parts.

Table 4-1. *The Parts List*

Part	Description	Price
2 LEDs	Basic 5mm LED	$0.35 each
3 Resistors (2x220ohm, 1x10k)	You can buy them in packs.	$0.25 each
Photocell	Sparkfun SEN-09088 or similar	$1.50
Arduino Board	Your best friend!	$30
Kinect Sensor	From Amazon or any other retailer	$140
Breadboard	We used Sparkfun PRT-00112, but any one will do.	$11.95

■ **Note** For every project in this book, you are expected to have a Kinect Device ($140) and an Arduino ($30), so these won't be included in the parts lists in the following chapters.

Before you start building, let's talk about serial communication. Serial is the base of any communication between your Arduino board and your computer, unless you are using a Bluetooth or Ethernet shield. Whenever you are uploading a program to the board, you are doing it through one of your computer's serial ports.

Serial Communication

Serial communication is a somehow "simple" communication protocol based on sequentially sending data bit by bit. All Arduino boards have one or more serial ports that load the sketches into the board and communicate with the computer at runtime. You are going to use serial communication to talk to the Arduino board from Processing and vice versa.

Of course, there is a library that allows you to control an Arduino board from a Processing sketch without having to write any Arduino code. But wait a minute; you're reading this book, so you must like coding! You're going to write your own serial communication protocol! Apart from the sheer pleasure of coding, writing your own protocols allows you to have a greater control over the data flows and the way the devices communicate.

■ **Caution** Digital pins 0 (TX) and 1 (RX) on the Arduino are used for serial communication. If you are using serial, you can't use them for digital input or output! One very common mistake is forgetting this and thus connecting a sensor or actuator to one of these pins, which will result in crazy values or behaviors. Watch out!

You're going to write a very simple serial communication routine that will allow you to turn an LED on and off from a Processing sketch. There are other serial communication examples in Processing and Arduino, so you may want to have a look at those, too.

Serial-Controlled LED

First, you need to write an Arduino sketch that will turn an LED on and off depending on the data in the serial buffer. You'll use the built-in LED at pin 13 for this example; it's the very tiny square light with an "L" next to it next to the Arduino logo (see Figure 4-2).

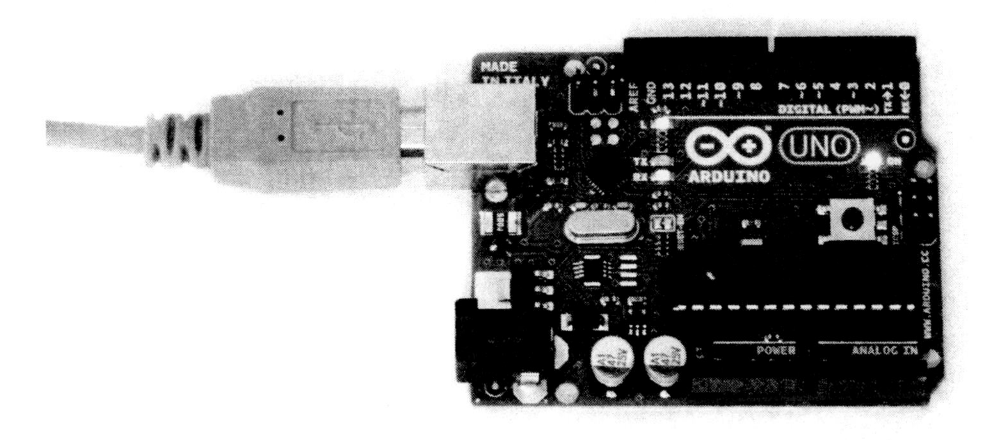

Figure 4-2. Arduino with serial received (RX) LED and built-in LED at pin 13 (L) on

Within `setup()`, you first initialize the serial communication and set pin 13 as an output pin. (Remember that pins are set to input by default, so you need to set them as output if you are planning to use them as such.)

```
void setup() {
  Serial.begin(9600);
  pinMode(13, OUTPUT);
}
```

In `loop()`, you check if there is any data in the serial buffer, and if so, read the most recent byte of information and store it in `input`. If `input` equals 1, turn the LED on; otherwise, turn it off.

```
void loop(){
  if (Serial.available()>0) {
      byte input=Serial.read();
      if(input == '1'){
      digitalWrite(13, HIGH);
    }else{
      digitalWrite(13, LOW);
    }
  }
}
```

■ **Note** The number '1' is within single quote marks both in the Processing sketch and in the Arduino code to indicate that you mean the *character* 1, represented by the binary value 00110001. If you wrote it without quotes in both places, it would still work because you are sending the *decimal value* 1, represented by the binary value 00000001. If you mix characters and decimal values, such as sending '1' and then writing `if(input == 1)` without quote marks, you are comparing two different values (the binary number 00110001 is not the same as 00000001), which would cause your `if` statement to always be wrong!

Now it's time to write a Processing sketch that will send a serial '1' message if the mouse is on the right side of the Processing frame, and a '0' if it is on the left side. First, import the Serial library into your sketch and create an object of the serial class.

```
import processing.serial.*;
Serial myPort;
```

Within `setup()`, set the size to 255x255, so no value of your mouse coordinate is higher than what can be contained in a byte of information. (You'll see why later on.) Then you get the name of the first serial port on your computer and initialize the `Serial` object to that port.

```
void setup()
{
  size(255, 255);
  String portName = Serial.list()[0];
  myPort = new Serial(this, portName, 9600);
}
```

In the `draw()` function, create a rectangle for visualization purposes and check which side of the screen your mouse X-coordinate is on. If it is on the left side of the screen, write a '1' character to the serial port, and a '0' if it is on the right side.

```
void draw() {
  noStroke();
  rect(0,0,width/2,height);

if (mouseX>width/2) {
    myPort.write('1');
  }
  else {
```

```
    myPort.write('0');
  }
}
```

If everything went as planned, the Arduino LED should now be controlled by the position of your mouse. This is great for controlling one device in binary mode (on/off), but what if you want to set the brightness of the LED to a certain level? You could send one byte to control 256 levels of brightness, thus creating a dimmer.

Arduino has two kinds of pins, analog and digital. You can emulate analog values by using PWM. If you have been wondering what these three letters next to some digital pins meant, now you know!

Pulse Width Modulation

Pulse width modulation is basically turning the power on and off so quickly that, because of our eye's persistence of vision, we perceive the light as being permanently on, but with less intensity. In this way, by changing the percentage of time that the electricity is on and off, we can trick our eyes into perceiving different intensities of continuous light (see Figure 4-3). This is very different from using a digital-to-analog converter (DAC), which instead creates a genuine varying voltage that we can use to get different intensities from an incandescent lamp.

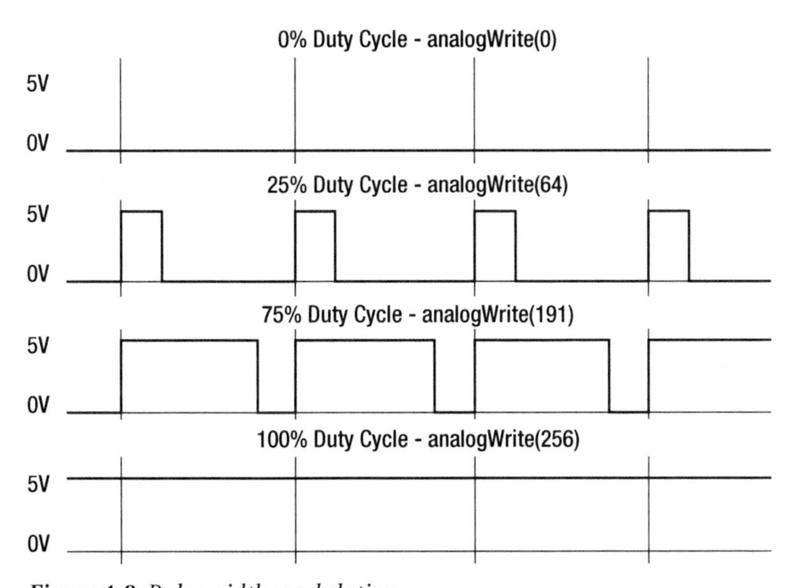

Figure 4-3. *Pulse width modulation*

PWM-Controlled LED

You're going to use pulse width modulation to create a dimmer for your light depending on the X-position of your mouse in the Processing sketch. Change the pin number to a PWM-enabled pin (we have chosen 11 here), read your input value as an integer, and send the value straight to your pin using `analogWrite`. Done! You have a serial driven dimmer!

Well, this is not quite true. When working with LEDs, you need to limit the current to a safe value for the LED, so let's add a resistor. (This is why you're building this circuit on a breadboard.) There are many

LED resistor calculators on the Internet, but trust me on this one—you need a 220-ohm resistor for this circuit, as indicated in the diagram in Figure 4-4.

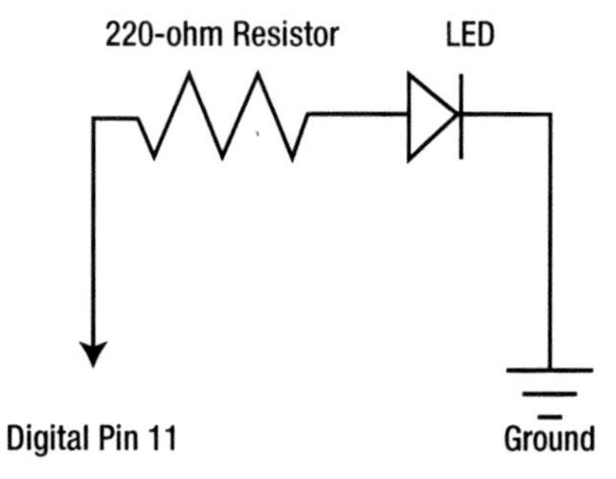

Figure 4-4. LED circuit

■ **Note** In the previous example you didn't use a resistor because pin 13 has a resistor attached to it, soldered to the board. You should try not to use this pin for input, as the internal resistor will pull the voltage level down, giving you a LOW signal at all times. If you absolutely need to use it as an input, you must use an external pull-down resistor. Pull-up and pull-down resistors are described in Chapter 1.

```
void setup() {
   Serial.begin(9600);
   pinMode(11, OUTPUT);
}
void loop(){
    if (Serial.available()) { // If data is available to read,
      int input=Serial.read();
      analogWrite(11, input);
    }
}
```

In the Processing sketch, you only need to replace the `draw()` function with the following one:

```
void draw(){
  // Create a gradient for visualisation
  for(int i = 0; i<width; i++){
    stroke(i);
    line(i,0,i,height);
  }
  // Send the value of the mouse's x-position
    myPort.write(mouseX);
}
```

Note that you are now sending and reading the X-coordinate as an integer, so you can use it to set the brightness directly on the `write()` function. Figure 4-5 shows the Processing sketch and the setup on the breadboard.

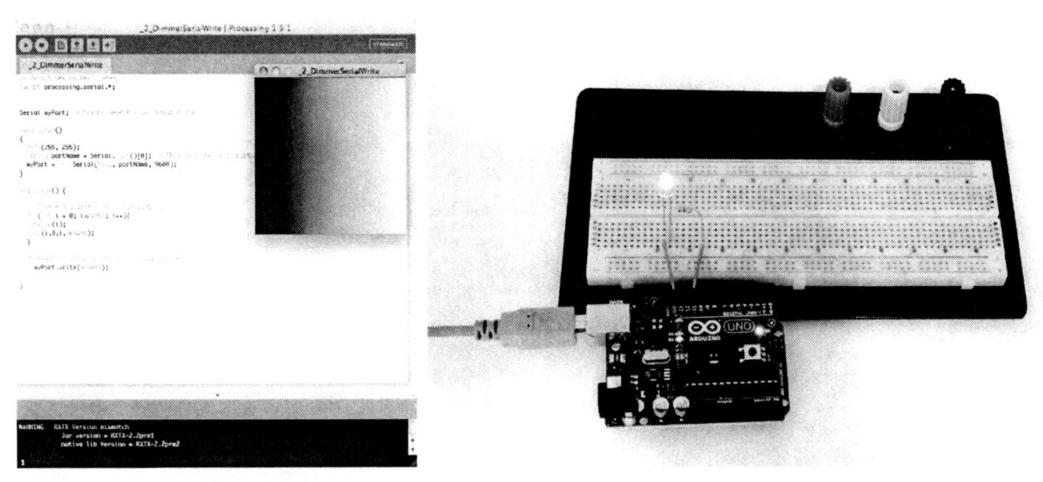

Figure 4-5. Processing sketch and LED mounted on breadboard

Writing Your Own Communication Protocol

Everything you did in the previous example works well because you're only controlling one LED and you can live with 256 values. But what if you want to control two or more devices and define a whole range for each? What if you want to pass on `float` values? Then you need to set up your own communication protocol, which you will do in this example.

You're going to use an *event trigger*, a signal to tell Arduino that the communication has started and that it can start reading the following bytes and use them to modify its behavior. Let's build on the previous example by adding another LED that will be controlled by the Y-coordinate of the mouse.

Serial-Controlled LEDs

You are now going to try to dim two LEDs by sliding your mouse through a Processing sketch. The X-position of the mouse will drive the brightness of one of the LEDs, while the Y-coordinate of the mouse will drive the other LED.

First, you need to declare the integers you're using in your Arduino sketch and define your pins as output.

```
int val, xVal, yVal;
void setup() {
  Serial.begin(9600);
  pinMode(10, OUTPUT);
  pinMode(11, OUTPUT);
}
```

In the loop, you check if enough data has been sent from the computer. You need two different values (one for each LED) and an event trigger character that indicates the start of the communication. Once you have identified the event trigger, you can read the next two values and send them straight to your LEDs.

```
void loop(){

  // check if enough data has been sent from the computer:
  if (Serial.available()>2) {
    // Read the first value. This indicates the beginning of the communication.
    val = Serial.read();
    // If the value is the event trigger character 'S'
    if(val == 'S'){
      // read the most recent byte, which is the x-value
      xVal = Serial.read();
     // Then read the y-value
      yVal = Serial.read();
    }
  }
  // And send those to the LEDS!
  analogWrite(10, xVal);
  analogWrite(11, yVal);
}
```

Load the Arduino sketch into your Arduino board. Now you need to implement the Processing sketch that will write the X and Y values to the serial port and talk to the code that is running in the Arduino. This is pretty similar to the previous example; you're only adding the event trigger character and two different values sequentially.

```
import processing.serial.*;
Serial myPort;

void setup()
{
  size(255, 255);
  String portName = Serial.list()[0]; // This gets the first port on your computer.
  myPort = new Serial(this, portName, 9600);
}

void draw() {
  // Create a gradient for visualisation
  for (int i = 0; i<width; i++) {
    for (int j = 0; j<height; j++) {
      color myCol = color(i,j,0);
      set(i, j, myCol);
      }
    }


// Send an event trigger character to indicate the beginning of the communication
    myPort.write('S');
    // Send the value of the mouse's x-position
    myPort.write(mouseX);
    // Send the value of the mouse's y-position
    myPort.write(mouseY);
}
```

That's it! You are now sending two different values from Processing to control two different devices with your Arduino board (see Figure 4-6). Of course, you can extend the same logic to three, four, or any higher number of devices attached to your Arduino board.

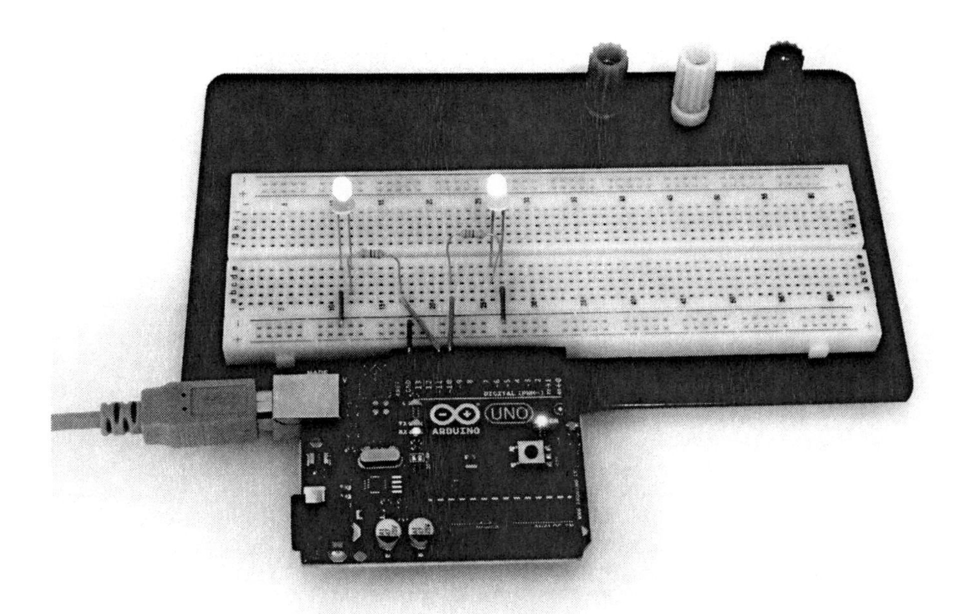

Figure 4-6. *Two LEDs mounted on the breadboard*

Now you know how to send serial data from your Processing sketch to your Arduino board, so what about telling your lights when to turn off without a mouse or a keyboard? It's time to bring in the Kinect!

Kinect-Controlled LEDs

You are going to take the previous example and change the mouse input to Kinect input. The aim is to have the brightness of two LEDs driven by the position of your hand. This is the first project in which you use your Kinect, so let's proceed one step at a time.

The Arduino code for this example is the same as that in the previous example, so if you skipped it, go back and copy the Arduino code and upload it to your board.

First, import the required libraries for the project and declare the variables that will be associated to the `Simple–OpenNI` and `Serial` objects.

```
import SimpleOpenNI.*;
import processing.serial.*;
SimpleOpenNI kinect;
Serial myPort;
```

Then define two `PVectors` to contain the position of the hand as defined by the Kinect (`handVec`), the position of the hand in projective screen coordinates (`mapHandVec`), and a color for the hand-tracking point.

```
PVector handVec = new PVector();
PVector mapHandVec = new PVector();
color handPointCol = color(255,0,0);
```

In setup(), you start by initializing the Simple-OpenNI object, passing the current PApplet (this) as a parameter. Then you enable all the functions you will be using. You need the depth map (to see yourself on screen), the hands, and the gestures. Use the Wave gesture to tell the Kinect when to start tracking your hand. Finally, set the sketch size according to Kinect's depth map size and initialize the Serial object as you did in the previous example.

```
void setup() {
  kinect = new SimpleOpenNI(this);
  // enable mirror
  kinect.setMirror(true);
  // enable depthMap generation, hands and gestures
  kinect.enableDepth();
  kinect.enableGesture();
  kinect.enableHands();
  // add focus gesture to initialise tracking
  kinect.addGesture("Wave");
  size(kinect.depthWidth(),kinect.depthHeight());
  String portName = Serial.list()[0]; // This gets the first port on your computer.
  myPort = new Serial(this, portName, 9600);
}
```

Now is the moment to add the Simple-OpenNI callbacks (the functions that the library invokes when it recognizes certain events). The first one is triggered when a gesture is recognized, in this case when you wave at the Kinect. You then remove the gesture from the capabilities, so Kinect doesn't keep looking for gestures. Once you know a gesture has been performed, you tell NITE to start tracking hands at the position where the gesture occurred.

```
void onRecognizeGesture(String strGesture, PVector idPosition, PVector endPosition)
{
  kinect.removeGesture(strGesture);
  kinect.startTrackingHands(endPosition);
}
```

This will trigger the creation of a Hand object. After creating the hand, change the color of your hand tracking point to green to indicate that the gesture was recognized.

```
void onCreateHands(int handId, PVector pos, float time)
{
  handVec = pos;
  handPointCol = color(0, 255, 0);
}
```

Finally, every time the position of the hand is updated, update the hand vector so you can work with it in the draw() function.

```
void onUpdateHands(int handId, PVector pos, float time)
{
  handVec = pos;
}
```

The first thing to do in the draw() function is to update the Simple-OpenNI object (kinect), which in turn calls the three previous functions. Once updated, the position of handVec is also updated, but handVec contains the real-scale 3D position of the hand, so you need to map its coordinates to your 2D screen coordinates. Luckily, Simple-OpenNI has a specific method for performing this operation (with a rather long name), which you will use to this effect.

```
void draw() {
  kinect.update();
  kinect.convertRealWorldToProjective(handVec,mapHandVec);
```

You now have all the data you need from the Kinect, so it's time to print the depth image to the screen and draw the hand-tracking point from the mapped coordinates.

```
  image(kinect.depthImage(), 0, 0);
  strokeWeight(10);
  stroke(handPointCol);
  point(mapHandVec.x, mapHandVec.y);
```

And lastly, send the event trigger character 'S' to the serial port, followed by the values of your hand position mapped to the limits of a serial message (0-255).

```
   // Send a marker to indicate the beginning of the communication
    myPort.write('S');
    // Send the value of the mouse's x-position
    myPort.write(int(255*mapHandVec.x/width));
    // Send the value of the mouse's y-position
    myPort.write(int(255*mapHandVec.y/height));
}
```

So, if you did things right, you should now have two LED lights on your breadboard that you can control with the vertical and horizontal movements of your hand. The Processing screen should look something like Figure 4-7.

Figure 4-7. Processing screen with hand tracking

Let's add one more feature to this project; you have been talking to Arduino from Processing, but serial communication works both ways, so you can use serial to send data back from the Arduino sensors into Processing. This generates a communication feedback loop that will be very useful in later projects.

Feedback from Arduino

Your sketch works and you are seeing your depth image on the screen. Let's imagine that you prefer seeing your RGB image on the screen whenever possible, but if the lighting level of the room is too low, you want the image to revert back to the depth image. You can do this by getting feedback from a light-sensing resistor connected to your Arduino board.

You're going to use a very simple and cheap photocell. This cell provides a full range of values between 0 and 1023 (depending on the resistor), so connect it to the analog input pin 0. The circuit for the photocell includes a 10k resistor and looks something like the diagram in Figure 4-8.

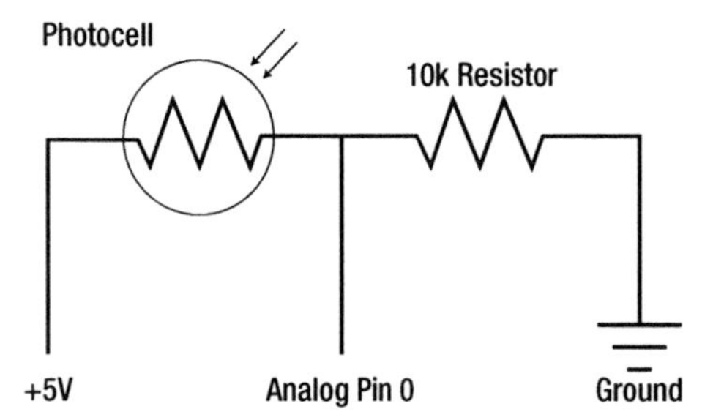

Figure 4-8. *Photocell circuit 1; voltage output increases with brightness*

This kind of circuit is called a *resistive divider*, which is a special case of voltage divider, a circuit that produces an output voltage (in your case, going to analog pin 0) that is a fraction of the input voltage (5V). The output voltage can be easily found applying Ohm's law.

$$V_{out} = \frac{R_2}{R_1 + R_2} \cdot V_{in}$$

R1 is the value of the resistor connected to Vin, and R2 is the value of the resistor connected to ground. The fixed resistor should have the same order of magnitude as the variable's resistor range. In your case, the photocell has a range of 1K (in bright light) to 10K (in the dark). You're using a 10K fixed resistor, which, applying the equation, will give you a voltage output range of 4.54V to 2.5V.

There is a catch to this circuit. If you switch the position of the photocell and the resistor (see Figure 4-9), it will still work but the values will be inverted; this means the numbers will grow as the light decreases. You can use it either way; just be aware of the input you are getting when you read the data in analog pin 0.

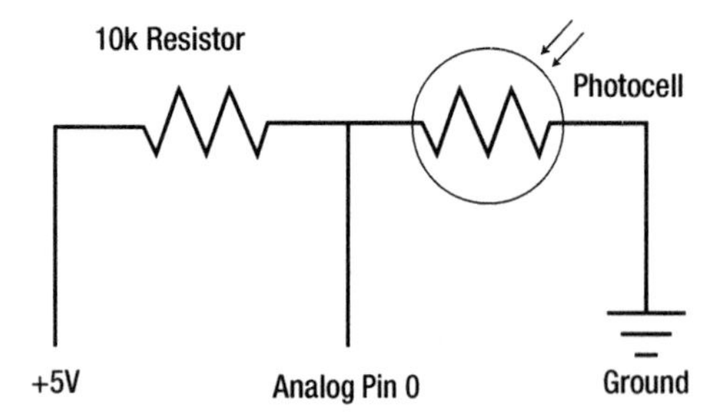

Figure 4-9. *Photocell circuit 2; voltage output decreases with brightness*

The full circuit is getting a little more complicated now, so we are including the full schematics that we produced with Fritzing, as covered in Chapter 1 (see Figures 4-10 and 4-11).

***Figure 4-10.** The physical project and Fritzing breadboard view*

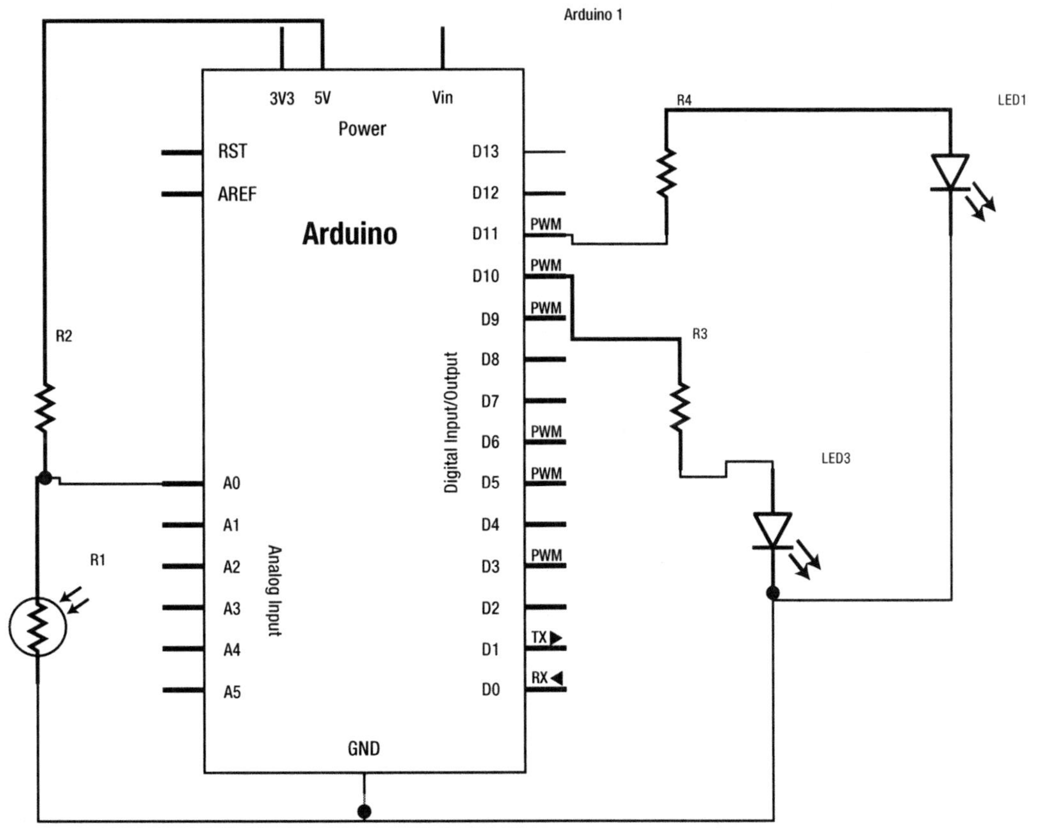

***Figure 4-11.** Project schematics produced with Fritzing*

In the Arduino sketch, you add a variable to store the value of the photocell and the pin number for your analog input. The setup() function will stay unchanged.

```
int val, xVal, yVal;
int sensorValue;
int sensorPin = 0;

void setup() {
  // initialize the serial communication:
  Serial.begin(9600);
  // initialize the serial communication:
  pinMode(10, OUTPUT);
  pinMode(11, OUTPUT);
}
```

Then add two lines at the beginning of the loop() to read the values of the resistor and send a message to the serial port. You can add a little delay later in order to allow the analog-to-digital converter to stabilize after the last reading.

```
void loop(){
  // Check Values from the Photocell
  sensorValue = analogRead(sensorPin);
  Serial.println(sensorValue);
  // wait a bit for the analog-to-digital converter
  // to stabilize after the last reading:
  delay(10);

  // check if data has been sent from the computer:
  if (Serial.available()>2) { // If data is available to read,
    val = Serial.read();
    if(val == 'S'){
      // read the most recent byte (which will be from 0 to 255):
      xVal = Serial.read();
      yVal = Serial.read();
    }
  }
  analogWrite(10, xVal);
  analogWrite(11, yVal);
}
```

In the Processing sketch, add a variable to store the values received.

```
import SimpleOpenNI.*;
import processing.serial.*;
SimpleOpenNI kinect;
Serial myPort;

PVector handVec = new PVector();
PVector mapHandVec = new PVector();
color handPointCol = color(255, 0, 0);
int photoVal;
```

Within setup(), enable the RGB image and, after initializing the port, specify that Processing doesn't generate a serialEvent() until a newline character has been received.

```
void setup() {
  kinect = new SimpleOpenNI(this);
  // disable mirror
  kinect.setMirror(true);

  // enable depthMap generation, hands and gestures
  kinect.enableDepth();
  kinect.enableRGB();
  kinect.enableGesture();
  kinect.enableHands();

  // add focus gesture to initialise tracking
  kinect.addGesture("Wave");

  size(kinect.depthWidth(), kinect.depthHeight());
  String portName = Serial.list()[0]; // This gets the first port on your computer.
  myPort = new Serial(this, portName, 9600);
  // don't generate a serialEvent() unless you get a newline character:
  myPort.bufferUntil('\n');
}
```

In the draw loop, create a conditional clause that will display RGB or the depth image on screen depending on the value of the photocell that you are acquiring inside the `serialEvent()` method described next. Then draw the hand point and send its values, as you did previously.

```
void draw() {
  kinect.update();
  kinect.convertRealWorldToProjective(handVec,mapHandVec);
  // Draw RGB or depthImageMap according to Photocell
  if (photoVal<500) {
    image(kinect.rgbImage(), 0, 0);
  } else {
    image(kinect.depthImage(), 0, 0);
  }

  strokeWeight(10);
  stroke(handPointCol);
  point(mapHandVec.x, mapHandVec.y);

  // Send a event trigger character to indicate the beginning of the communication
  myPort.write('S');
  // Send the value of the mouse's x-position
  myPort.write(int(255*mapHandVec.x/width));
  // Send the value of the mouse's y-position
  myPort.write(int(255*mapHandVec.y/height));
}
```

Next, add a `serialEvent()` method, which is called by Processing when you receive the newline character (remember you sent a `Serial.printline` from Arduino, which includes a newline character at the end of the message). Read the message, trim any white spaces, and convert it to an integer.

```
void serialEvent (Serial myPort) {
  // get the ASCII string:
  String inString = myPort.readStringUntil('\n');
  if (inString != null) {
```

```
    // trim off any whitespace:
    inString = trim(inString);
    // convert to an integer
    photoVal = int(inString);
  }
  else {
    println("No data to display");
  }
}
```

The NITE callbacks stay the same, as in the previous example. You just add some print line commands to keep track of the process on the console.

```
void onCreateHands(int handId, PVector pos, float time)
{
  println("onCreateHands - handId: " + handId + ", pos: " + pos + ", time:" + time);
  handVec = pos;
  handPointCol = color(0, 255, 0);
}

void onUpdateHands(int handId, PVector pos, float time)
{
  println("onUpdateHandsCb - handId: " + handId + ", pos: " + pos + ", time:" + time);
  handVec = pos;
}

void onRecognizeGesture(String strGesture, PVector idPosition, PVector endPosition)
{
  println("onRecognizeGesture - strGesture: " + strGesture + ", idPosition: " + idPosition +
", endPosition:" +   endPosition);
  kinect.removeGesture(strGesture);
  kinect.startTrackingHands(endPosition);
}
```

In this way, whenever the lighting level drops under a certain threshold, your image is automatically changed to the depth image, which is independent of the ambient light.

Summary

This chapter was all about making Kinect and Arduino talk through Processing, and you now have a working example. You are indeed gathering information from the physical world with your Kinect device and sending that information from Processing to Arduino through serial communication. You are also gathering lighting level information through a photocell connected to the Arduino board and modifying the Processing visualization according to it.

During this process, you used Kinect's hand-tracking capabilities for the first time, learned how to use serial communication, and wrote your own protocols. You have also learned how to use photocells, voltage dividers, and Arduino PWM-enabled pins. This is probably enough for an introductory project; the next chapter has even more exciting stuff!

CHAPTER 5

Kinect Remote Control

by Ciriaco Castro and Enrique Ramos

Imagine controlling all your home appliances remotely just using body movements and gestures. There are a number of machines and devices that can be controlled electrically in an automated house, such as television sets, DVD players, radios, lamps, and the heating system. In order to avoid the problem of tinkering with mains power and the obvious risks of opening up expensive devices, this chapter will focus on hacking an external element that can control a broad range of devices: a general-purpose remote control. Hacking a remote control is a really simple job; you just open the remote and connect it to an Arduino board. You then use the Kinect to perform gesture recognition in order to activate different functions.

In this project you will learn how to use NITE's gesture recognition capabilities and you'll even implement a simple gesture recognition routine yourself. You will send the data gathered to Arduino using serial communication, so you will ultimately be able to control the remote control with your body movements. This application can be used on pretty much any infrared-based remote control just by connecting the Arduino pins to the correct buttons on the remote.

Parts List for Kinect Remote Control

All you need for this project is a general-purpose remote control (the guts of which you can see in Figure 5-1), your Kinect, an Arduino board, a prototyping shield, some relays, and a bunch of cables for connecting them. Table 5-1 has more details on the required parts.

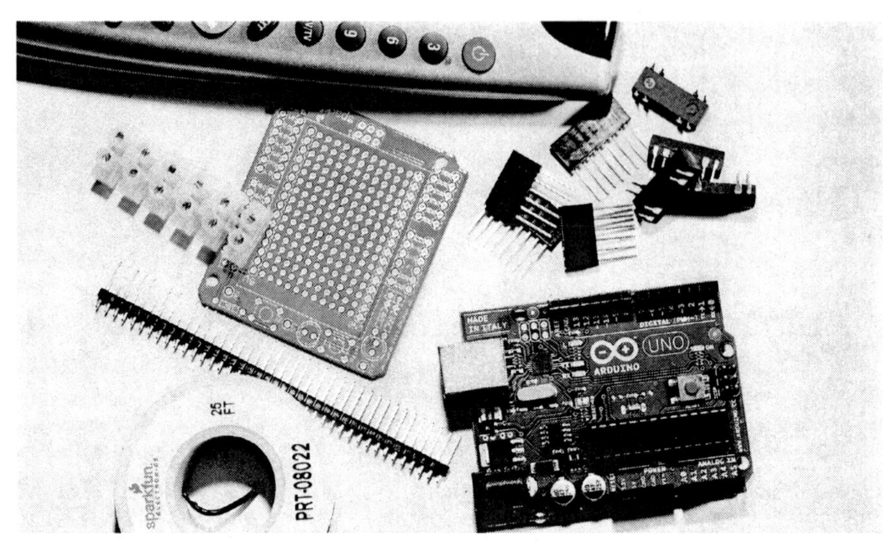

Figure 5-1. Parts of the remote control

Table 5-1. The Parts List

Part	Description	Price
1 universal remote control	From Amazon or any other retailer	$11.99
1 prototype shield	We used Sparkfun DEV-07914, but any one will do.	$16.95
5 breakaway headers-straight	For connecting your Arduino to the prototype Shield	$1.50
1 connector strip	Screw connector for joining cables	$0.95
4 5V reed relay	Similar to Sparkfun COM-00100	$1.95/each

You'll start by hacking the remote control, connecting it to Arduino, and testing it to make sure it works properly. Then you'll develop the Kinect sketch that will track your hands and gestures. Once everything is neat and tidy, you will connect the Kinect to Arduino using serial communication, as you did in the previous chapter.

There is a wide range of remote controls available and they all look quite different from each another, but as long as they are infrared-based they all work pretty much in the same way.

Hacking a Remote Control

Remote controls are completely inactive until you push a button. When you hit one of the buttons, the remote will send a signal to a receiver device, which should interpret the signal and perform the action you intend it to perform. Some remote controls use radio signals to communicate with the receiver devices, but most communicate via infrared light. You'll be focusing on infrared remotes.

When you hit a button on your remote, you are actually closing one of the open circuits inside it. At that moment, the remote starts to send a signal that goes to the receiver device. This carrier signal consists of a series of precisely spaced infrared pulses constituting a code that is processed by the receiver device. Once decrypted, the receiver triggers different functions according to the signal received. Apart from sending out pulses that encode the instructions for the receiver, the remote control sends a short code that identifies the device for which the message is intended.

The first thing you need to do is pair the universal remote control with the device that you want to control. Put the batteries in and follow the instructions. Pay attention to how long you have to press a button for a command to be transmitted because this information is useful for refining the Arduino code that controls the remote. Once you have tested the device (by changing channels and volume), it's time to move on to the most interesting part: opening up the remote (see Figure 5-2)!

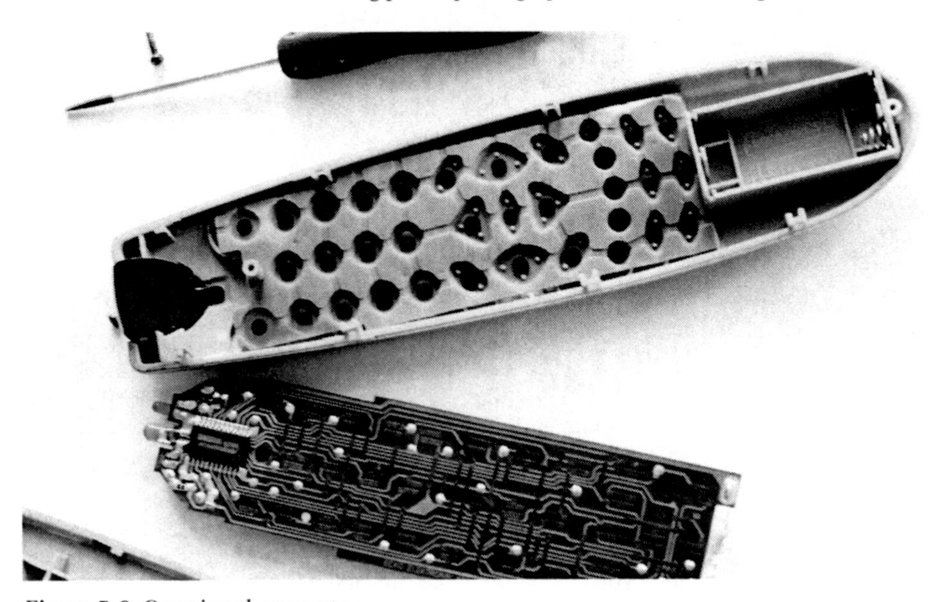

Figure 5-2. Opening the remote

Remove the batteries and the screws so you can see the printed circuit board. Observe that the buttons are made of plastic and have a conductive material on the inner side. You have to locate the solder pads, which are the areas where you can solder your cables (see Figure 5-3). When you connect them back, you bridge the circuit in the same way as you do by holding a button.

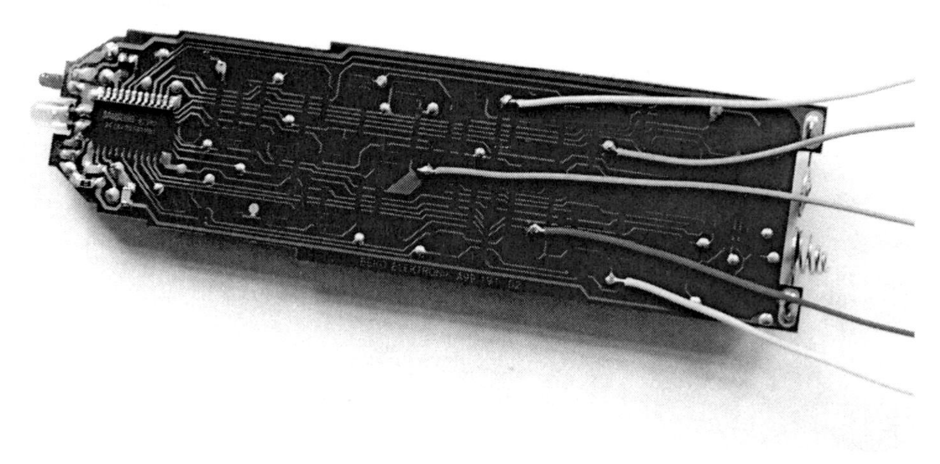

Figure 5-3. Wires soldered to the remote pads

Normally two pins connect when the button is pressed, but some remotes have four to make the connection stronger. Sometimes it is not easy to find which solder pads you want to connect in order to transmit the right command. There are two main ways of finding them. The most straightforward method is by putting the batteries back in the remote and using a short cable to try to connect the right pads. You can sit in front of the TV, try the connections you think are right, and check what happens! A second, more scientific option is by using a multimeter to measure the resistance between the pads. If you are testing the right pads, you will see a change from high values of resistance to very low values when you press a button.

In this project, you need to find the right pads for sending the commands Channel+, Channel -, Volume +, and Volume -. Once you find them, you can connect the cables so that they are ready to be soldered to the prototype shield. Use cables of different colors so you don't lose track of them!

Connecting the Remote to Arduino

We're going to explain the basics of how to build a prototype shield and how to connect it to Arduino. Here you will find enough information for building it, but if you want a deeper explanation and information on how to build other Arduino applications, look to the book *Practical Arduino* by Jonathan Oxer and Hugh Blemings (Apress, 2009).

The way that the remote control circuit operates is by closing a circuit. When you hold a button, it works as a mechanical switch. You are going to perform the same function by using relays. A relay is an electrically operated switch that controls a circuit with a low-power signal; both circuits remain electrically isolated. In your case, a low current coming from Arduino controls the flow of current in the remote control. 5v relays require about 20 mA to operate, meaning that you can connect them directly to your Arduino outputs (Figure 5-4).

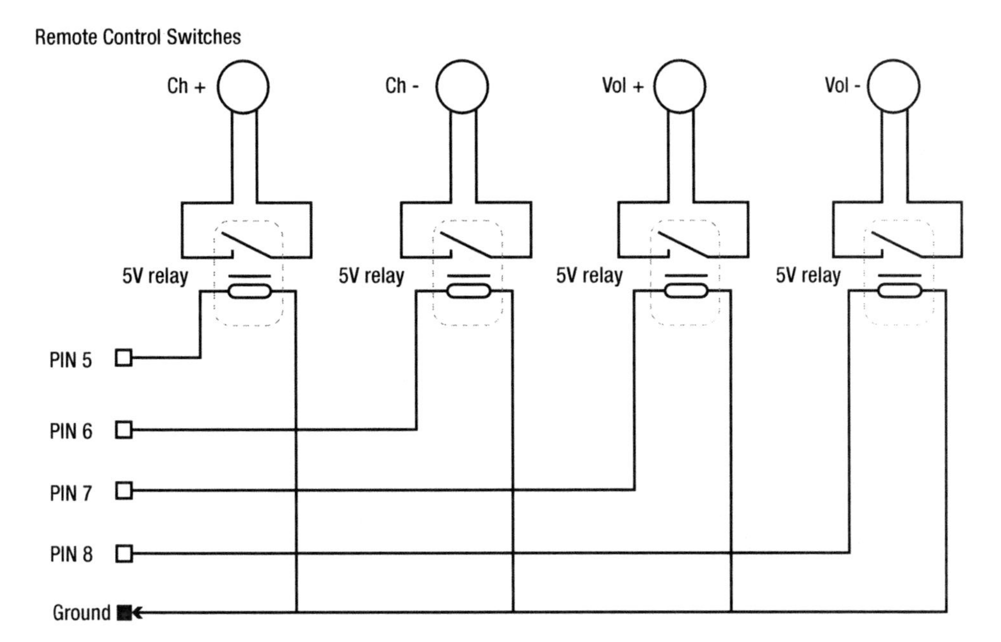

Figure 5-4. Remote control switches testing a relay

First, you're going to do a test on a breadboard in order to understand how to connect a relay to your Arduino board: you're going to control a LED through a relay.

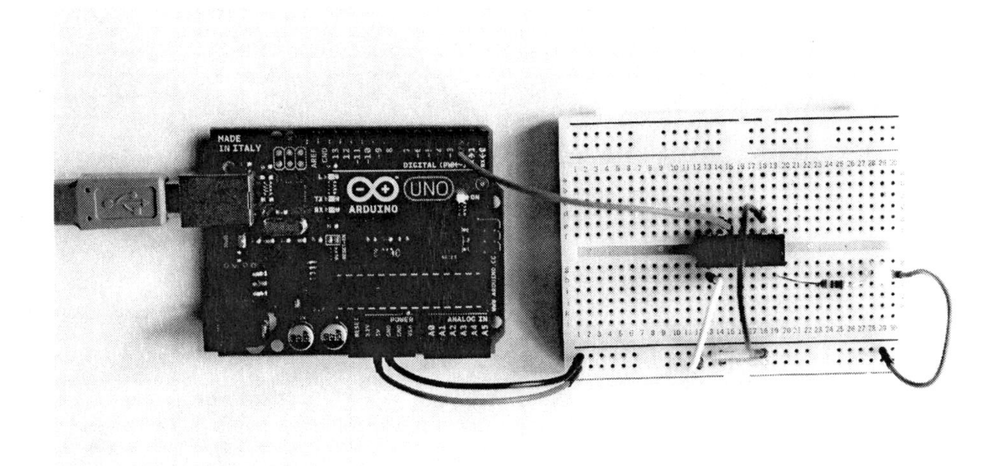

Figure 5-5. Testing the relay on a breadboard

Relays have their input (coil connection) in the central pair of pins and the output on the outer pair of pins. First you insert the relay in the breadboard, making sure it spans unconnected rows. Connect power (5V) and ground to the breadboard. Then insert the LED into the breadboard, connecting the short leg to ground. After that, connect the relay. Input pins are connected to pin 2 of the Arduino board and ground. For the output, connect power in one end and a resistor in the other that connects to the longer leg of the LED. This constitutes the basic circuit, as shown in Figures 5-5 and 5-6. Then you upload into Arduino the example Blink, changing the output pin to pin 2.

```
void setup() {
  pinMode(2, OUTPUT);    // Set the digital pin as an output.
}
void loop() {
  digitalWrite(2, HIGH);          // set the LED on
  delay(1000);                    // wait for a second
  digitalWrite(2, LOW);           // set the LED off
  delay(1000);                    // wait for a second
}
```

Your Arduino sends a signal to the relay, and the relay allows the current to pass during the same time that your signal is High (which is 1 second, according to the code). You will hear a little noise, meaning the relay is switching on and off.

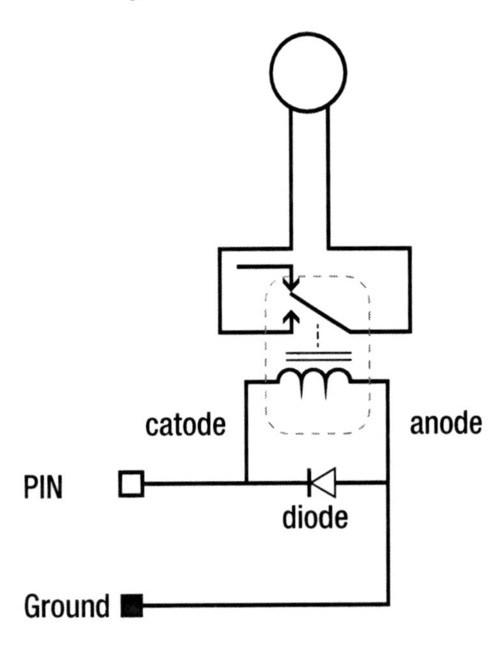

Figure 5-6. Schematics of the relay circuit

■ **Tip** It's necessary to introduce a diode in order to protect your circuit. This is due to the way a relay operates; it generates a magnetic field to mechanically close the circuit. When the relay is turned off, the magnetic field stops and generates a reverse spike. This is potentially dangerous for your circuits, as you are receiving a negative voltage on the same pin where you were supplying positive voltage. If the spike is big enough, it can damage the circuit. In order to solve this problem, you can connect a diode across the relay coil, protecting your Arduino. Diodes work only in one direction, stopping the negative voltage from going to your Arduino. To connect them, you need to connect the cathode lead (the end with the stripe) towards the Arduino pin output and the anode lead to the ground. It's important to take this into consideration when you are using high voltage. However, due to the range you are using in this exercise (very low power), it's unlikely that the relay will damage your Arduino, so you're not going to incorporate them in your prototype shield.

Assembling the Prototype Shield

Now that you understand the principles behind a relay, you can start building stuff! The first thing is to mount four relays on the prototype shield. Observing the shield, you can identify one line for power and other line for ground. There should be enough space for assembling four relays (Figures 5-7 and 5-8).

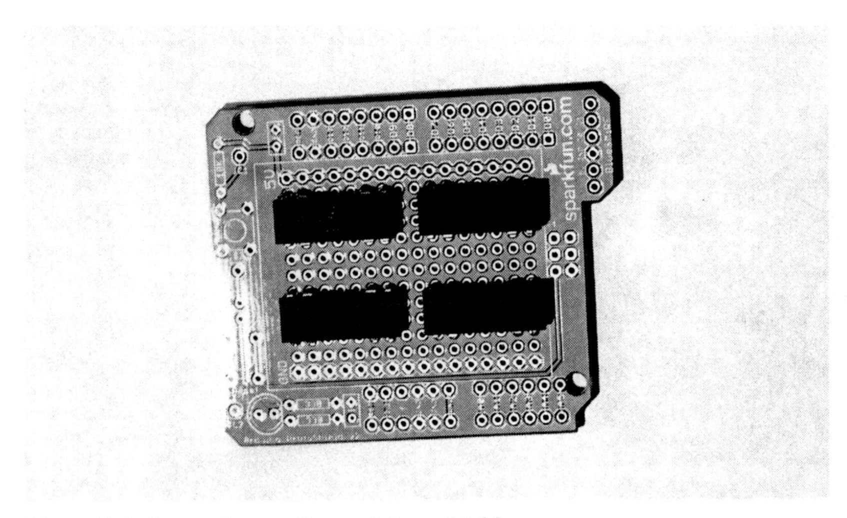

***Figure 5-7.** Four relays on the prototype shield*

Figure 5-8. Back of prototype shield

Start by fitting the relays and soldering the necessary connectors (two legs of inputs and two legs of outputs). Then, using short cables, connect one of the coils of each relay together and to the ground (see Figure 5-9).

Figure 5-9. Relays connected to Arduino shield

You can also incorporate the breakaway headers to connect the prototype shield to the Arduino (see Figure 5-9). Once the headers are fitted, turn the prototype shield around and connect the other coil to the matching Arduino digital pin. In this example, you're using pins 5, 6, 7, and 8, but you can use any of them; just make sure they're coordinated with the code (see Figure 5-10)!

Figure 5-10. Breakaway headers

You have the relays' inputs connected to Arduino, so now you connect the outputs; just connecting a pair of cables to each output connection will do (see Figure 5-11). Make sure you leave enough length to connect them to the remote control. You can screw one connector in each cable to make it easy to connect them with the cables coming from the remote control.

Figure 5-11. Finished shield

You're almost done. Attach the prototype shield to your Arduino board and connect each pair of cables with the corresponding ones on the remote control. Table 5-2 lists the correct connections and Figure 5-12 shows the end result.

Table 5-2. Remote Control Connections

Pair of Cables from Relay	Pair of Cables on Remote
Arduino pin 5	Channel +
Arduino pin 6	Channel -
Arduino pin 7	Volume +
Arduino pin 8	Volume -

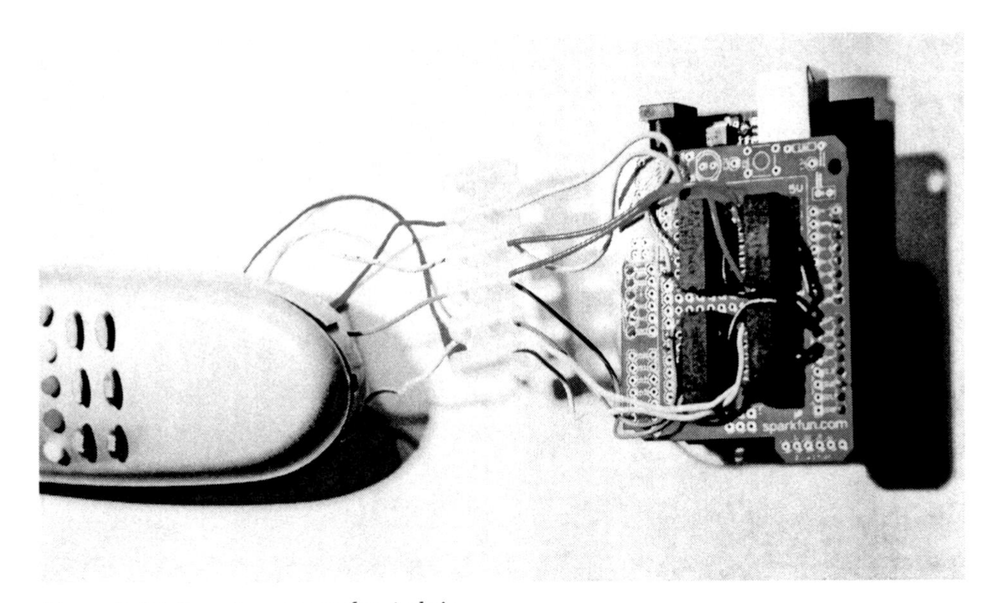

Figure 5-12. Remote connected to Arduino

Now your prototype shield is ready to be tested! If your remote control runs on 3V, instead of using its own batteries you can run it from Arduino. Just remove the batteries and solder a pair of wires in the positive and ground terminals, connecting them to the 3.3V pin and ground respectively (+ goes to 3.3V!). This allows the remote control to take its power from the Arduino so you don't have to worry about batteries running down.

Testing the Circuit

Now that you have the prototype shield ready and connected to the remote control, it's time to program your Arduino in order to do the first tests! Once you're sure that it's working and ready to receive data, you can develop the Processing application that will drive the remote.

Let's start from the basic Arduino code for testing. You begin by defining a set of variables, assigning a symbolic name to each pin (5,6,7,8). You also define a pulse that is the time that the pin is going to be High, equivalent to the time that you hold a button in the remote control.

```
int ChannelPlusPin = 5;
int ChannelLessPin = 6;
int VolumePlusPin = 7;
int VolumeLessPin = 8;
int pulse = 250;
```

After defining the variables, start the `setup()` function. Describe each pin as outputs and start the serial communication.

```
void setup(){  pinMode(ChannelPlusPin, OUTPUT);
  pinMode(ChannelLessPin, OUTPUT);
  pinMode(VolumePlusPin, OUTPUT);
  pinMode(VolumeLessPin, OUTPUT);
  Serial.begin(9600);// Start serial communication at 9600 bps
}
```

Before starting the `loop()`, define an external function to perform a series of commands. This function is called from the `loop()`, modifying the state of each pin to High (on), waiting some time (pulse previously defined), and returning Low (off). The function takes two arguments for the pin number and the pulse. The `Serial.print` commands write text to the serial port (ON, name of the pin used, and OFF) so you can be aware of which function has been called at every moment.

```
void updatePin (int pin, int pulse){
  Serial.println("RECEIVED");
  Serial.println(pin);
  digitalWrite(pin, HIGH);
  delayMicroseconds(pulse);
  digitalWrite(pin, LOW);
  Serial.println("OFF");
}
```

The main `loop()` function is where all the action happens. Arduino is listening for available serial data. If there is data, it will be stored inside a `char` variable; according to the value, it will call the `updatePin()` function to the pin assigned. If the message is "`1`", Arduino will modify the state of the pin `ChannelPlus` (Pin 5), activating it for a period of time. In other words, this will turn on the relay connected to this pin, activating the remote control and changing the channel to the next one. After the pulse time has elapsed, the pin will be brought back to low voltage, so the relay will be turned off and the mechanism will come to a steady state, waiting for new data to come in.

```
void loop(){
   if (Serial.available()) {
    char val=Serial.read();
    if(val == '1') {
      updatePin(ChannelPlusPin, pulse);
    } else if(val == '2') {
     updatePin(ChannelLessPin, pulse);
    } else if(val == '3') {
    updatePin(VolumePlusPin, pulse);
    } else if(val == '4') {
   updatePin(VolumeLessPin, pulse);
    }
  }
}
```

This code is written for four commands, so you're using just four pins; feel free to add more pins and more commands in order to have a full set of instructions for your remote control.

Once you have finished writing your code, plug the Arduino into the USB port, select the correct Arduino (in our case Arduino Uno), verify the sketch, and upload it!

In physical computing it's a good idea to get accustomed to frequently checking if everything is connected correctly. So take a moment now to check the following:

- The Arduino is connected with a USB cable to the computer.
- The remote control has working batteries.
- The Arduino is plugged into your prototype shield.
- The TV is on.
- The remote control is pointing towards the TV.

For testing purposes, use the serial communication console to introduce your values by typing on the keyboard (see Figure 5-13). Double-check that serial communication is at the same baud rate as the one specified in the code (9600 in our case).

Figure 5-13. *Using the Serial Monitor to input data*

Enter a value of `1` in the text area. You should read a response from your Arduino saying that the data has been received and the pin number being used, followed by an off message. If everything worked as expected, the remote should have changed the channel up to the next one; you should be able to send the other commands by manually introducing other values as 2, 3, or 4.

Kinect Hand Tracking and Gesture Recognition

You now have a hacked remote control that is connected to an Arduino board and is just waiting for data to stream in. Your next goal is to easily communicate the commands "next channel," "previous channel," "volume up," and "volume down" to the remote using hand gestures.

We have figured out that the easiest way to tell the Kinect that we want to change the channel is by doing a quick movement with our hand to the left or right, so you're going to implement a right or left "swipe" gesture for that purpose. For the volume, we could have done the same with "up" and "down", but it is not that easy to "swipe" upwards or downwards (try it!), and it's nicer to have a continuous control over the volume, so we have decided that the good old-fashioned potentiometer will do the deal. You will use the circle detector included with NITE to allow you to smoothly change the volume.

Libraries and Setup

Start by importing the core Serial and OpenGL libraries and the Simple-OpenNI library to your sketch; also, declare the variables for the Serial and Simple-OpenNI objects.

```
import SimpleOpenNI.*;
import processing.opengl.*;
import processing.serial.*;

SimpleOpenNI kinect;
Serial myPort;
```

In the previous project you used NITE through Simple-OpenNI callbacks, but you didn't declare the NITE objects explicitly. In this project, you're going to use two NITE functions: hand tracking and gesture recognition for the waving, hand raising, and circle detection. For this purpose, you declare an `XnVSessionManager` object to deal with NITE gesture recognition, an `XnVPointControl` object to keep track of the hand points, and an `XnVCircleDetector` to trigger events when it recognizes a circle being drawn with a hand.

■ **Note** All classes, objects, and methods starting with "`XnV`" refer to NITE commands.

```
XnVSessionManager sessionManager;
XnVPointControl pointControl;
XnVCircleDetector circleDetector;
```

This project requires that you display text on screen, so you need to declare a text font. In this project, we used `SansSerif-12`, so you need to create the font first by going to Tools menu > Create Font.

```
PFont font;
```

The sketch tracks your hand as it moves on screen, and then it triggers events whenever it recognizes a "channel change" or "volume change" gesture, which each have their own timing. This requires the main `draw()` loop to run in three different modes. It remembers these modes in the variable called `mode`.

```
int mode = 0;
```

Now declare the variables for the hand tracking and volume control functions. For hand tracking, you need a signal that a hand is being tracked (handsTrackFlag), variables for hand positions (real position and projected on screen), and the arrayList of PVectors containing previous positions.

```
boolean handsTrackFlag = true;
PVector screenHandVec = new PVector();
PVector handVec = new PVector();
ArrayList<PVector>    handVecList = new ArrayList<PVector>();
int handVecListSize = 30;
```

For volume control and channel, you declare previous and current rotations of the hand in the circle, a radius, an angle for the volume control gizmo, and its center in absolute and relative coordinates. For channel change, you need a changeChannel integer to define the direction of the change and a channelTime timer to keep track of the channel change gizmo.

```
float rot;
float prevRot;
float rad;
float angle;
PVector centerVec = new PVector();
PVector screenCenterVec = new PVector();
int changeChannel;
int channelTime;
```

In setup(), initialize all the objects and then enable the functions that you expect to be using through the execution of the program. You need the depth map, gestures, and hands enabled in the kinect object.

```
void setup(){
  // Simple-openni object
  kinect = new SimpleOpenNI(this);
  kinect.setMirror(true);

  // enable depthMap generation, hands + gestures
  kinect.enableDepth();
  kinect.enableGesture();
  kinect.enableHands();
```

This is important: initialize the NITE session manager by adding the two gestures you'll be working with, and then initialize the hand point controller and circle detector. Call the methods RegisterPointCreate, RegisterPointDestroy, and RegisterPointUpdate in the point control object (passing the current applet "this" as a parameter) so the XnVPointControl object will invoke your callback functions. Now you have a control in the processes triggered at the creation, destruction, and update intervals of a hand point.

In the same way, call the RegisterCircle and RegisterNoCircle in the circle detector so it invokes your callbacks for circle detection. Finally, add the listeners to the NITE session so they get initialized when you start your session.

```
  // setup NITE
  sessionManager = kinect.createSessionManager("Wave", "RaiseHand");
  // Setup NITE.s Hand Point Control
  pointControl = new XnVPointControl();
  pointControl.RegisterPointCreate(this);
  pointControl.RegisterPointDestroy(this);
  pointControl.RegisterPointUpdate(this);
```

```
  // Setup NITE's Circle Detector
  circleDetector = new XnVCircleDetector();
  circleDetector.RegisterCircle(this);
  circleDetector.RegisterNoCircle(this);

  // Add the two of them to the session
  sessionManager.AddListener(circleDetector);
  sessionManager.AddListener(pointControl);
```

Wow, that was a little bit difficult! Don't worry if you didn't quite understand what was going on with these functions; it will all become clearer when you get to the callback functions. The only things you have left to do in your setup() function are to set the size of the sketch to match the depthMap and to initialize the font and the serial communication.

```
  size(kinect.depthWidth(), kinect.depthHeight());
  smooth();
  font = loadFont("SansSerif-12.vlw");

  String portName = Serial.list()[0]; // This gets the first port on your computer.
  myPort = new Serial(this, portName, 9600);
}
```

Now, before you get into the draw() loop, you need to write the callbacks from NITE so all the previous declaring and calling of weird functions makes some sense.

NITE Callbacks

What you have done by setting up these objects is take advantage of NITE's gesture recognition framework. Now you can include a series of functions (the ones you specified!) that will be invoked when specific events are triggered by NITE.

Regarding hand recognition, you call functions when the hand points are created, destroyed, and updated (see Figure 5-14). Upon creation of a hand point, you set the flag to true so you know that a hand is being tracked, set the handVec PVector to the position of the hand by using the parameter coming from NITE's call of the function, clear the list, and add the first point.

```
void onPointCreate(XnVHandPointContext pContext){
  println("onPointCreate:");
  handsTrackFlag = true;
  handVec.set(pContext.getPtPosition().getX(),
                         pContext.getPtPosition().getY(),
                         pContext.getPtPosition().getZ());
  handVecList.clear();
  handVecList.add(handVec.get());
}
```

On hand destruction, set the flag to false.

```
void onPointDestroy(int nID){
  println("PointDestroy: " + nID);
  handsTrackFlag = false;
}
```

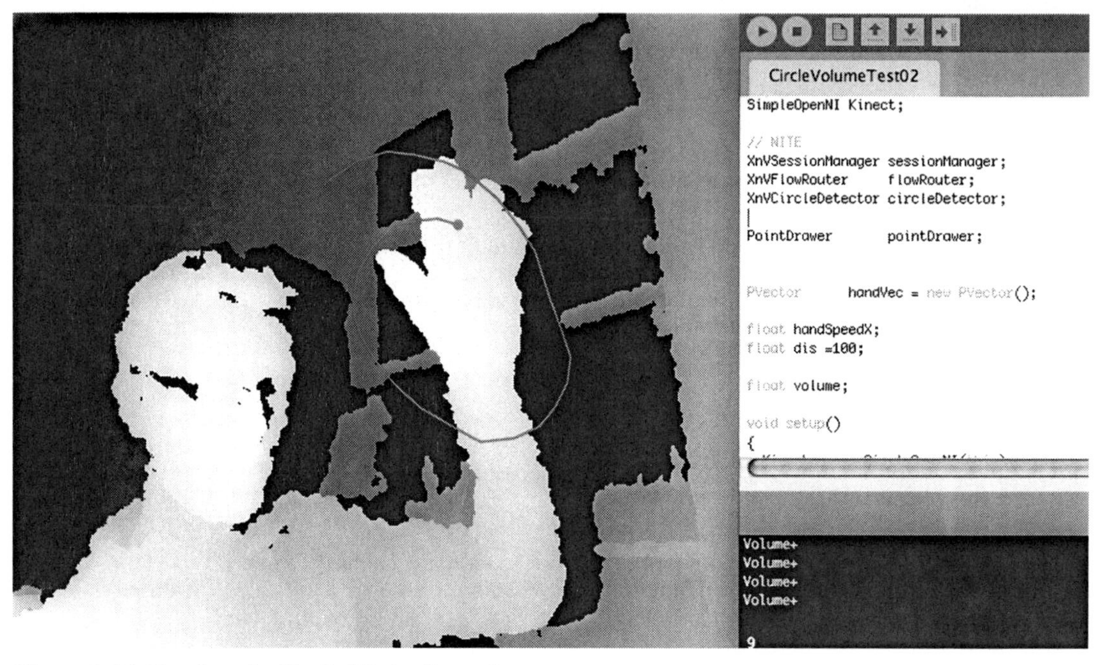

Figure 5-14. Hand tracked by XnVPointControl

And whenever the hand point is updated, you update the hand vector and add another point to the list. If the list is larger than the size specified, get rid of its oldest element.

```
void onPointUpdate(XnVHandPointContext pContext){
  handVec.set(pContext.getPtPosition().getX(),
                     pContext.getPtPosition().getY(),
                     pContext.getPtPosition().getZ());
  handVecList.add(0, handVec.get());
  // remove the last point
  if (handVecList.size() >= handVecListSize)  {
    handVecList.remove(handVecList.size()-1);
  }
}
```

The circle detector has its own callback functions as well: one for the circle "being there" and one when there is no circle. Simple.

When there is a circle, you get your rotation from the parameter `fTimes`. This parameter gives you the number of rotations your hand has made around the circle, so a 90 degree rotation will mean `fTimes` is equal to 0.25. This is very useful when you want to control the volume with a constant rotation of your hand. You need to extract your angle as well, which you also get from `fTimes` with a little function. You need the circle's center as well, so you extract it from the circle object with `circle.getPtCenter()`. But be careful! The center you are given doesn't include a z-coordinate! You get the z-coordinate from the hand vector, and then you convert to projective to get the projection of the circle on screen for visualization. You then extract the radius of the circle and set your sketch mode to "1", or volume control mode.

```
void onCircle(float fTimes, boolean bConfident, XnVCircle circle)
{
```

```
  println("onCircle: " + fTimes + " , bConfident=" + bConfident);
  rot = fTimes;
  angle = (fTimes % 1.0f) * 2 * PI - PI/2;
  centerVec.set(circle.getPtCenter().getX(), circle.getPtCenter().getY(), handVec.z);
  kinect.convertRealWorldToProjective(centerVec, screenCenterVec);
  rad = circle.getFRadius();
  mode = 1;
}
```

If there is no circle, you just tell your program to go back to mode "0", or waiting mode.

```
void onNoCircle(float fTimes, int reason)
{
  println("onNoCircle: " + fTimes + " , reason= " + reason);
  mode = 0;
}
```

Draw Loop and Other Functions

The previous callback functions give you all the data you need from your Kinect device. Now you transform the data into clear signals to be sent to the Arduino, and you implement a simple interface to understand on screen what is happening in your code.

The main `draw()` loop, as usual, updates the `kinect` object, and draws the depth map on screen so you can follow what the Kinect is seeing. Then the loop runs into a switch statement that breaks the loop into three different paths.

```
void draw(){
  background(0);
  kinect.update();
  kinect.update(sessionManager);
  image(kinect.depthImage(), 0, 0);  // draw depthImageMap
```

The switch structure chooses one of the cases' blocks according to the value of the mode variable. While mode is `0` (waiting mode), you keep checking the x-component of the hand to identify the swipe gesture. Then you draw the hand position and tail.

If mode is `1` (volume control mode), you call the function `volumeControl()`, which will draw the gizmo and send the volume signal to Arduino.

If mode is `2` (channel change mode), you get stuck in a loop that sends the channel change signal to Arduino, draws the gizmo, and avoids any other signal to be attended to for a number of frames.

```
  switch(mode) {
    case 0:
    checkSpeed();
    if (handsTrackFlag) { drawHand(); }
    break;

    case 1:
    volumeControl();
    break;

    case 2:
    channelChange(changeChannel);
    channelTime++;
if (channelTime > 10) {
```

```
      channelTime = 0;
      mode = 0;
    }
    break;
  }
}
```

The checkSpeed() function is where you have implemented your simple swipe gesture recognition. If you think about it, what is different in that gesture from the normal behavior of the hand? Its horizontal speed! Actually, just by analyzing the current horizontal speed of the hand, you can detect the channel change gesture. And as you can extract the current hand position and the previous one from your handVecList, you can measure the difference in their x-coordinates to set a threshold that defines when the hand is in normal movement and when it is swiping. Set a threshold of 50mm between frames (this may need to be fine-tuned for faster or slower computers).

If you recognize a gesture, you send the program to channel change mode and then set the channelChange variable to 1 for right direction and -1 for left direction.

```
void checkSpeed() {

  if (handVecList.size() > 1) {
    PVector vel = PVector.sub(handVecList.get(0), handVecList.get(1));
    if (vel.x > 50) {
      mode = 2;
      changeChannel = 1;
    }
     else if (vel.x < -50) {
      changeChannel = -1;
      mode = 2;
    }

  }
}
```

When the main loop is in channel change mode, it calls channelChange and passes in a parameter defining the side of the gesture. If it's the first loop (channelTime is equal to 0), you send the channel change signal to the Arduino. In any case, you draw an arrow with some text to indicate that you are changing the channel (see Figure 5-15). You limit the sending of the signal to the first loop to make sure you're not sending repeated signals while the arrow is displayed on screen.

```
void channelChange(int sign) {
  String channelChange;
  pushStyle();
  if (sign==1) {
    stroke(255, 0, 0);
    fill(255, 0, 0);
    // Send the signal only if it's the first loop
    if (channelTime == 0)myPort.write(1);
    textAlign(LEFT);
    channelChange = "Next Channel";
  }
  else{
    stroke(0, 255, 0);
    fill(0, 255, 0);
    // Send the signal only if it's the first loop
    if (channelTime == 0)myPort.write(2);
```

```
    textAlign(RIGHT);
    channelChange = "Previous Channel";
  }

  // Draw the arrow on screen
  strokeWeight(10);
  pushMatrix();
  translate(width/2,height/2);
  line(0,0,sign*200,0);
  triangle(sign*200,20,sign*200,-20,sign*250,0);
  textFont(font,20);
  text(channelChange,0,40);
  popMatrix();
  popStyle();

}
```

Figure 5-15. Next channel and previous channel gestures

When you enter volume control mode in the main draw loop, you call the volumeControl method. This function draws the circle that NITE has detected by making use of the radius, center, and angle variables updated on every circle update.

You need to check if the present rotation of the hand is bigger than the previous one (meaning you are rotating clockwise) or smaller (counter-clockwise rotation). In the first case, you send the "volume up" signal to the Arduino, and in the second one, "volume down." You don't need to worry that you are sending repeated signals because that is exactly what you do when you hold the volume button down on your remote! You also display the circle and the current rotation of the hand on screen, as shown in Figure 5-16.

```
void volumeControl() {
  String volumeText = "You Can Now Change the Volume";
  fill(150);
  ellipse(screenCenterVec.x, screenCenterVec.y, 2*rad, 2*rad);
  fill(255);

  if (rot>prevRot) {
```

```
    fill(0, 0, 255);
    volumeText = "Volume Level Up";
    myPort.write(3);
  }
  else {
    fill(0, 255, 0);
    volumeText = "Volume Level Down";
    myPort.write(4);
  }

  prevRot = rot;
  text(volumeText, screenCenterVec.x, screenCenterVec.y);
  line(screenCenterVec.x, screenCenterVec.y, screenCenterVec.x+rad*cos(angle),
        screenCenterVec.y+rad*sin(angle));
}
```

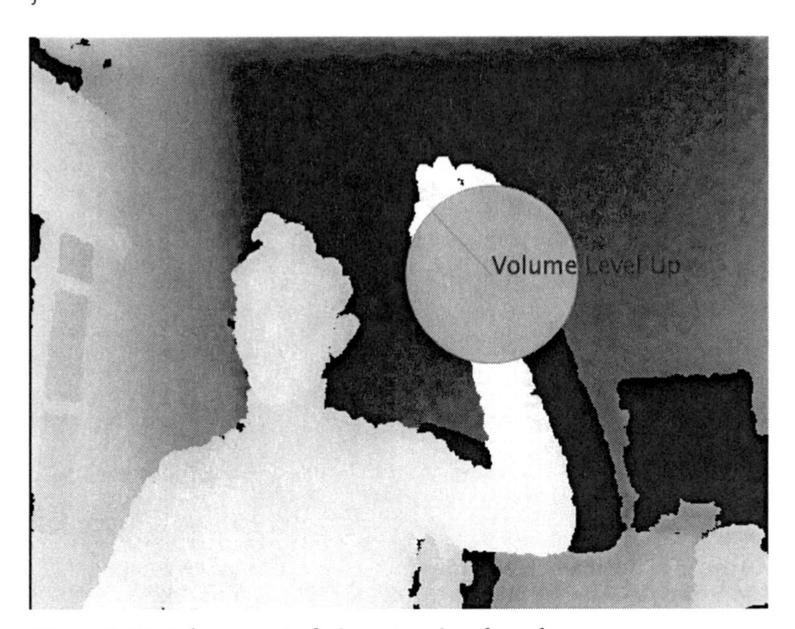

Figure 5-16. Volume control gizmo turning the volume up

Finally, add a drawHand function very similar to the one you implemented in the previous chapter (thus we won't go into the details again here).

```
void drawHand() {

  stroke(255, 0, 0);
  pushStyle();
  strokeWeight(6);
  kinect.convertRealWorldToProjective(handVec, screenHandVec);
  point(screenHandVec.x, screenHandVec.y);
  popStyle();

  noFill();
```

```
  Iterator itr = handVecList.iterator();

  beginShape();
  while ( itr.hasNext ())
  {
    PVector p = (PVector) itr.next();
    PVector sp = new PVector();
    kinect.convertRealWorldToProjective(p, sp);
    vertex(sp.x, sp.y);
  }
  endShape();

}
```

To sum up, once the sketch is up and running you can wave your hand in front of the Kinect (saying hello with your hand) and a red dot will appear in the center of your hand. This means NITE has recognized your hand and it's tracking it. You'll also see a long tail of the 30 last positions of your hand.

If you swipe your hand horizontally (much in the same way as Tom Cruise in the movie *Minority Report*), you'll see the channel change signal on screen. If you do a circular movement with your hand, you'll see the volume control gizmo. If you want to stop the volume control, you need only move your hand away from the virtual circle that you have drawn in space.

It's fun to see how awkward it is at the beginning, but you quickly get used to the different movements. If you find that the swipe gesture is too sensitive or too narrow, you can change the threshold until you feel more comfortable with it.

You have to run this code with the Arduino connected to your computer; otherwise you'll get a serial error. If you want to run it without an Arduino, you can comment out all the calls to the `serial` object.

Connecting the Processing Sketch to Arduino

So now you have a Processing sketch that recognizes several different gestures and an Arduino programmed to control a remote. And you have cunningly prepared your Processing sketch to send the same serial data that you are using to control your Arduino.

So that's it! If you connect both the Kinect and the Arduino to your computer, upload the previous code to your Arduino, and run the Processing sketch, you should be able to stand back, wave at your Kinect, and start controlling your TV with hand gestures! Play with it, and get to ready impress your guests with a session of channel zapping without a remote. They'll surely be most impressed.

Summary

This chapter led you through the process of opening a remote control, examining its inner workings, and connecting it to your Arduino board in order to control it from your computer. Then you used NITE's gesture recognition capabilities, complemented with your own implementation of a gesture recognition routine, to alter the state of the remote by using body gestures.

As a result, you have built a system that allows you to control a home appliance (a TV set in this case) by waving, swiping, and circling your hand in space without the need of any electronic device in your hand.

This example aims to constitute a base for hacking any other appliances controlled by a remote. This is only the principle. Once you have opened a TV remote control, you can investigate other connections and see which commands are sent. Most infrared remote devices work by the same principles, so you can also hack a remote-control car or helicopter and control these devices with the movement of your hands or your body! You'll see more of this in the following chapters so stay tuned!

CHAPTER 6

Kinect Networked Puppet

by Enrique Ramos and Ciriaco Castro

In this chapter, you are going to make use of one of the most amazing capabilities of the Kinect: skeleton tracking. This feature will allow you to build a gesture-based puppeteering system that you will use to control a simple, servo-driven puppet.

Natural interaction applied to puppetry has an important outcome: the puppeteer (you) and the puppet don't need to be tied by a physical connection anymore. In the second part of the chapter, you will extend your project and implement network connectivity that will permit you to control the puppet remotely from any point in the world, as shown in Figure 6-1.

This project will introduce you to the ins and outs of skeleton tracking so you can detect the user's joints and limbs movements in space. The user data will be mapped to servo angles and transmitted via the Internet to a remote receiver program, which will in turn send them to the Arduino board controlling the servos of the robotic puppet.

Throughout the following pages, you will also be learning how to drive servos from Arduino and how to communicate over networks with a few lines of code using Processing's Net library.

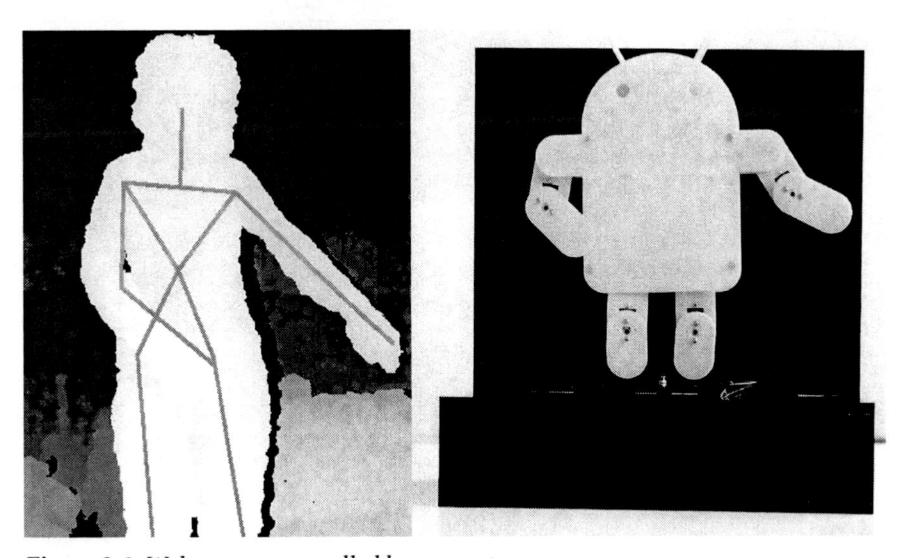

***Figure 6-1.** Web puppet controlled by a remote user*

Table 6-1. The Parts List

Part	Description	Price
8 small servos	Hi Tec HS 55	$9.99 each
1 big servo	Hi Tec HS-311	$7.99
Strip board	94 x 53 mm copper	$16.95
Extension lead cables	1 for each servo	$ 1.75 each
23 breakaway headers-straight	You should have a bunch of these already somewhere in your lab, or bedroom…	$ 1.50
1 green Perspex board	Colored acrylic 400 x 225 mm	$ 15.50
1 black Perspex board	Colored acrylic 450 x 650 mm	$ 19.50

The Puppet

We have set out to use only open source tools in this book, so we're going to use an open source design for the puppet. We have thus chosen the Android Robot by Google, the symbol for their open source mobile phone OS, which is licensed under the Creative Commons Attribution. The design will need to be slightly altered to add some joints to allow a higher range of movements. You will also design and build a stage were the robot will stand.

■ **Note** Android Robot can be found at `http://www.android.com/branding.html`. Parts of the design of this chapter's robot are modifications on work created and shared by Google and used according to terms described in the Creative Commons 3.0 Attribution License.

Servos

The movements of your puppet will be based on the rotation of nine small servos, so now is a good moment to understand the theory behind servo control. Servos are DC motors that include a feedback mechanism that allows you to keep track of the position of the motor at every moment. The way you control a servo is by sending it pulses with a specific duration. Generally, the range of a servo goes from 1 millisecond for 0 degrees to 2 milliseconds for 180 degrees. There are servos that can rotate 360 degrees, and you can even hack servos to allow for continuous rotation.

You need to refresh your control signals every 20ms because the servo expects to get an update every 20ms, or 50 times per second. Figure 6-2 shows the normal pattern for servo control.

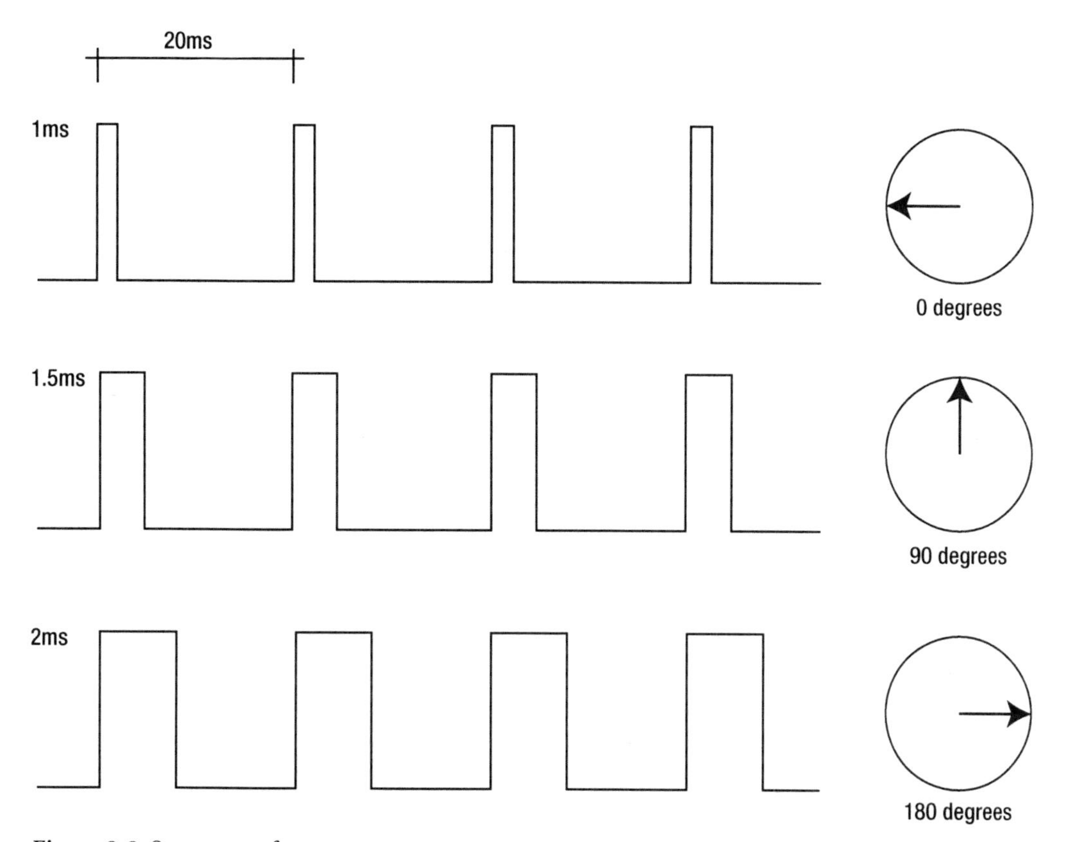

Figure 6-2. Servo wave form

In Chapter 3, you learned about pulse width modulation (PWM) and you used it to dim an LED lamp. The way you control servos is very similar to PWM, but in this case, you will be dealing with the pulsing yourself instead of sending an analog value and allowing Arduino to break it down to pulses.

There is a Servo library included with Arduino, but you are going to write your own servo control functions. This will save you from a lot of problems when working with a number of servos. Let's see the differences in the two approaches to servo control with a simple example.

For the sake of this example, you will connect a servo to the digital pin 5 and control it with a potentiometer connected to analog pin 0. The following is the code you need for that using the Servo library. First, import the Servo library and declare the servo object.

```
#include <Servo.h>
Servo servo1;
```

In `setup()`, use the `Servo.attach()` function to set the servo to pin 5.

```
void setup() {
  servo1.attach(5);
}
```

And in `loop()`, read the potentiometer value and map its values to a range of 0-180. Finally, write the angle value to the servo and add a delay to allow the servo to reach the new position.

```
void loop()
{
  int angle = analogRead(0);
  angle = map(angle, 0, 1023, 0, 180);
  servo1.write(angle);
  delay(15);
}
```

If you were using your own servo control instead, this would look something like the following:

```
int servoPin = 5;
unsigned long previousMillis = 0;
long interval = 20;

void setup() {
  pinMode (servoPin, OUTPUT);
}

void loop() {
  int pot = analogRead(0);
  int servoPulse = map(pot,0,1023,500,2500);
  // Update servo only if 20 milliseconds have elapsed since last update
  unsigned long currentMillis = millis();
  if(currentMillis - previousMillis > interval) {
    previousMillis = currentMillis;
    updateServo(servoPin, servoPulse);
  }
}

void updateServo (int pin, int pulse){
  digitalWrite(pin, HIGH);
  delayMicroseconds(pulse);
  digitalWrite(pin, LOW);
}
```

This requires a little explanation. The updating of the servos follows a very similar logic to the “blink without delay” Arduino example. (This is one of the examples included with Arduino IDE and is used to control time-depending events without the use of a delay. If you have never seen this sketch, go to the Arduino examples and have a good look at it. You will find it within 2.Digital in the Arduino examples.) You need to send a pulse to the servo every 20ms but you don’t want to add delays to your code, so you’re not interfering with other time-dependent routines included in the future project. When you have been running for longer than the specified interval, you update the servo. Note that the servo sends a HIGH pulse and then there is a delay. This one is inevitable because you need to send a very precise pulse, and anyway it’s such a short delay (in microseconds) that you can live with it.

If you want to control several servos at the same time, you only need to add some more pins and update all of them at the same time! Let’s add another servo and make it move just through half its rotational span (90 degrees).

```
int servo1Pin = 5;
int servo2Pin = 6;
unsigned long previousMillis = 0;
long interval = 20;
```

```
void setup() {
  pinMode (servo1Pin, OUTPUT);
  pinMode (servo2Pin, OUTPUT);
}

void loop() {
  int pot = analogRead(0); // Read the potentiometer
  int servo1Pulse = map(pot,0,1023,1000,2000);
  int servo2Pulse = map(pot,0,1023,1500,2000);

  // Update servo only if 20 milliseconds have elapsed since last update
  unsigned long currentMillis = millis();
  if(currentMillis - previousMillis > interval) {
    previousMillis = currentMillis;
    updateServo(servo1Pin, servo1Pulse);
    updateServo(servo2Pin, servo2Pulse);
  }
}

void updateServo (int pin, int pulse){
  digitalWrite(pin, HIGH);
  delayMicroseconds(pulse);
  digitalWrite(pin, LOW);
}
```

This control technique is the one you will be using throughout the rest of the chapter and in further chapters that make use of servos. Try to familiarize yourself with the dynamics of servo control with a couple of servos only. When you are ready, you can move on to the following sections where you will be controlling up to nine servos at the same time.

Building the Stage

Now that you know what a servo is and how it works, let's move on to an easier bit: the stage. Basically you want a background and some support from which to hang the puppet. You're using black Perspex so your robot will stand out. The stage consists of a hollow base to hide all the mechanisms plus a backdrop, as you can see in Figure 6-4.

Figure 6-3 shows the stage parts as per the Illustrator file provided (C06_Stage_Parts.ai). The two small rectangular pieces have been added in order to support your Arduino and give stability to the stage. Note that the front is slightly raised to hide the gap between the robot's feet and the stage.

The two round holes on the central element permit the servo cables to reach the Arduino board at the base. The rectangular hole is for the servo that rotates the robot.

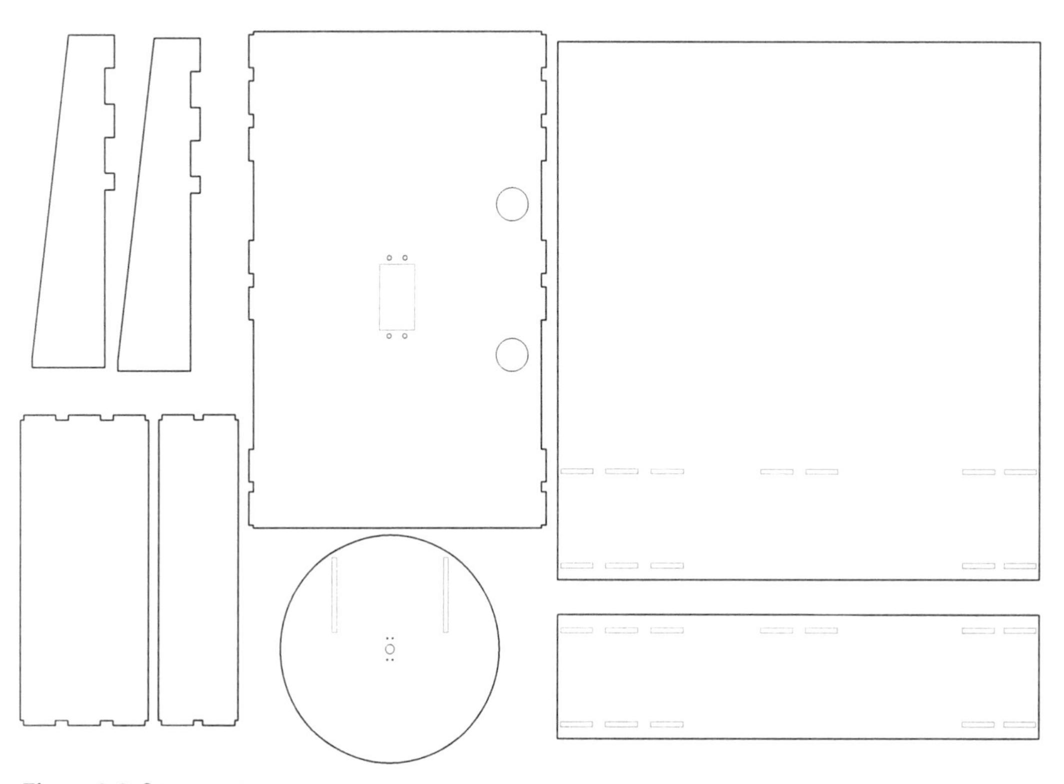

***Figure 6-3.** Stage parts*

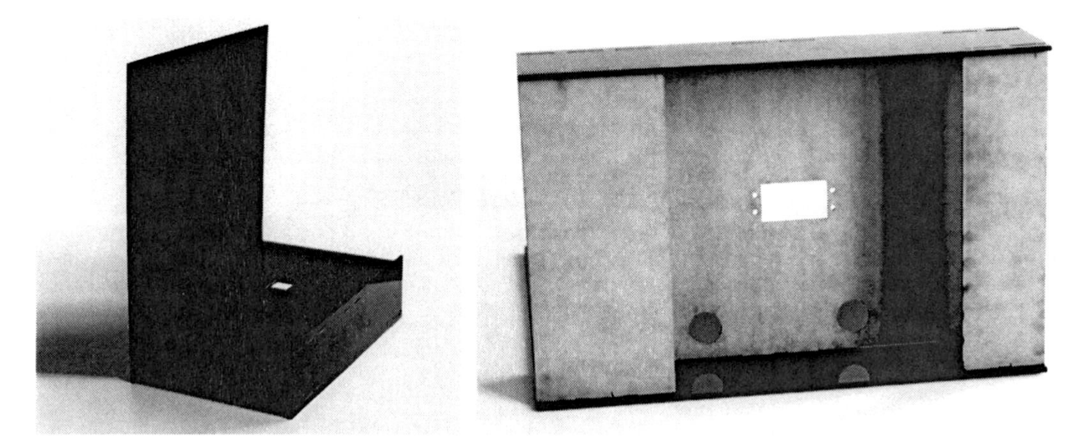

***Figure 6-4.** Assembled stage*

The puppet rotation servo is attached to the base with four screws, as shown in Figure 6-5.

***Figure 6-5.** Servo fixed to the base*

The robot is attached to two supports that are also made of black Perspex so they don't stand out from the background, as shown in Figure 6-6. These supports are connected to a disc, thereby allowing rotation. This disc will be screwed to a servo circular horn, and then attached to the rotation servo that you have previously fixed to the base.

***Figure 6-6.** Servo arm fixed to disc (left), and robot supports (right)*

Building the Puppet

The puppet is made of laser-cut green Perspex. If you are thinking of using the same design, you can use the Illustrator vector files included with this book (`C06_Puppet_Parts.ai`) and shown in Figure 6-7.

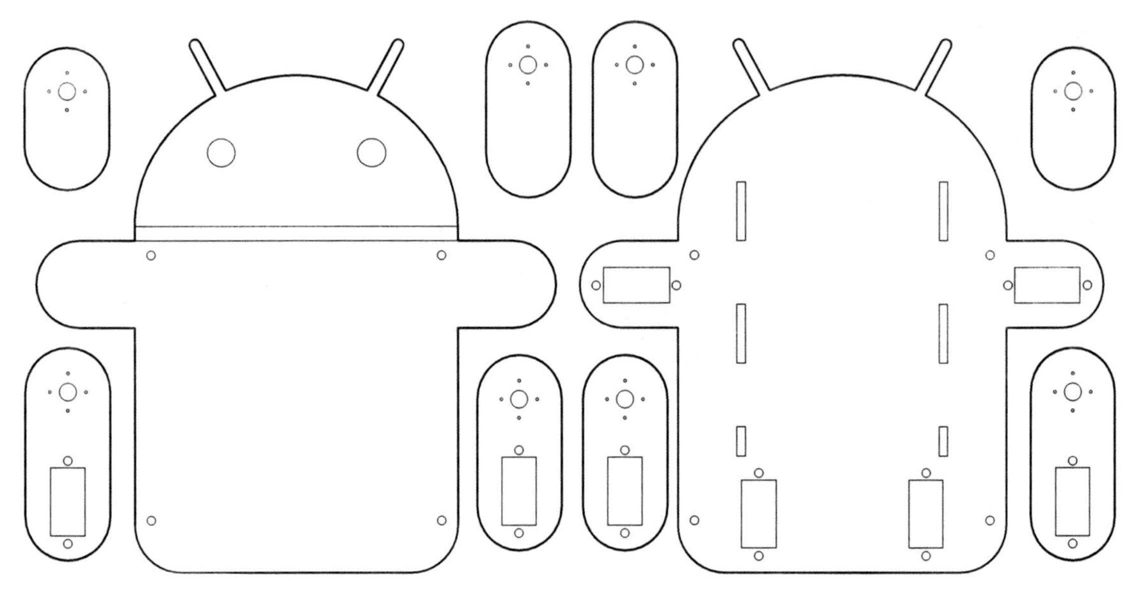

***Figure 6-7.** Puppet pieces*

There is a front body that covers up all the mechanisms and a back body that connects all the servos and is fixed to the support. Both bodies are connected by four bolts passing through the holes provided for that purpose, which you can see in Figure 6-8.

***Figure 6-8.** Assembling servos to the back body*

Now you can start to assemble the legs and arms, screwing the double servo arm, and fixing a second servo, as shown in Figure 6-9.

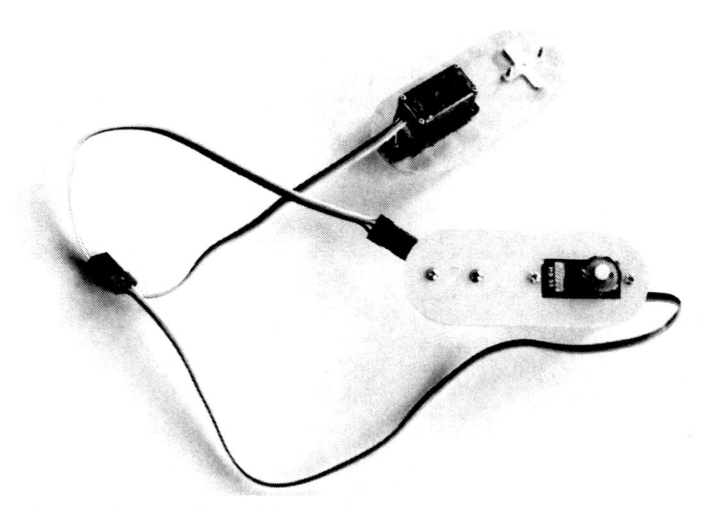

Figure 6-9. *Arms/legs with servomotors*

Before screwing and fixing all arms to the corresponding servos, you should set all servos to the middle position (1500) so later you can map them in an easy way. The complete arm will look like that in Figure 6-10.

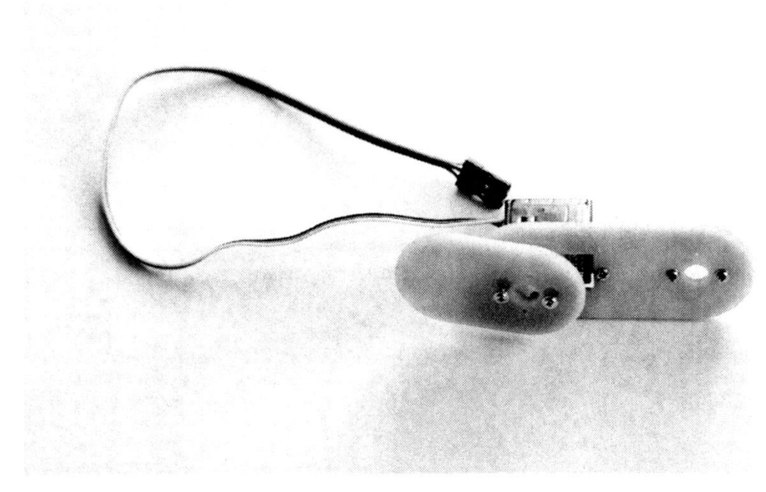

Figure 6-10. *Arm assembled*

Repeat the same process with the other arm and legs. Once you have all these parts assembled, attach the arms and legs to the back body piece and fix the puppet onto the stage, as shown in Figure 6-11. You're now ready to build the circuit!

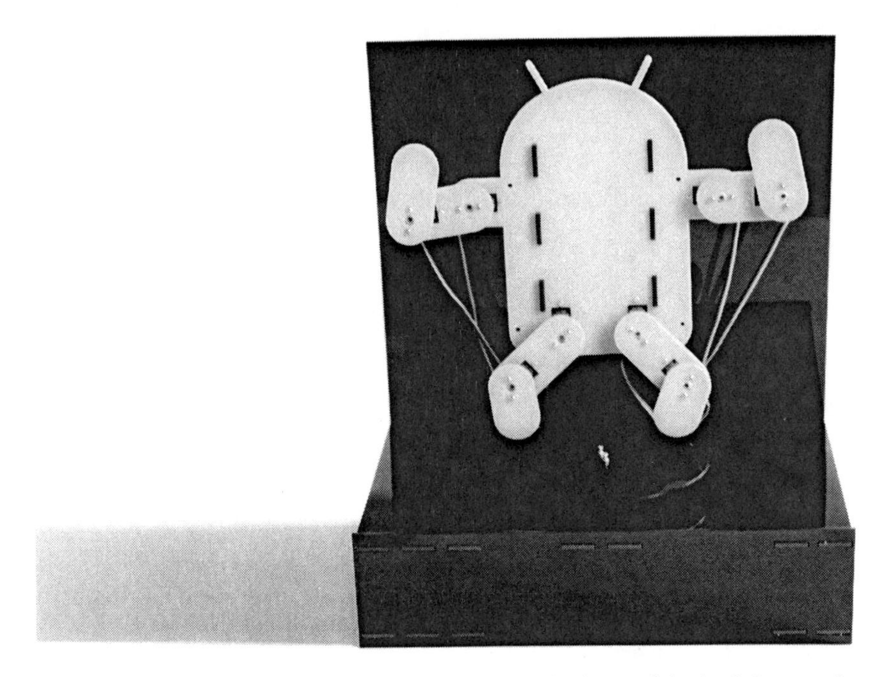

Figure 6-11. *The assembled puppet on the stage. The front of the body has not been attached yet.*

The front body piece is the last thing that you assemble once everything is double-checked. Now you have the puppet up and running.

Building the Circuit

The circuit that you are going to prepare is going to be plugged into the Arduino board and is designed to control nine servomotors. You first run all servos using just the power of your USB connection, but you then add an external power source, as all servos working at the same time will exceed the power of the USB (Figure 6-12). You will use Arduino pins 3, 4, 5, 6, 7, 8, 9, 10, and 11, following the diagram in Figure 6-13.

Build your circuit on a strip board. Start by measuring the amount of strips that you need and soldering the breakaway headers. Make sure that the Arduino USB cable is at one extreme of the board so the cable can be plugged and unplugged without interfering with the breakaway headers.

Figure 6-12. *Circuit components*

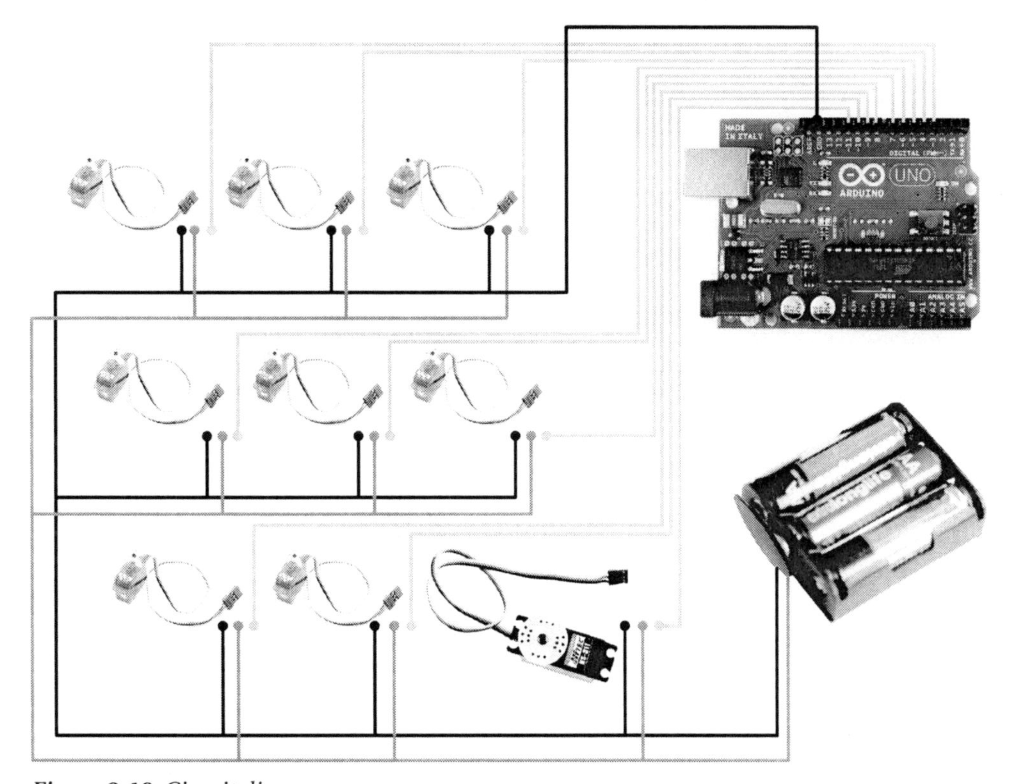

Figure 6-13. *Circuit diagram*

Double-check the direction of the copper strips on the back of the strip board, and start scratching the strips so you are sure that you don't connect the two rows of pins on your Arduino.

You can use a potentiometer to check the resistance between strips to make sure that there is no connection. After that, you are ready to start soldering the breakaway header corresponding to Arduino pins 3, 4, 5, 6, 7, 8, 9, 10, and 11. Then you solder a breakaway header for ground and 5V.

After soldering, double-check that they are placed correctly by plugging the shield into your Arduino board and making sure that it attaches easily (Figure 6-14). You might need to bend some of the headers with a nipper to get the shield properly connected.

Figure 6-14. *Strip board and Arduino*

Now start soldering the breakaway headers for your servos. Before doing this, look at the servo cables. Servos have three cables: black for ground, red (middle one) for power, and yellow or brown for the signal. The signal comes from the Arduino pin; power and ground can come from Arduino 5V and ground or from an external power source. The important thing is that the circuit ground is connected to Arduino ground; otherwise you won't be able to control your servos, and they will behave in strange ways.

Due to the geometry of the strip board, you should arrange the headers in two columns, leaving a group of four headers together (2 for arms, 2 for legs) and one for the rotation of the robot. The circuit will look like that in Figure 6-15.

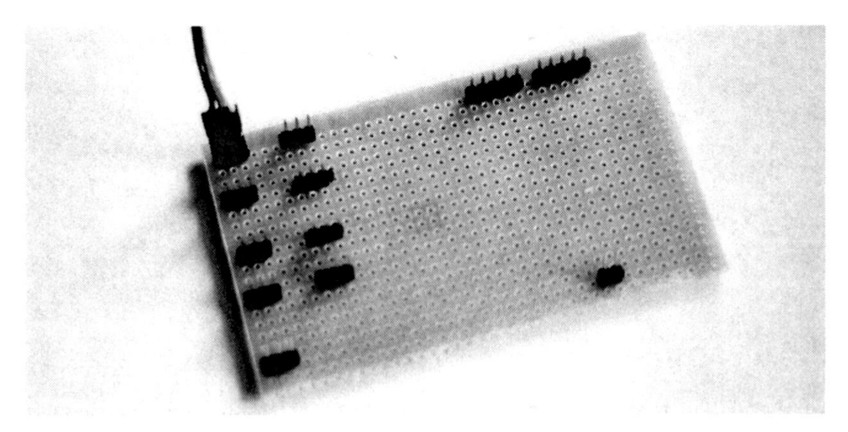

Figure 6-15. *Strip board and headers*

After completing the soldering, it's time to scratch the connection between the signals (yellow cable of the servo motors). With a little patience and a scalpel, remove the copper that connects the line (see Figure 6-16). After you have removed all the material, check that the connection is completely broken by using a multimeter.

Figure 6-16. *Scratched strip board*

The next step is to connect power and ground. Solder a cable that connects the 5V pin to the power column (middle pin of the header) and a small cable that connects the power of the other column (Figure 6-17). Repeat the same process with the ground (one cable form Arduino ground to the ground column, and a small cable connecting the two columns ground between them).

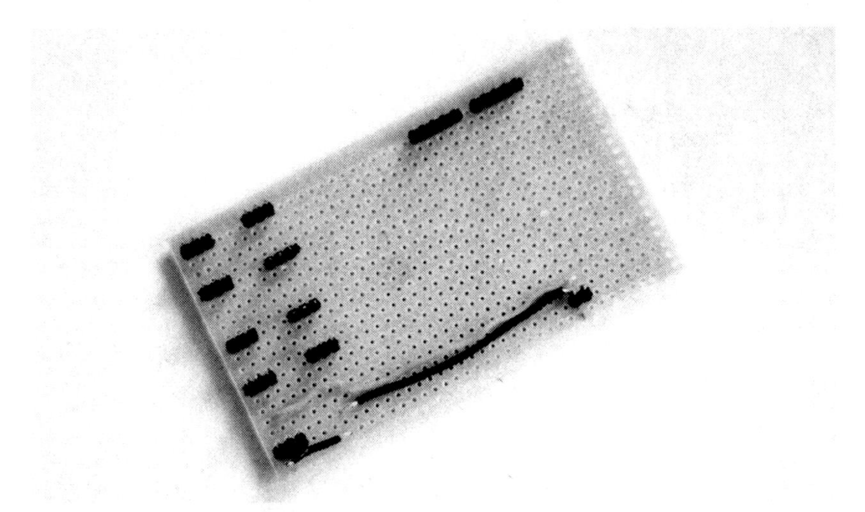

Figure 6-17. *Strip board, power, and ground*

Now, connect all the signals to the corresponding pin. You need to solder one cable connecting the Arduino pin strip to the header signal. It is important that you do this in a tidy and orderly manner (see Figures 6-18 and 6-19) so it's easy for you to remember which pin number connects to each servo.

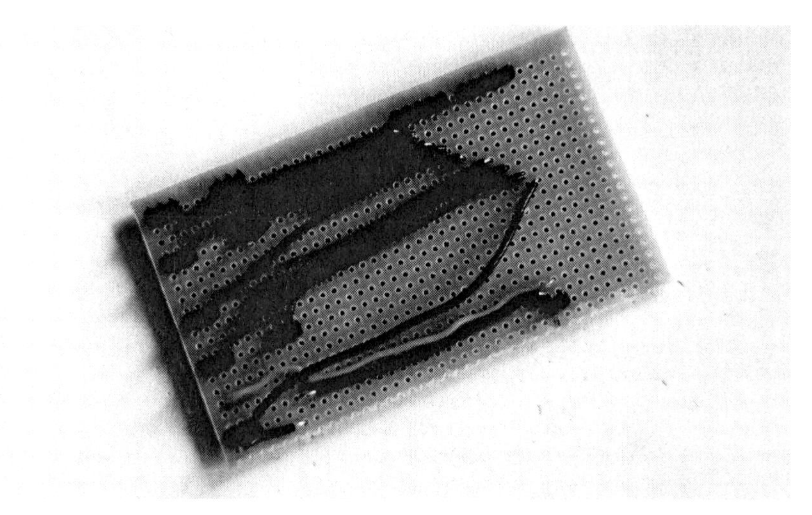

Figure 6-18. Strip board, all pins connected front

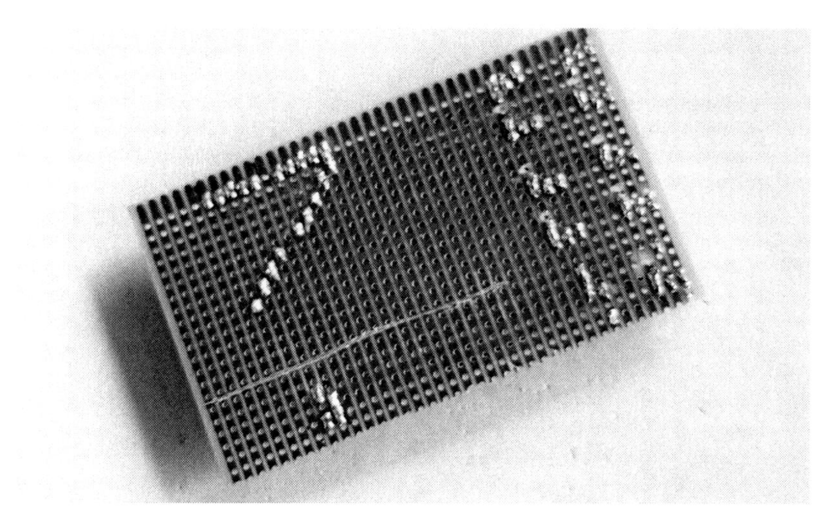

Figure 6-19. Strip board, all pins connected back

Then you can plug in your Arduino (Figure 6-20) and do a first test.

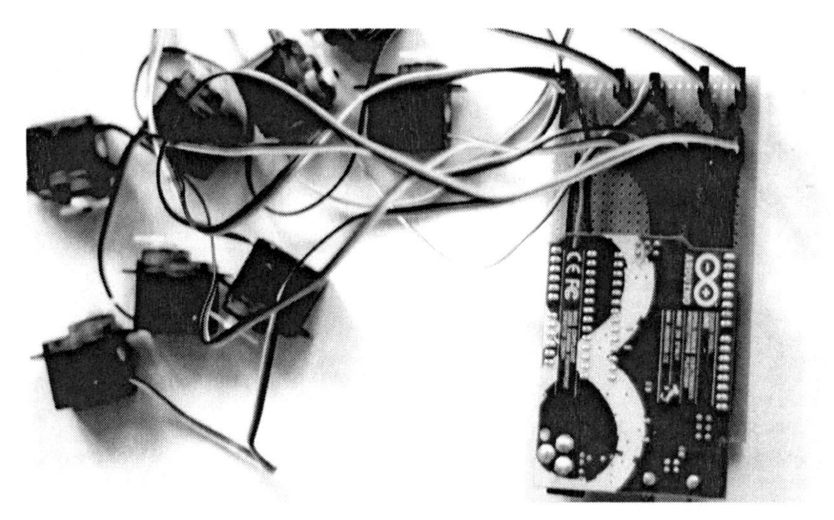

Figure 6-20. Strip board, connected for test

Testing the Servos

The code for this first test is quite similar to the one in the servo introduction, but instead of using a potentiometer you are going to keep the servo moving from 0 to 180 degrees and back to zero. This way you can check if all of the servos work properly or if you have to debug the circuit (making sure that the soldering is right!).

The code calls all the servo pins as outputs and updates each pin with a pulse that will change between values 500-2500 (0-180 degrees). Once it arrives at 2500, it will decrease to 500.

```
int servo3Pin = 3;
int servo4Pin = 4;
int servo5Pin = 5;
int servo6Pin = 6;
int servo7Pin = 7;
int servo8Pin = 8;
int servo9Pin = 9;
int servo10Pin = 10;
int servo11Pin = 11;

int servoPulse = 1500;
int speedServo = 50;

unsigned long previousMillis = 0;
long interval = 20;

void setup() {
  pinMode (servo3Pin, OUTPUT);
  pinMode (servo4Pin, OUTPUT);
  pinMode (servo5Pin, OUTPUT);
  pinMode (servo6Pin, OUTPUT);
```

```
  pinMode (servo7Pin, OUTPUT);
  pinMode (servo8Pin, OUTPUT);
  pinMode (servo9Pin, OUTPUT);
  pinMode (servo10Pin, OUTPUT);
  pinMode (servo11Pin, OUTPUT);
}

void loop() {

  unsigned long currentMillis = millis();
  if(currentMillis - previousMillis > interval) {
    previousMillis = currentMillis;

    updateServo(servo3Pin, servoPulse);
    updateServo(servo4Pin, servoPulse);
    updateServo(servo5Pin, servoPulse);
    updateServo(servo6Pin, servoPulse);
    updateServo(servo7Pin, servoPulse);
    updateServo(servo8Pin, servoPulse);
    updateServo(servo9Pin, servoPulse);
    updateServo(servo10Pin, servoPulse);
    updateServo(servo11Pin, servoPulse);

     servoPulse += speedServo;
     if(servoPulse > 2500 || servoPulse <500){
     speedServo *= -1;
    }
  }
}

void updateServo (int pin, int pulse){
  digitalWrite(pin, HIGH);
  delayMicroseconds(pulse);
  digitalWrite(pin, LOW);
}
```

Once you upload the code, all servos should start to move. If one servo doesn't move, check the circuit, the connections, and the Arduino pin (in case the fault is in the code).

Obviously, the Arduino 5V power isn't enough for running all nine servos at the same time. You can see that the Arduino power LED starts to fade out and blink; also, the servos aren't moving properly and instead make a hiss, like they are struggling. To avoid this, add an external power source to your circuit.

The first thing is to cut the cable that connects with the Arduino 5V pin (you don't want to fry your Arduino!). Then you connect a 3-battery clip to your circuit (see Figure 6-21). Solder the red cable power to the servo power stripe and the ground to the servo ground stripe. Note that ground is still connected to Arduino ground.

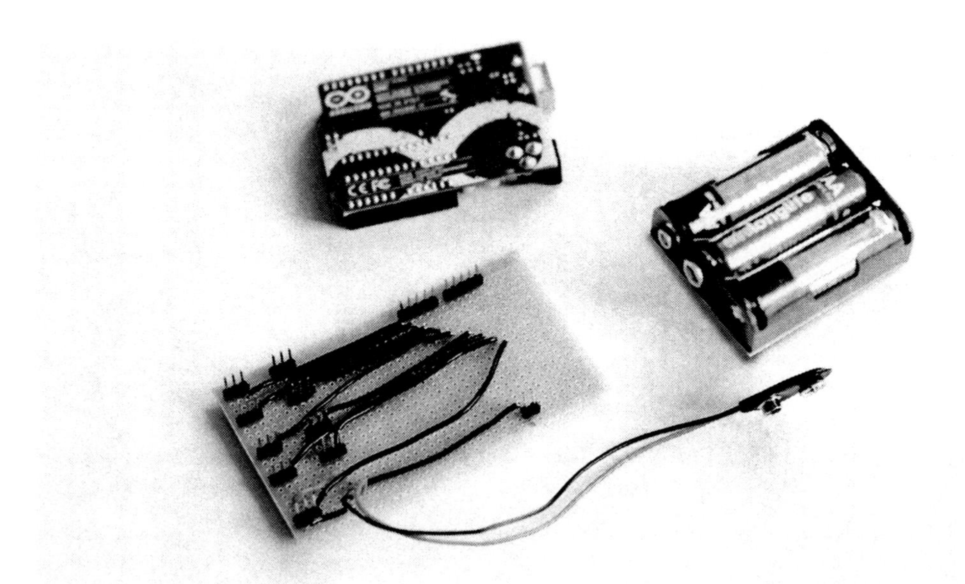

Figure 6-21. Circuit with battery clip

So now you can have a pack of four AA batteries in a battery holder connected to the battery clip. This should be enough power to move all your servos. Another option is to connect a power adapter plugged into a mains socket, but make sure that the output voltage is 5V. Later in the book, we will explain another way of getting a constant 5V current using a computer power supply.

Setting the Servos to the Starting Position

Reconnect your Arduino and the batteries and check that all servos move smoothly. You're almost ready to start coding. The last thing you need to do is set all servos to a starting position, so you can fix all of the servo arms.

You're going to send all shoulder and leg servos to 500 (0 degrees) or 2500 (180 degrees) depending if they are on the left side or right side, and the elbows to 1500 (90 degrees), allowing the movement in two directions. The last thing is to position the servo that will rotate the puppet to 1500 (90 degrees). Figure 6-22 shows all the angles.

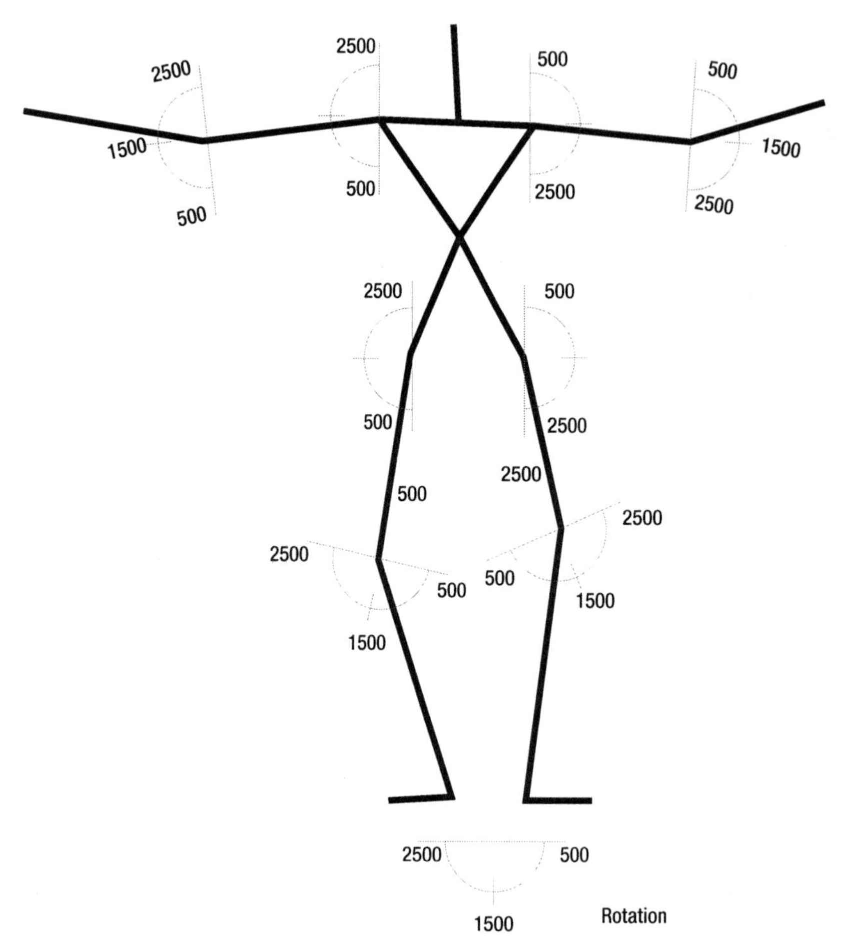

Figure 6-22. Puppet servo angles

The code is similar to the previous; you're just commenting or deleting some lines (the ones that change the servo pulse) and adding new servo pulses for the 90 and 180 degrees position.

```
int servo3Pin = 3;
int servo4Pin = 4;
int servo5Pin = 5;
int servo6Pin = 6;
int servo7Pin = 7;
int servo8Pin = 8;
int servo9Pin = 9;
int servo10Pin = 10;
int servo11Pin = 11;

int servoPulse = 500;
int servoPulse2 = 1500;
```

```
int servoPulse3 = 2500;
int speedServo = 50;

unsigned long previousMillis = 0;
long interval = 20;

void setup() {
  pinMode (servo3Pin, OUTPUT);
  pinMode (servo4Pin, OUTPUT);
  pinMode (servo5Pin, OUTPUT);
  pinMode (servo6Pin, OUTPUT);
  pinMode (servo7Pin, OUTPUT);
  pinMode (servo8Pin, OUTPUT);
  pinMode (servo9Pin, OUTPUT);
  pinMode (servo10Pin, OUTPUT);
  pinMode (servo11Pin, OUTPUT);
}

void loop() {
unsigned long currentMillis = millis();
if(currentMillis - previousMillis > interval) {
    previousMillis = currentMillis;
    updateServo(servo3Pin, servoPulse);//lef Shoulder
    updateServo(servo4Pin, servoPulse2);//left Elbow
    updateServo(servo5Pin, servoPulse);//left Hip
    updateServo(servo6Pin, servoPulse2);//left Knee
    updateServo(servo7Pin, servoPulse3); //right Shoulder
    updateServo(servo8Pin, servoPulse2);//rigt Elbow
    updateServo(servo9Pin, servoPulse3);// right Hip
    updateServo(servo10Pin, servoPulse2);//right Knee
    updateServo(servo11Pin, servoPulse2);//move it to the central position
  }
}

void updateServo (int pin, int pulse){
  digitalWrite(pin, HIGH);
  delayMicroseconds(pulse);
  digitalWrite(pin, LOW);
}
```

That servo position will be the starting position of the puppet. Once this position has been fixed, you know the orientation of all limbs so you can screw in your servo arms. Then you cover everything with the front body piece and use some cable ties to hold the cables in position (see Figure 6-23). Just make sure you keep enough length to allow the movement!

Figure 6-23. *Puppet completed*

Skeleton Tracking On-Screen

The first thing you need to get sorted for this project is tracking your skeleton with Kinect. We briefly described skeletonization in Chapter 3, but here you are going to delve deeply into it, so hang on!

You will start off by tracking the skeleton on-screen, and then you will attach the virtual puppet to the physical one you have just built. Simple-OpenNI provides a nice interface for skeleton tracking and a couple of callback functions to which you can add your own code. Start by importing and initializing Simple-OpenNI.

```
import SimpleOpenNI.*;
SimpleOpenNI kinect;
```

Set the mirroring to On so it's easy for you to relate to your on-screen image, and enable the depth image and the User. Pass the parameter `SimpleOpenNI.SKEL_PROFILE_ALL` to this function to enable all the joints. Set your sketch size to the depth map dimensions.

```
public void setup() {
  kinect = new SimpleOpenNI(this);
  kinect.setMirror(true);
  kinect.enableDepth();
  kinect.enableUser(SimpleOpenNI.SKEL_PROFILE_ALL);
  size(kinect.depthWidth(), kinect.depthHeight());
}
```

In the `draw()` loop, simply update the Kinect data, draw the depth map, and, if Kinect is tracking a skeleton, call the function `drawSkeleton()` to print your skeleton on screen.

```
public void draw() {
  kinect.update();
  image(kinect.depthImage(), 0, 0);
  if (kinect.isTrackingSkeleton(1)) {
    drawSkeleton(1);
  }
}
```

We took this drawSkeleton() function from one of the examples in Simple-OpenNI. It takes an integer as a parameter, which is the user ID (it will be always 1 if you are just tracking a single skeleton). This function runs through all the necessary steps to link the skeleton joints with lines, defining the skeleton that you will see on screen. The function makes use of the public method drawLimb() from the Simple-OpenNI class.

```
void drawSkeleton(int userId) {
  pushStyle();
  stroke(255,0,0);
  strokeWeight(3);
  kinect.drawLimb(userId, SimpleOpenNI.SKEL_HEAD, SimpleOpenNI.SKEL_NECK);
  kinect.drawLimb(userId, SimpleOpenNI.SKEL_NECK, SimpleOpenNI.SKEL_LEFT_SHOULDER);
  kinect.drawLimb(userId, SimpleOpenNI.SKEL_LEFT_SHOULDER, SimpleOpenNI.SKEL_LEFT_ELBOW);
  kinect.drawLimb(userId, SimpleOpenNI.SKEL_LEFT_ELBOW, SimpleOpenNI.SKEL_LEFT_HAND);

  kinect.drawLimb(userId, SimpleOpenNI.SKEL_NECK, SimpleOpenNI.SKEL_RIGHT_SHOULDER);
  kinect.drawLimb(userId, SimpleOpenNI.SKEL_RIGHT_SHOULDER, SimpleOpenNI.SKEL_RIGHT_ELBOW);
  kinect.drawLimb(userId, SimpleOpenNI.SKEL_RIGHT_ELBOW, SimpleOpenNI.SKEL_RIGHT_HAND);
  kinect.drawLimb(userId, SimpleOpenNI.SKEL_LEFT_SHOULDER, SimpleOpenNI.SKEL_TORSO);
  kinect.drawLimb(userId, SimpleOpenNI.SKEL_RIGHT_SHOULDER, SimpleOpenNI.SKEL_TORSO);
  kinect.drawLimb(userId, SimpleOpenNI.SKEL_TORSO, SimpleOpenNI.SKEL_LEFT_HIP);
  kinect.drawLimb(userId, SimpleOpenNI.SKEL_LEFT_HIP, SimpleOpenNI.SKEL_LEFT_KNEE);
  kinect.drawLimb(userId, SimpleOpenNI.SKEL_LEFT_KNEE, SimpleOpenNI.SKEL_LEFT_FOOT);
  kinect.drawLimb(userId, SimpleOpenNI.SKEL_TORSO, SimpleOpenNI.SKEL_RIGHT_HIP);
  kinect.drawLimb(userId, SimpleOpenNI.SKEL_RIGHT_HIP, SimpleOpenNI.SKEL_RIGHT_KNEE);
  kinect.drawLimb(userId, SimpleOpenNI.SKEL_RIGHT_KNEE, SimpleOpenNI.SKEL_RIGHT_FOOT);
  popStyle();
}
```

Simple-OpenNI Events

Use the Simple-OpenNI callback functions to trigger the pose detection and skeleton tracking capabilities. The function onNewUser() is called when a user is detected. You know the user is there, but you haven't calibrated their skeleton yet, so you start the pose detection.

```
public void onNewUser(int userId) {
  println("onNewUser - userId: " + userId);
  if (kinect.isTrackingSkeleton(1))  return;
  println("  start pose detection");
  kinect.startPoseDetection("Psi", userId);
}
```

When the user is lost, you simply print it on the console.

```
public void onLostUser(int userId) {
  println("onLostUser - userId: " + userId);
}
```

When you detect a pose, you can stop detecting poses (you already have one!) and request a skeleton calibration to NITE.

```
public void onStartPose(String pose, int userId) {
  println("onStartPose - userId: " + userId + ", pose: " + pose);
  println(" stop pose detection");
  kinect.stopPoseDetection(userId);
  kinect.requestCalibrationSkeleton(userId, true);
}
```

You also display messages when the pose is finished and when the calibration has started.

```
public void onEndPose(String pose, int userId) {
  println("onEndPose - userId: " + userId + ", pose: " + pose);
}
public void onStartCalibration(int userId) {
  println("onStartCalibration - userId: " + userId);
}
```

If the calibration is finished and it was successful, you start tracking the skeleton. If it wasn't, you restart the pose detection routine.

```
public void onEndCalibration(int userId, boolean successfull) {
  println("onEndCalibration - userId: " + userId + ", successfull: " + successfull);
  if (successfull) {
    println("  User calibrated !!!");
    kinect.startTrackingSkeleton(userId);
  }
  else {
    println("  Failed to calibrate user !!!");
    println("  Start pose detection");
    kinect.startPoseDetection("Psi", userId);
  }
}
```

Run this sketch and stand in front of your Kinect in the start pose. After a few seconds, your skeleton should appear on screen, and follow your every movement until you disappear from screen (see Figure 6-24).

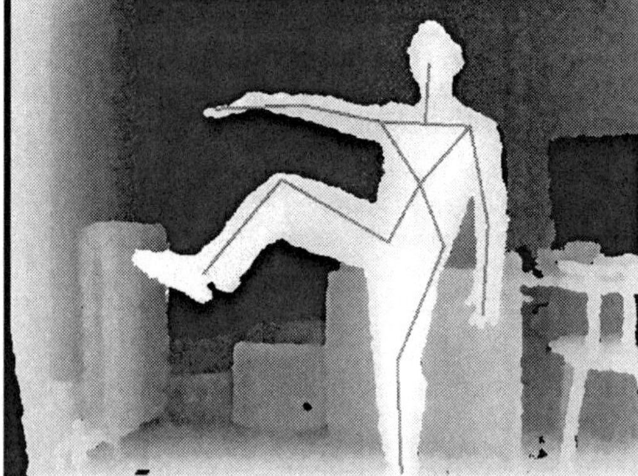

Figure 6-24. *Start position and calibrated user*

Angle Calculation

With the previous program, you were able to track your skeleton and display it on screen. That is the base of your puppet control. Now you need to implement a routine that translates your poses to the angles of eight servos that will ultimately drive the puppet.

The eight servos are placed on the shoulders, elbows, hips, and knees of your puppet. If you can find out the current bending angles of your joints, you can map those angles to the servos and thus make the puppet move like you. You implement this with the cunning use of geometry.

Taking the previous code as a base, you include a series of routines that will use the skeleton to calculate all the joints' angles. First, add some PVectors at the beginning of the sketch (outside of any function) to store the position of all your joints.

```
// Left Arm Vectors
PVector lHand = new PVector();
PVector lElbow = new PVector();
PVector lShoulder = new PVector();
// Left Leg Vectors
PVector lFoot = new PVector();
PVector lKnee = new PVector();
PVector lHip = new PVector();
// Right Arm Vectors
PVector rHand = new PVector();
PVector rElbow = new PVector();
PVector rShoulder = new PVector();
// Right Leg Vectors
PVector rFoot = new PVector();
PVector rKnee = new PVector();
PVector rHip  = new PVector();
```

You also need a PVector array to contain the angles of your joints.

```
float[] angles = new float[9];
```

You're interested in nine angles, one for each servo on your puppet.

- `angles [0]` Rotation of the body
- `angles [1]` Left elbow
- `angles [2]` Left shoulder
- `angles [3]` Left knee
- `angles [4]` Left Hip
- `angles [5]` Right elbow
- `angles [6]` Right shoulder
- `angles [7]` Right knee
- `angles [8]` Right Hip

You will be wrapping all these calculations into a new function called `updateAngles()` that you call from your main `draw()` function only in the case you are tracking a skeleton. Add this function to the if statement at the end of the `draw()` function, and then print out the values of the `angles` array.

```
  if (kinect.isTrackingSkeleton(1)) {
    updateAngles();
   println(angles);
    drawSkeleton(1);
}
```

You need a function that calculates the angle described by the lines joining three points, so implement it next; call it the angle() function.

```
float angle(PVector a, PVector b, PVector c) {

  float angle01 = atan2(a.y - b.y, a.x - b.x);
  float angle02 = atan2(b.y - c.y, b.x - c.x);
  float ang = angle02 - angle01;
  return ang;
}
```

The function updateAngles() performs the calculation of all these angles and stores them into the previously defined angles[] array. The first step is storing each joint position in one PVector. The method getJointPosition() helps you with this process. It takes three parameters: the first one is the ID of the skeleton, the second one is the joint you want to extract the coordinates from, and the third one is the PVector that is set to those coordinates.

```
void updateAngles() {
  // Left Arm
  kinect.getJointPositionSkeleton(1, SimpleOpenNI.SKEL_LEFT_HAND, lHand);
  kinect.getJointPositionSkeleton(1, SimpleOpenNI.SKEL_LEFT_ELBOW, lElbow);
  kinect.getJointPositionSkeleton(1, SimpleOpenNI.SKEL_LEFT_SHOULDER, lShoulder);
  // Left Leg
  kinect.getJointPositionSkeleton(1, SimpleOpenNI.SKEL_LEFT_FOOT, lFoot);
  kinect.getJointPositionSkeleton(1, SimpleOpenNI.SKEL_LEFT_KNEE, lKnee);
  kinect.getJointPositionSkeleton(1, SimpleOpenNI.SKEL_LEFT_HIP, lHip);
  // Right Arm
  kinect.getJointPositionSkeleton(1, SimpleOpenNI.SKEL_RIGHT_HAND, rHand);
  kinect.getJointPositionSkeleton(1, SimpleOpenNI.SKEL_RIGHT_ELBOW, rElbow);
  kinect.getJointPositionSkeleton(1, SimpleOpenNI.SKEL_RIGHT_SHOULDER, rShoulder);
  // Right Leg
  kinect.getJointPositionSkeleton(1, SimpleOpenNI.SKEL_RIGHT_FOOT, rFoot);
  kinect.getJointPositionSkeleton(1, SimpleOpenNI.SKEL_RIGHT_KNEE, rKnee);
  kinect.getJointPositionSkeleton(1, SimpleOpenNI.SKEL_RIGHT_HIP, rHip);
```

Now you have all your joints nicely stored in the PVectors you defined for that purpose. These are three-dimensional vectors of real-world coordinates. You transform them to projective coordinates so you get the second angles (remember your puppet is pretty flat!), but first you need to extract your body rotation angle (see Figure 6-25), which is the only rotation of the plane that you need.

```
  angles[0] = atan2(PVector.sub(rShoulder, lShoulder).z,
  PVector.sub(rShoulder, lShoulder).x);
```

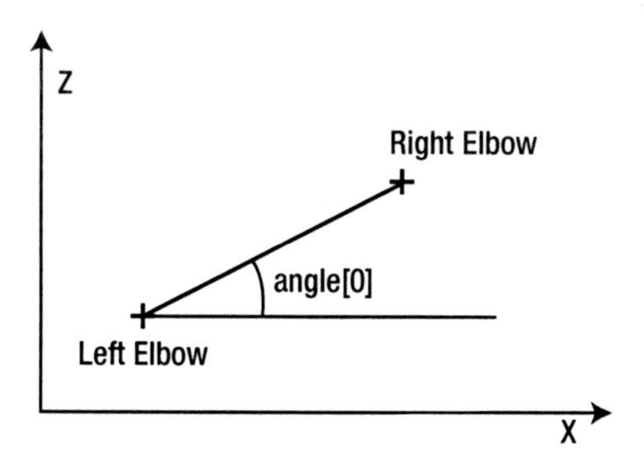

Figure 6-25. Rotation of the body

After this angle has been calculated, you can transform all the joints to projective coordinates (on screen coordinates).

```
kinect.convertRealWorldToProjective(rFoot, rFoot);
kinect.convertRealWorldToProjective(rKnee, rKnee);
kinect.convertRealWorldToProjective(rHip, rHip);
kinect.convertRealWorldToProjective(lFoot, lFoot);
kinect.convertRealWorldToProjective(lKnee, lKnee);
kinect.convertRealWorldToProjective(lHip, lHip);
kinect.convertRealWorldToProjective(lHand, lHand);
kinect.convertRealWorldToProjective(lElbow, lElbow);
kinect.convertRealWorldToProjective(lShoulder, lShoulder);
kinect.convertRealWorldToProjective(rHand, rHand);
kinect.convertRealWorldToProjective(rElbow, rElbow);
kinect.convertRealWorldToProjective(rShoulder, rShoulder);
```

And finally, you use the angle function you implemented previously to compute all the necessary angles for the control of the puppet.

```
  // Left-Side Angles
  angles[1] = angle(lShoulder, lElbow, lHand);
  angles[2] = angle(rShoulder, lShoulder, lElbow);
  angles[3] = angle(lHip, lKnee, lFoot);
  angles[4] = angle(new PVector(lHip.x, 0), lHip, lKnee);
  // Right-Side Angles
  angles[5] = angle(rHand, rElbow, rShoulder);
  angles[6] = angle(rElbow, rShoulder, lShoulder );
  angles[7] = angle(rFoot, rKnee, rHip);
  angles[8] = angle(rKnee, rHip, new PVector(rHip.x, 0));
}
```

Now you have all the data you need to control your puppet! If you run the angle calculation sketch and start the skeleton tracking, you will see the values of your joints' angles appear on the console. You could now connect the serial cable to your computer, add a serial communication protocol, and start driving the puppet straight away. But let's build some flexibility into this application. You want to be able to control the puppet remotely through the Internet, and to do this you need to learn some basics of network communication.

Network Communication

In previous chapters you used Serial communication to send messages from your computer to the Arduino board. Now you are going to follow an analogous process to let two machines communicate over the Internet.

Luckily, Processing includes a core library called Network that will help you go about the network communication with just a few lines of code. Start Processing and go to `examples/Libraries/Network` to open the SharedCanvasServer and SharedCanvasClient examples. (We won't explain the whole code because you are going to do that with your own project later, but we will use these sketches to illustrate the concepts.)

If you run the SharedCanvasServer sketch, you get a frame in which you can draw lines with your mouse. Not very impressive. If you open the SharedCanvasClient, you get the same but Processing throws a "Connection refused" error at you in the console. If you open the server sketch first, and then the client sketch, you have two canvases and no errors. Now, if you paint in one of them, the lines appear on the other one as well, as shown in Figure 6-26. You have two Processing sketches talking to each other; that's a good start!

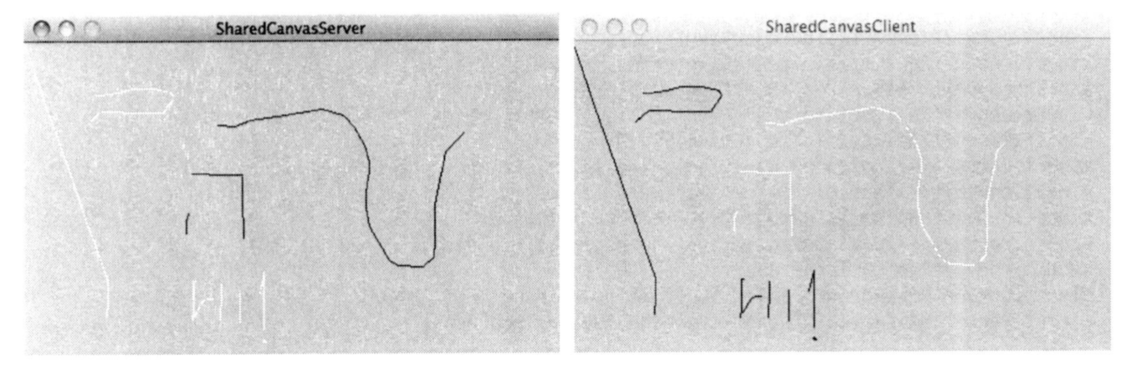

Figure 6-26. *Shared canvas*

Communicating Within a Local Network

If you try the same over two computers, though, it won't work. If you look into the `setup` functions of the server sketch, you find the line in which you are starting the server.

```
s = new Server(this, 12345);
```

And in the client sketch, you'll find the line in which you start the client.

```
c = new Client(this, "127.0.0.1", 12345);
```

The client is trying to connect to the computer with IP address 127.0.0.1 through port 12345. Well, this address happens to be the *localhost* address of the computer. It worked well as long as you had both sketches running on the same computer, but if you want to make them work on different machines, you need to provide the client with the IP address of the server. You can find the internal IP address of your computer easily. On a Mac, open Network Preferences; it's on the top right. You should see something like "*AirPort is connected to xxxxxx and has the IP address 192.168.0.101.*" On a PC, go to the Start menu. In "Search programs and files" write `cmd` and press Enter. You have opened a console. Now type `ipconfig` in the console and press Enter again. Many text lines will appear, one of which says something like "*IPv4 Addres..:192.168.0.103*". This is your IP address.

If you go now to your client sketch and substitute the 127.0.0.1 with the IP address of the machine in which you are running the server (the other computer) and run both sketches (remember to run the server first), it should work smoothly.

```
c = new Client(this, "192.168.0.101", 12345);
```

Now you have two Processing sketches running on different computers talking to each other. But the ultimate goal is to be able to communicate over the Internet. Can this code make that happen? Can I go to my friend's house with the client sketch, run it, and be connected to my machine at home? Well, unfortunately not yet. The IP addresses you've been using are internal IP addresses. They are good as long as both computers are connected to the same local network. The local IP address is known only by the router, so if you go beyond the reach of your local network, these addresses won't make sense any more. What you need is the external IP address of your router.

Communicating over the Internet

The easiest way to find your external IP address is using a web site. If you go to `www.whatsmyip.com` or similar, you will find a line saying "Your IP Address Is: `24.28.51.56`" or a similar number. This is you external IP address, the one you need to include in your Client sketch if you want to connect to your computer from a computer outside of your local network.

But you need to do just one more thing on the server side. The client is trying to connect to your computer using the external IP; now you need to tell the router that any incoming connection addressed to the port 12345 needs to be send to your machine. This is called *port forwarding*. If you are an online gamer, you probably know about port forwarding because you need to set up socket connections when you want to play online games. What you are trying to establish here is a synchronous connection, or socket, which is a bi-directional open communication channel between your machine and another machine over the Internet.

Port forwarding depends on the router you are using at home, but in general, if you type `192.168.0.1` in your browser, it will lead you to your router configuration menu. Provided that you have the user name and password (by default, it is often "admin"), you will find the port forwarding setting somewhere in the menu. You need to specify the IP address of the computer you are forwarding to (your server computer), and the port that the communication is sent to (12345 in our example), both in TCP and UDP. Dig your router's instructions manual from the dark place you hid it five years ago (or find the manual on the Internet, in the likely case you have no remote clue where your hard copy could be) if you need some help.

You should now be in a position where you can run your server sketch on your computer (where you forwarded the port) and have one of your geeky friends run the client sketch with your external address at his home computer. The two sketches should be communicating smoothly across neighborhoods, countries, and even continents!

Server Applet: Sending the Angles over a Network

So now that you know how to communicate across networks, you are going to add this functionality to your angle calculation sketch so you can send the angles over the network for another program to retrieve them from a remote location.

All your helper functions and callbacks stay the same, but you add some lines to the `setup()` and draw() functions. You first import the Network library and declare a Server object. Then you initialize all the PVectors and the angles array as per your previous code.

```
import processing.net.*;
import SimpleOpenNI.*;
```

```
SimpleOpenNI kinect;
Server s; // Server Object

// Left Arm Vectors
PVector lHand = new PVector();
PVector lElbow = new PVector();
PVector lShoulder = new PVector();
// Left Leg Vectors
PVector lFoot = new PVector();
PVector lKnee = new PVector();
PVector lHip = new PVector();
// Right Arm Vectors
PVector rHand = new PVector();
PVector rElbow = new PVector();
PVector rShoulder = new PVector();
// Right Leg Vectors
PVector rFoot = new PVector();
PVector rKnee = new PVector();
PVector rHip = new PVector();

// Articulation Angles
float[] angles = new float[9];
```

In the setup() function, you initialize the server object, passing the port you are communicating through as a parameter.

```
public void setup() {
  kinect = new SimpleOpenNI(this);
  kinect.setMirror(true);
  kinect.enableDepth();
  kinect.enableUser(SimpleOpenNI.SKEL_PROFILE_ALL);

  frameRate(10);
  size(kinect.depthWidth(), kinect.depthHeight());

  s = new Server(this, 12345); // Start a simple server on a port
}
```

And finally, at the end of the draw() loop and only in case you are tracking a skeleton, you write a long string of characters to your port. The string is composed of all the joint angles separated by white spaces, so you can split the string at the other end of the line.

```
public void draw() {
  kinect.update();  // update the Kinect
  image(kinect.depthImage(), 0, 0);  // draw depthImageMap
if (kinect.isTrackingSkeleton(1)) {
    updateAngles();
    drawSkeleton(1);

    // Write the angles to the socket
    s.write(angles[0] + " " + angles[1] + " " + angles[2] + " "
        + angles[3] + " " + angles[4] + " " + angles[5] + " "
        + angles[6] + " " + angles[7] + " " + angles[8] + "\n");
  }
}
```

Client Applet: Controlling the Puppet

The server applet is finished and ready to start sending data over the Internet. Now you need a client sketch to receive the angle data and send it to Arduino to get translated into servo positions that will become movements of your puppet.

You're going to implement a virtual puppet resembling the physical puppet (Figure 6-27), so check that everything is working properly before sending data to the servos. You will implement the serial communication at the same time, so once everything is working you only need to check that the serial boolean is set to true for the data to be sent to Arduino.

Figure 6-27. Client applet

After importing the Net and Serial libraries, create a boolean variable called `serial` that you set to `false` if you haven't plugged the serial cable yet and want to try the sketch. You are going to be displaying text, so you also need to create and initialize a font.

```
import processing.net.*;
import processing.serial.*;

Serial myPort;
boolean serial = true;
PFont font;
```

You need to declare a client object called `c`, declare a String to contain the message coming from the server applet, and a `data[]` array to store the angle values once you have extracted them from the incoming string. You are going to use a PShape to draw a version of your physical puppet on screen so you can test all the movements virtually first.

■ **Note** PShape is a Processing datatype for storing SVG shapes. You can create SVG files with Adobe Illustrator, and then show them in Processing using PShapes. For more information, refer to the Processing help.

```
Client c;
String input;
float data[] = new float[9];
PShape s;
```

In the setup() function, initialize the client object. Remember to replace the IP address with the external IP of the server computer if you are communicating over the Internet or the internal IP if you are communicating within the same local network. You will need to add the file Android.svg (provided with the chapter images) to your sketch folder so it can be loaded into the PShapes.

```
void setup()
{
  size(640, 700);
  background(255);
  stroke(0);
  frameRate(10);

  // Connect to the server's IP address and port
  c = new Client(this, "127.0.0.1", 12345); // Replace with your server's IP and port
  font = loadFont("SansSerif-14.vlw");
  textFont(font);
  textAlign(CENTER);
  s = loadShape("Android.svg");
  shapeMode(CENTER);
  smooth();

  if (serial) {
    String portName = Serial.list()[0]; // This gets the first port on your computer.
    myPort = new Serial(this, portName, 9600);
  }
}
```

Within the draw() loop, check for incoming data from the server; in a positive case, read is as a string. Then trim off the newline character and split the resulting string into as many floats as there are sub-strings separated by white spaces in your incoming message. Store these floats (the angles of your limbs sent from the server sketch) into the data[] array.

Run a little validation test, making sure that all the values acquired fall within the acceptable range of -PI/2 to PI/2.

```
void draw()
{
  background(0);
  // Receive data from server
  if (c.available() > 0) {
    input = c.readString();
    input = input.substring(0, input.indexOf("\n")); // Only up to the newline
    data = float(split(input, ' ')); // Split values into an array

    for (int i = 0 ; i < data.length; i++) {
```

```
      if(data[i] >  PI/2) { data[i] = PI/2; }
      if(data[i] < -PI/2) { data[i] = PI/2; }
    }
  }
```

Once the data is nicely stored in an array, draw your puppet on screen and add the limbs in the desired position. You first draw the shape on screen in a specific position, which you have figured out by testing. Then use the function drawLimb() to draw the robot's limbs using the angles in the data array as parameters.

```
 shape(s, 300, 100, 400, 400);
  drawLimb(150, 210, PI, data[2], data[1], 50);
  drawLimb(477, 210, 0, -data[6], -data[5], 50);
  drawLimb(228, 385, PI/2, data[4], data[3], 60);
  drawLimb(405, 385, PI/2, -data[8], -data[7], 60);
```

Then draw an array of circles showing the incoming angles in an abstract way so you can debug inconsistencies and check that you are receiving reasonable values.

```
  stroke(200);
  fill(200);
  for (int i = 0; i < data.length; i++) {
    pushMatrix();
    translate(50+i*65, height/1.2);
    noFill();
    ellipse(0, 0, 60, 60);
    text("Servo " + i + "\n" + round(degrees(data[i])), 0, 55);
    rotate(data[i]);
    line(0, 0, 30, 0);
    popMatrix();
  }
```

And finally, if your boolean serial is true, use the sendSerialData() function to communicate the angles to Arduino.

```
  if (serial)sendSerialData();
}
```

The drawLimb() function is designed to draw an arm or leg on screen starting at the position specified by the parameters x and y. The angle0 variable specifies the angle that you have to add to the servo angle to accurately display its movement on screen. angle1 and angle2 determine the servo angles of the first and second joint angles on the limb. Limbsize is used as the length of the limb.

```
void drawLimb(int x, int y, float angle0, float angle1, float angle2, float limbSize) {
  pushStyle();
  strokeCap(ROUND);
  strokeWeight(62);
  stroke(134, 189, 66);
  pushMatrix();
  translate(x, y);
  rotate(angle0);
  rotate(angle1);
  line(0, 0, limbSize, 0);
  translate(limbSize, 0);
  rotate(angle2);
  line(0, 0, limbSize, 0);
  popMatrix();
```

```
  popStyle();
}
```

The function sendSerialData() works similarly to the ones used in previous chapters, sending a triggering character to Arduino and then writing a series of integer data values to the serial port. These values are the angles mapped to a range of 0 to 250, expressed as integers.

```
void sendSerialData() {
  myPort.write('S');
  for (int i=0;i<data.length;i++) {
    int serialAngle = (int)map(data[i], -PI/2, PI/2, 0, 255);
    myPort.write(serialAngle);
  }
}
```

Final Arduino Code

If you run the server applet and then the client applet, you should be able now to control your virtual puppet by dancing in front of your Kinect, as shown in Figure 6-28. Once you have all of your Processing code ready, it's time to plug in your Arduino and type the last bits of code.

Figure 6-28. Server and client applets communicating over a network

You're going to store the values coming from the client Processing sketch, and remap these values from 0-255 to the range of the servo (500 to 2500). It is important to match the data coming from the Processing sketch with the right pin of Arduino, and map it in the right direction (clockwise or counterclockwise) in order to match the movements. The rest of the functions are those you have been using since the beginning of this chapter.

```
float temp1, temp2, temp3, temp4, temp5, temp6, temp7, temp8, temp9;
// Joint Servos
int servo3Pin = 3; //shoulder 1
int servo4Pin = 4; // arm 1
int servo5Pin = 5; // shoulder 2
```

```
int servo6Pin = 6; // arm 2
int servo7Pin = 7; // hip 1
int servo8Pin = 8; // leg 1
int servo9Pin = 9; // hip 2
int servo10Pin = 10; //leg 2
int servo11Pin = 11; //rotation

//initial pulse values
int pulse1 = 500;
int pulse2 = 1500;
int pulse3 = 2500;
int pulse4 = 1500;
int pulse5 = 500;
int pulse6 = 2500;
int pulse7 = 2500;
int pulse8 = 500;
int pulse9 = 1500;

int speedServo = 0;
unsigned long previousMillis = 0;
long interval = 20;

void setup() {
  pinMode (servo3Pin, OUTPUT);
  pinMode (servo4Pin, OUTPUT);
  pinMode (servo5Pin, OUTPUT);
  pinMode (servo6Pin, OUTPUT);
  pinMode (servo7Pin, OUTPUT);
  pinMode (servo8Pin, OUTPUT);
  pinMode (servo9Pin, OUTPUT);
  pinMode (servo10Pin, OUTPUT);
  pinMode (servo11Pin, OUTPUT);

  Serial.begin(9600);
}

void loop() {
  if (Serial.available() > 18) {
    char led = Serial.read();
    if (led == 'S'){
      temp1 = Serial.read();
      temp2 = Serial.read();
      temp3 = Serial.read();
      temp4 = Serial.read();
      temp5 = Serial.read();
      temp6 = Serial.read();
      temp7 = Serial.read();
      temp8 = Serial.read();
      temp9 = Serial.read();
    }
  }
```

Next, remap the angles from the incoming range (0-255) to the servo range (500-2500).

```
pulse9 = (int)map(temp1,0,255,2500,500);        //rotation
pulse2 = (int)map(temp2,0,255,500,2500);        //leftElbow
pulse1 = (int)map(temp3,0,255,500,2500);        //left Shoulder
pulse4 = (int)map(temp4,0,255,2500,500);        //left Knee
pulse3 = (int)map(temp5,0,255,500,2500);        //left Hip
pulse6 = (int)map(temp6,0,255,2500,500);        //right Elbow
pulse5 = (int)map(temp7,0,255,2500,500);        //right Shoulder
pulse8 = (int)map(temp8,0,255,500,2500);        //right Knee
pulse7 = (int)map(temp9,0,255,2500,500);        //right Hip
```

And finally, move the servo to the new position.

```
  unsigned long currentMillis = millis();
  if(currentMillis - previousMillis > interval) {
    previousMillis = currentMillis;

    updateServo(servo3Pin, pulse1);
    updateServo(servo4Pin, pulse2);
    updateServo(servo5Pin, pulse3);
    updateServo(servo6Pin, pulse4);
    updateServo(servo7Pin, pulse5);
    updateServo(servo8Pin, pulse6);
    updateServo(servo9Pin, pulse7);
    updateServo(servo10Pin,pulse8);
    updateServo(servo11Pin, pulse9);
  }
}

void updateServo (int pin, int pulse){
  digitalWrite(pin, HIGH);
  delayMicroseconds(pulse);
  digitalWrite(pin, LOW);
}
```

Now upload your code into Arduino, call your most remote Kinect-enabled friend, and ask him to run the server code and dance in front of his Kinect. Then run your client applet, sit down, and enjoy the performance!

Summary

In this chapter, you learned how to use NITE's skeleton tracking capabilities from Processing and how to use the data acquired to control a physical puppet. You were introduced to servos and how to control them from Arduino; you also learned how to make use of network communication, both within a local network and beyond, spanning across the Internet.

The outcome of the chapter is a system with which you can control remote devices using natural interaction from home. In this example, you used it to control a puppet that literally copies your body gestures, but the concept can be endlessly extended. You could develop it to control more complicated devices, using other types of natural interaction algorithms and adapting it to your particular needs.

CHAPTER 7

Mood Lamps

by Ciriaco Castro and Enrique Ramos

Light greatly affects the spaces we inhabit. We all try control the lighting levels and quality in our homes by installing specific lighting fixtures, wall lamps, spotlights, etc. to create the desired ambience. In this chapter, you are going to learn how to build your own wireless RGB LED lamp and how to install a lighting control system that lets you modify the state of several lamps via body gestures. This lighting system will also be aware of your position in the room and will respond by increasing the brightness of the lamp closest to you.

You will learn about the RGB color mode so you can build lamps like the one in Figure 7-1 that will cast a wide spectrum of colors using four LED lights: three colored (red, green, and blue) and one white. You will use wireless communication via two XBee modules, and you will work with a smaller version of your board, the Arduino Nano. The use of spatial body gesture recognition will take you to a deeper exploration of three-dimensional NITE's skeleton tracking capabilities. Finally, you will have a first encounter with data storage and retrieval from external files, so you can save and load the layout of your lamps. Table 7-1 lists the required parts for this chapter.

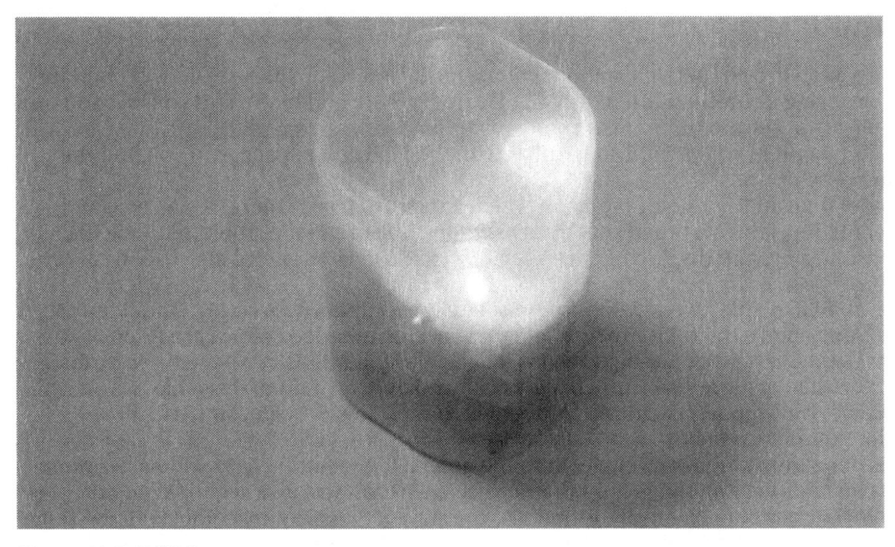

Figure 7-1. RGB lamp

Table 7-1. The Parts List

Part	Description	Price
1 Arduino Nano	Little brother in the Arduino family	$33.99
2 XBee 1mW antenna series1	Wireless module	$22.95 each
XBee Explorer Regulated	From SparkFun or similar	$9.95
XBee Explorer dongle	From SparkFun or similar	$24.95
Strip board	94 x 53 mm copper	$3.95
23 breakaway headers-straight	You should have a bunch of these already from previous chapters.	$1.50
1 ultra bright white LED	From SparkFun or similar	$1.50
1 ultra bright red LED	From SparkFun or similar	$1.50
1 ultra bright green LED	From SparkFun or similar	$1.50
1 ultra bright blue LED	From SparkFun or similar	$1.50
1 translucent pencil holder	From your favorite stationery store	$2.99

RGB Color Space

Throughout this chapter, we will frequently talk about colors, hues, and color spaces. You will build the whole spectrum of colors using a combination of three LEDs (red, green, and blue), and you will add one extra white LED for intensity at certain moments. If you are familiar with any graphics software, you will easily recognize the RGB palette. If you are not so familiar with the RGB color space, you will find the following introduction pretty useful.

The RGB color model is an additive color model in which a series of basic colors (red, green, and blue) are added together to reproduce a broad spectrum of colors. The original purpose of the RGB color model was sensing, representing, and displaying images in electronic devices such as television sets and computers screens.

To form a color on an RGB display, you need three colored lights superimposed. By modifying the intensity of each color, you achieve the right mixture to represent the intended color. Zero intensity of each component forms the darkest color (no light means black), and full intensity on the three forms white. (Note that it's difficult to achieve real white by physically mixing the light of three LEDs. You can buy RGB LED lights that are quite good at achieving this effect, but for this project you'll use three different lights for a more dramatic effect.)

If one of the lamps has a much higher intensity than the others, the resulting light will have a hue similar to this color. If two lamps are shining with full intensity, the result will be a secondary color (cyan, magenta, or yellow). Secondary colors are formed by the sum of two primary colors with the same intensity: cyan is green + blue, magenta is red + blue, and yellow is red + green, as shown in Figure 7-2.

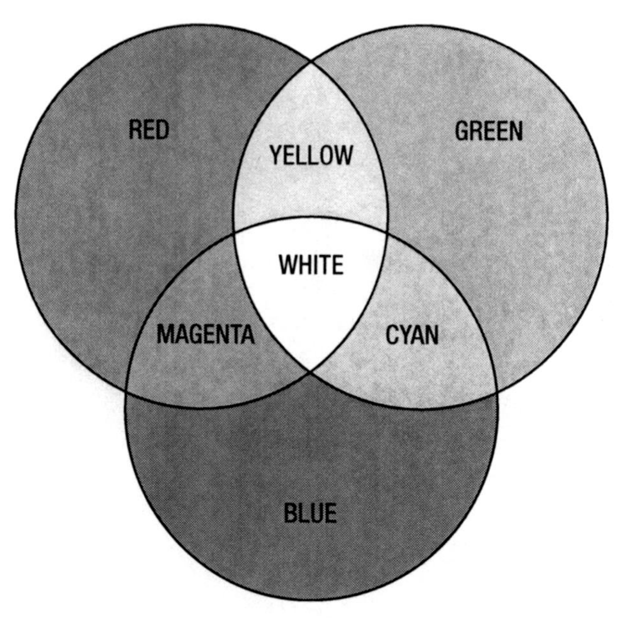

Figure 7-2. RGB chart

Common applications of the RGB color model are the display of colors on television sets and computer monitors. Each pixel of the screen is composed by three small and very close but still separate RGB light sources. If you stand close enough to the screen, you can distinguish the source's colors (this means really close). Otherwise, the separate sources are indistinguishable, which tricks the eye into seeing a given solid color.

In computing, the component values are often stored as integer numbers in the range of 0 to 255, and this is the way you will work with color intensities in Arduino and Processing: going from values 0 (light off) to 255 (full intensity).

Color changes the quality of an ambience. By modifying lighting effects and hues, you can alter your perception of a space. The Chinese theory of Feng Shui takes the use of color to a more profound level, associating energies and moods according to the right combination of colors. You can apply different colors to a room according to different situations or moods, whether by following the directions of millenary Feng Shui or just your personal taste.

Arduino Nano

Due to the size of your lamp, the use of an Arduino Nano is more appropriate for this project. An Arduino Nano, shown in Figure 7-3, is a small and complete Arduino board designed and produced by Gravitec. It has some minor differences from the Arduino Uno you used in previous projects. Instead of having a standard USB socket, it has a Mini-B USB. The pins are the same as on the Arduino Uno, just male instead of female.

Figure 7-3. Arduino Nano

As you saw in Chapter 1, each of the 14 digital pins on the Nano can be used as input and output. You also have an LED on pin 13, and you can use digital pin 0 (RX) for receiving and 1(TX) for transmitting serial communication. Remember to select the correct Arduino board in the Arduino IDE.

For powering the Arduino Nano you can use the Mini-B USB connection, a 6-20V unregulated external power supply in pin 30, or a 5V regulated external power supply in pin 27. This is important because you will use a wireless connection, so both the wireless module and the Arduino board will be externally powered.

Building the Circuit

Let's build your circle step by step. First, you will build the circuit that allows the control of the four LEDs from Arduino, and then you will make the whole installation wireless by adding a battery pack and an XBee radio module. Figure 7-4 shows the whole setup.

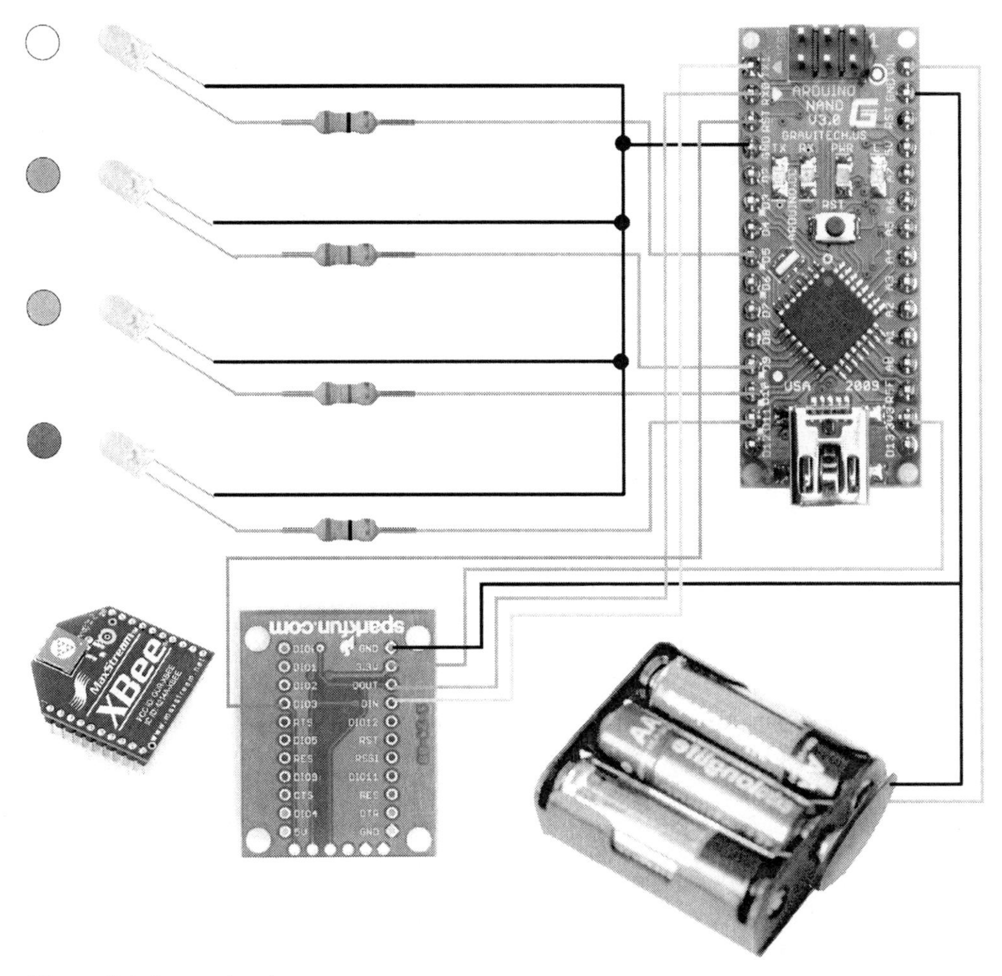

Figure 7-4. Lamp circuit

Get started by connecting the four LEDs (white, red, green, and blue) to your Arduino Nano, building the circuit using a strip board. The components are shown in Figure 7-5.

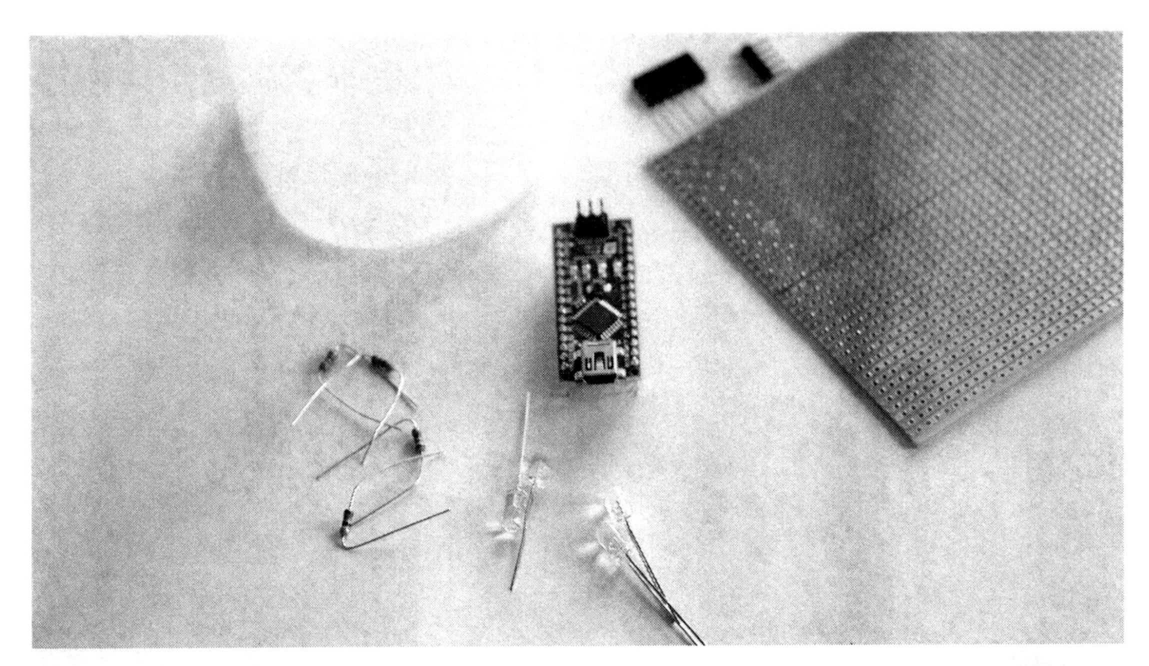

Figure 7-5. *Components*

As usual, you start by cutting the strip board into a small enough size so you can insert the circuit inside your pencil holder but large enough to connect all the components. For the moment, you are keeping it simple—just controlling the LEDs using your Mini USB connection—but you will increase the amount of components later when you add the wireless XBee module.

Once you have cut the piece, scratch the board before soldering the pins that will hold your Arduino Nano, as shown in Figure 7-6. Double-check the direction of the copper strips on the back of the board so you don't connect the two rows of pins to your Arduino.

Figure 7-6. *Scratched strip board*

After that, position your breakout headers and solder them to the strip board. You could actually place the Arduino Nano and solder it to your strip board if you want to make a permanent installation, but let's use the breakout headers so you can easily detach the Arduino board for other projects.

In principle, you don't need to connect all the pins to the board; you can add more later if you need them. First, solder the breakout headers that correspond to the analog pins, ground, and serial communication, and some of the digital pins on the other side to give some stability to the Arduino Nano, as shown in Figure 7-7.

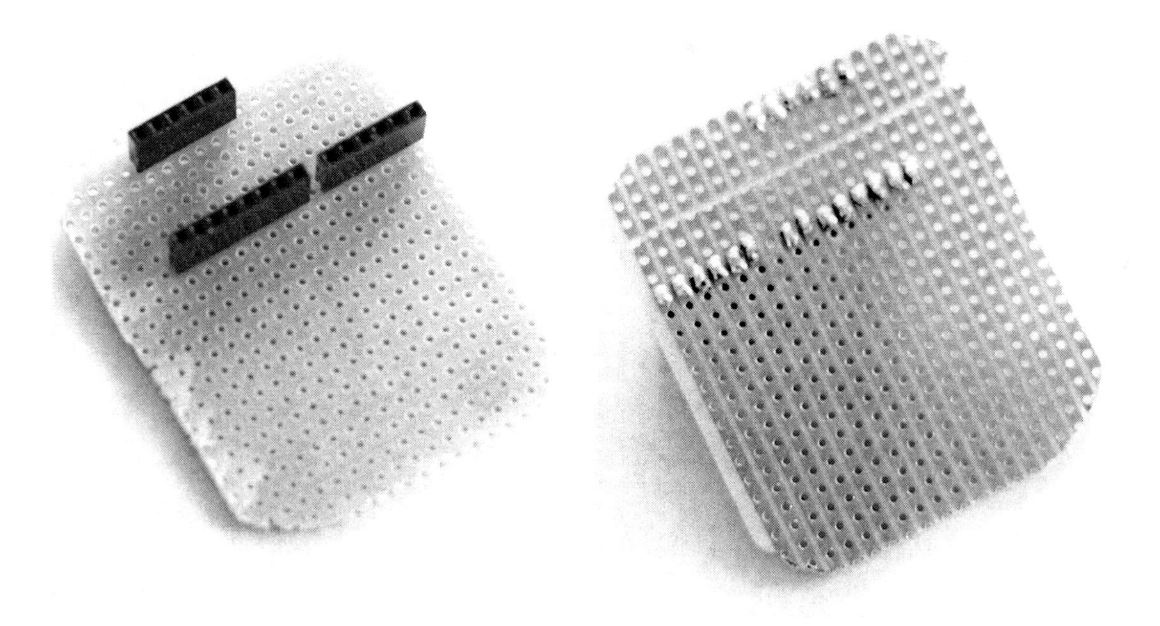

Figure 7-7. *Breakout headers (detail)*

Resistors

The next step is to solder the resistors for the LEDs. In Chapter 4, we asked you to trust our choice of a resistor. Now that you are a bit more experienced, you will learn how to choose the right resistor according to the kind of LED you are using.

LEDs are semiconductors. The flow in an LED is an exponential function of the voltage across it, meaning that a small change in voltage can produce a huge change in current. Using a resistor, you will manage to have current and voltage linearly related. This means that a change in voltage will now produce a proportional change in current, thus limiting the current in the LED to a safe value.

■ **Note** In order to know which resistor is adequate, you can use an online resistor calculator, like `http://led.linear1.org/1led.wiz`, in which you can introduce the supply voltage, diode-forward voltage, and diode-rated current to obtain a suitable resistance value and the equivalent resistor color code.

Understanding Resistor Color Codes

Resistors have a series of color stripes that denote their resistance value. Normally, resistors have a four-band code. The first two bands represent the most significant digits of the resistance value. Colors are assigned to the numbers between 0 and 9, and the color bands basically translate the numbers into a visible code. Table 7-2 provides the code.

***Table 7-2.** Resistor Color Code*

Black	Brown	Red	Orange	Yellow	Green	Blue	Violet	Gray	White
0	1	2	3	4	5	6	7	8	9

So if a resistor has brown and red as the first two bands, the most significant digits will be 1 and 2 (12). The third band is the multiplier using the power of ten (or how many zeros to add to your most significant values), using the same assigned value for each color as in the previous step. For example, if you have a resistor that has brown, red and red, that means 12 (significant digits brown =1, red =2) x 100 (multiplier, red =2 meaning 10^2) = 1200Ω (1.2kΩ).

The last band is called the tolerance band (the deviation from the specified value) and it's usually spaced away from the others. Gold means 5% of tolerance and silver means 10% of tolerance; if there's no tolerance band that means a 20% of tolerance.

The values of your LEDs from the distributor's specifications are as follows:

- White LED: Supply voltage 5V, forward voltage 3.3V, rated current 30mA, meaning a resistor of 68 ohms (color code: blue, grey, black).
- Red LED: Supply voltage 5V, forward voltage 1.85V, rated current 30mA, meaning a resistor of 120 ohms (color code: brown, red, brown).
- Green LED: Supply voltage 5V, forward voltage 3.3V, rated current 25mA, meaning a resistor of 150 ohms (color code: brown, green, brown).
- Blue LED: Supply voltage 5V, forward voltage 3.3V, rated current 30mA, meaning a resistor of 68 ohms (color code: blue, grey, black).

Now that you know your resistors, you can solder them to your board. You're going to use pins 5, 9, 10, and 11 for your LEDs (white, red, green, and blue), and scratch the strip board to separate all grounds, as shown in Figures 7-8 and 7-9.

Figure 7-8. *Resistors soldered on board*

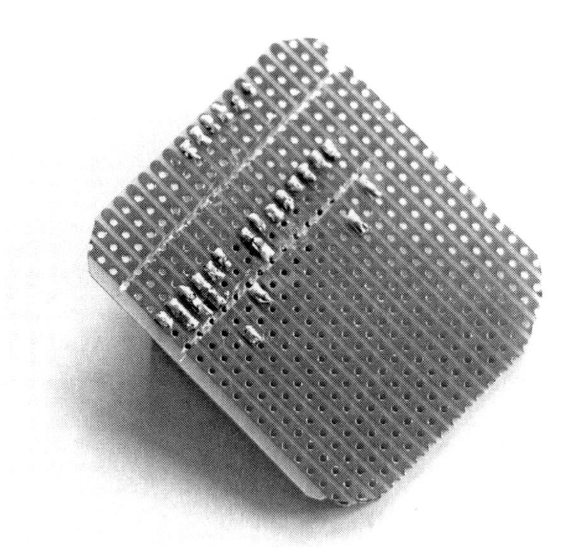

Figure 7-9. *Resistors soldered on board (back)*

The next step is to connect all grounds together and back to Arduino's ground, and then solder the LEDs to your circuits, as shown in Figure 7-10 and 7-11. Make sure that the right color is connected to the right pin; also remember that the shorter legs on the LEDs (cathode) connect to ground and the longer legs connect to the resistors.

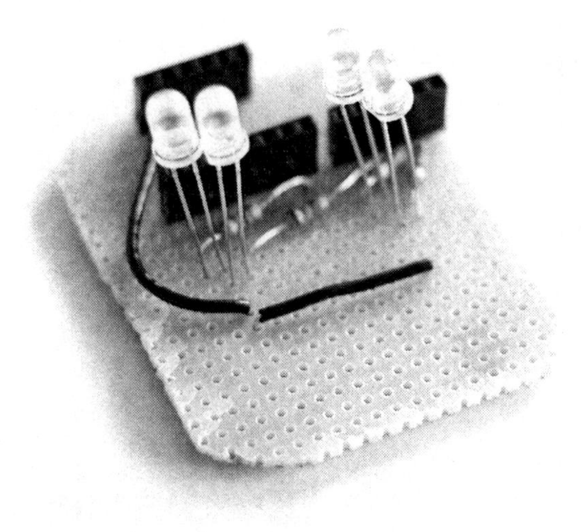

Figure 7-10. *LEDs on circuit*

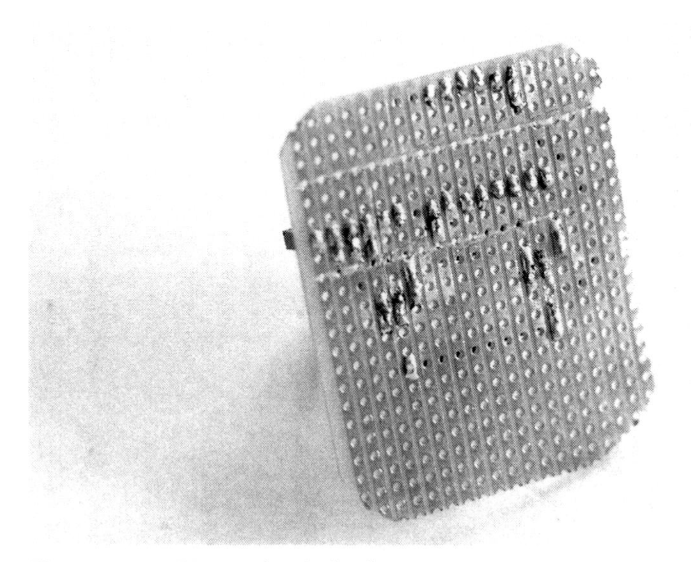

Figure 7-11. *LEDs on circuit (back)*

Testing the Circuit

Now that everything is soldered, you can use a simple piece of code to test the circuit. This code programs the LEDs to dim at different speeds so you can appreciate the richness and variety of the colors resulting from the mix. The code is based on the Fade example from Arduino; it's been modified to drive more than one LED.

```
int whiteLED = 5;
int redLED = 9;
int greenLED = 10;
int blueLED = 11;

int Wbr = 0;
int Rbr = 0;
int Gbr = 0;
int Bbr = 0;

int Wspeed = 0;
int Rspeed = 2;
int Gspeed = 3;
int Bspeed = 5;

void setup()  {
  pinMode(whiteLED, OUTPUT);
  pinMode(redLED, OUTPUT);
  pinMode(greenLED, OUTPUT);
  pinMode(blueLED, OUTPUT);
}

void loop()  {
  analogWrite(whiteLED, Wbr);
  analogWrite(redLED, Rbr);
  analogWrite(greenLED, Gbr);
  analogWrite(blueLED, Bbr);

  Wbr +=  Wspeed;
  if (Wbr == 0 || Wbr == 255) {
    Wspeed *= -1 ;
  }
  Rbr +=  Rspeed;
  if (Rbr == 0 || Rbr == 255) {
    Rspeed *= -1 ;
  }
  Gbr +=  Gspeed;
  if (Gbr == 0 || Gbr == 255) {
    Gspeed *= -1 ;
  }
  Bbr +=  Bspeed;
  if (Bbr == 0 || Bbr == 255) {
    Bspeed *= -1 ;
  }

  delay(30);
}
```

Figure 7-12. RGB lamp test

Run this code and watch your lamp change colors smoothly (Figure 7-12). Each LED changes its intensity at a different speed. You can tweak the code and change the speed of each lamp at will. The lamp looks good now, but you still have a Mini USB cable attached to it. The USB cable brings power and data to the Arduino, so removing it means you need to plug an external power source and a wireless module to the board, as shown in Figure 7-13.

Figure 7-13. XBee and battery clip

To substitute the USB power on your freestanding module, connect a battery clip to your circuit. But first you need to solder a new header connecting pin 30 in your Arduino (Vin, a 6-20V unregulated external power supply). Then connect the red cable in the battery clip (power) to this new pin and the black cable (ground) to the other grounds (Figure 7-14).

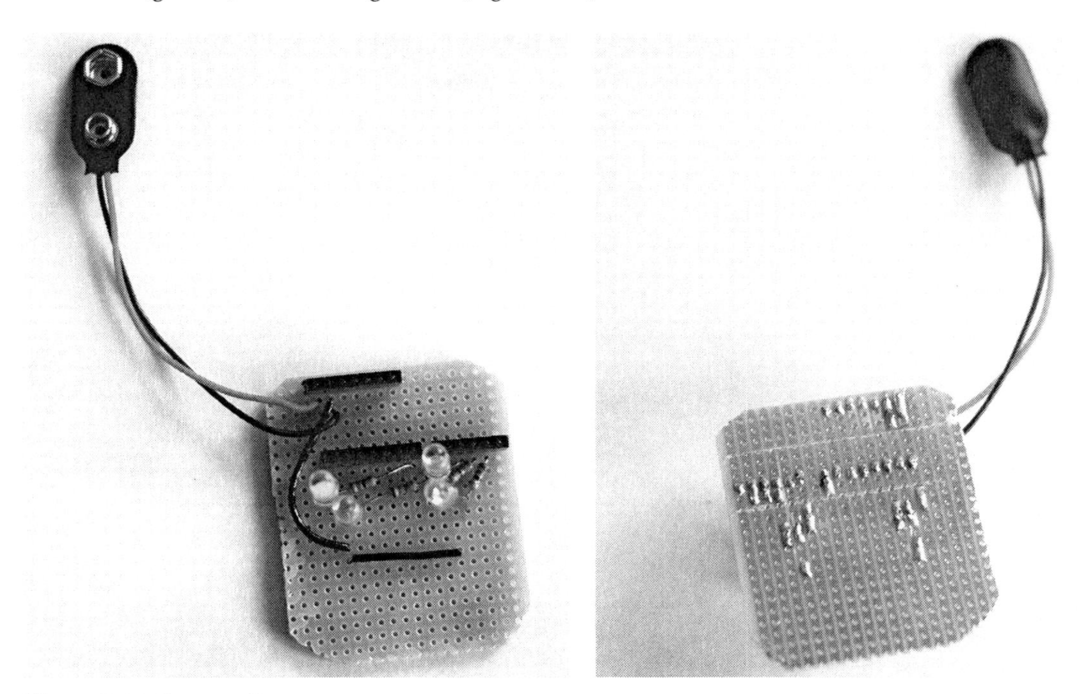

Figure 7-14. *Battery clip*

Connect the battery holder to the battery clip and observe that the lights behave the same way as when the mini USB cable is connected. The code was already uploaded from the previous test and Arduino just followed the routine without receiving any external data from the computer.

XBee Wireless Module

You're now going to add an XBee to the circuit. XBee is a brand name from Digi International for a family of radio modules. These wireless modules allow your Arduino to communicate wirelessly with your computer, avoiding the inconvenience of having cables sticking out of your beautiful lamps.

XBee pins are spaced 2 mm apart, which means that XBee can't be directly connected to a breadboard. The XBee modules must be first connected to an XBee Explorer USB (broadcaster) and to an XBee Explorer Regulated (receiver).

The first step is to install the drivers in order for your computer to recognize the XBee module. The XBee Explorer uses a FTDI chip to convert the serial transmission to USB signals. The FTDI drivers can be found at `http://www.ftdichip.com/FTDrivers.htm`. Select the right platform and install the drivers. After installing the drivers and plugging in your XBee Explorer mini USB, you should be able to see your device using Terminal (`ls /dev/tty.*`) if you are using a Mac or by going to `Start/Control panel/Hardware and Sound/Device Manager/ Universal Serial Bus controllers` if you are a Windows user.

Once you have your broadcaster installed, you need to connect your receiver to the Arduino board. You could plug your XBee into a shield that connects the module directly to Arduino, but for this project you are going to use the normal XBee Explorer so you can analyze the connections.

XBee has many pins but it takes only five of them to create a working wireless connection with Arduino via its built-in serial communication protocol. The five connections that you have to make between the Arduino board and the XBee Explorer Regulated are described in Table 7-3 and shown in Figures 7-15 through 7-17.

Table 7-3. *XBee Explorer Regulated to Arduino Connections*

XBee Explorer Regulated Pins	Arduino Board Pins
XBee 3.3V	Arduino 3.3V
XBee GND	Arduino GND
XBee DIN	Arduino TX (Digital Pin 1)
XBee DOUT	Arduino RX (Digital Pin 0)
XBee DIO3	Arduino RESET

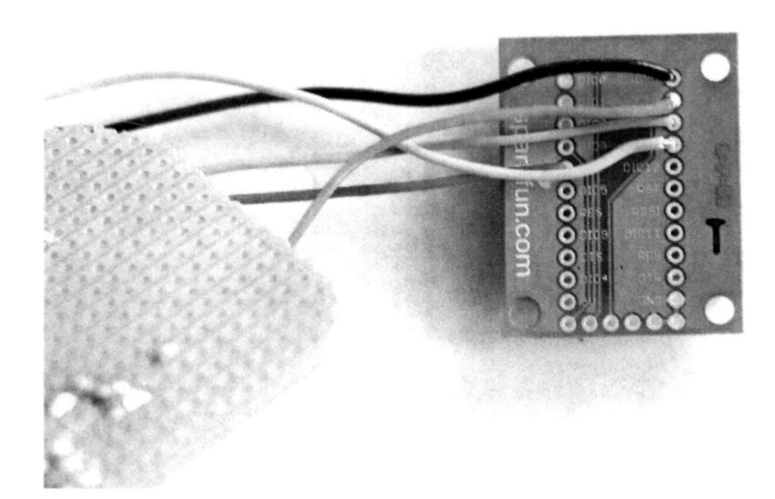

Figure 7-15. *XBee Explorer connected to Arduino*

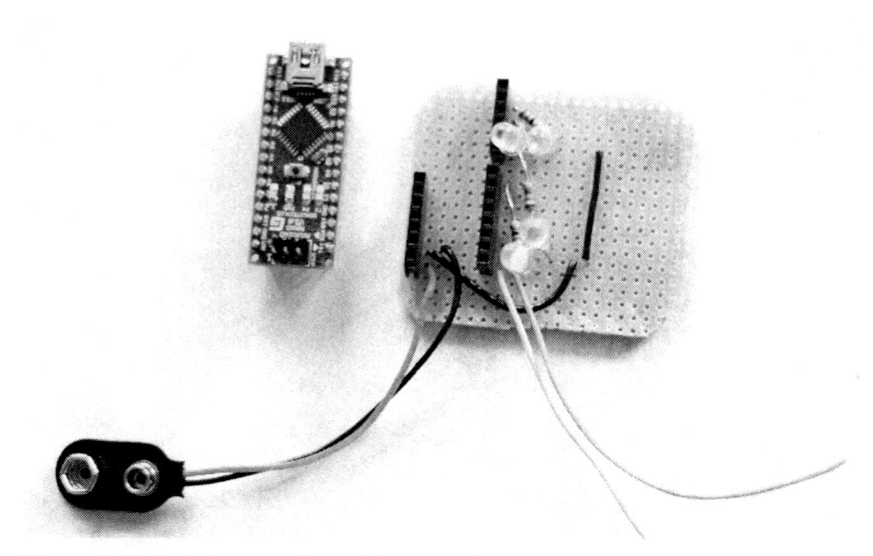

Figure 7-16. Strip board with serial connections

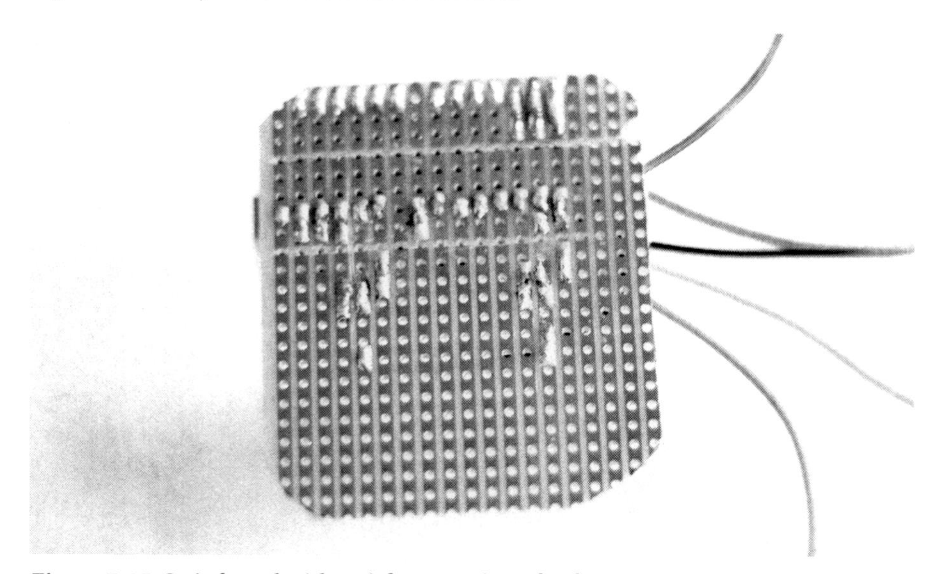

Figure 7-17. Strip board with serial connections (back)

The last step is to power your XBee, so pick a cable for power from the 3V Arduino pin to the XBee Explorer's VCC or 3.3V, and connect Arduino's ground to the XBee Explorer's ground.

Once you have the circuit assembled (Figure 7-18), pile it so it fits it inside your plastic pencil holder, building your lamp (Figure 7-19).

Figure 7-18. Final circuit

■ **Tip** A piece of advice: perforate a hole on top of the pencil holder so if the components get stuck inside, you can push them out.

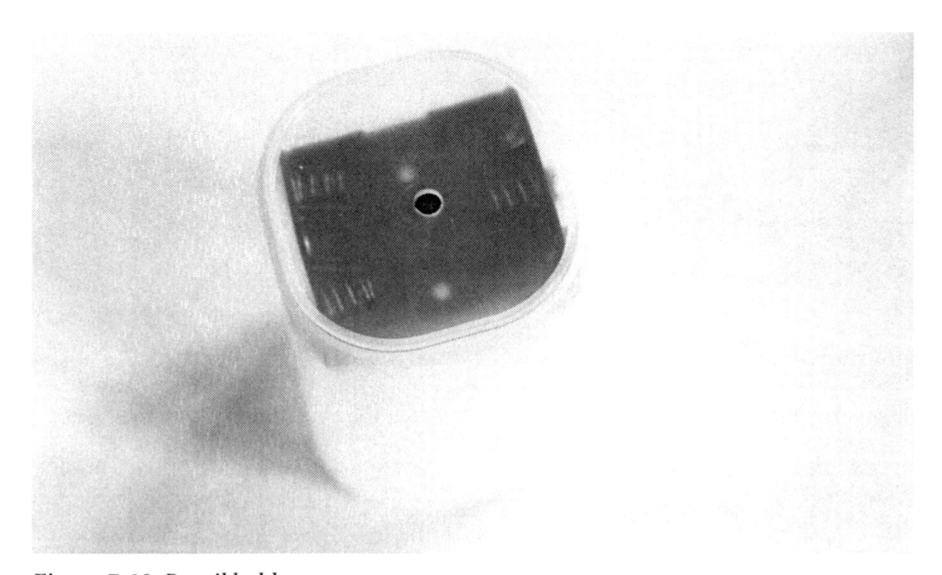

Figure 7-19. Pencil holder

After assembling the lamp, the next step is to use the Kinect to control the light's colors. You will learn how to control several lights at the same time, implementing code that will work on any number of lamps, from one to as many as you can build. If you feel patient, build a couple more of them!

Arduino Programming

Now it's time to implement and upload the Arduino code to the board, so you can move on to programming the user-lamp interaction in Processing. The following Arduino code declares the LED pin variables and four integers (Wval, Rval, Gval, and Bval) that store the brightness level of each LED. You are building several lamps, so you will declare an integer called lampNumber that determines which lamp it is. You need to change this number before uploading the code to every lamp's Arduino, starting from zero and counting up.

```
int whiteLED = 5;
int redLED = 9;
int greenLED = 10;
int blueLED = 11;

int WVal, RVal, GVal, BVal;
int lampNumber = 0;
```

Within the setup() function, start the serial communication and set the four pins as output.

```
void setup() {
  Serial.begin(9600);
  pinMode(whiteLED, OUTPUT);
  pinMode(redLED, OUTPUT);
  pinMode(greenLED, OUTPUT);
  pinMode(blueLED, OUTPUT);
}
```

Within the loop() function, check if you have enough data in the serial buffer and read the first value. If this value is the event trigger 'S', read the next piece of data, which you interpret as the number of the lamp for which the communication is intended. If it's equal to this lamp's lampNumber, read the following four numbers and apply their values directly to the four LEDs connected to the Arduino.

```
void loop(){
if (Serial.available()>5) {   char val = Serial.read();
    if(val == 'S'){

      int messageNumber = Serial.read();
      if(messageNumber == lampNumber){
        WVal = Serial.read();
        RVal = Serial.read();
        GVal = Serial.read();
        BVal = Serial.read();
      }
    }
  }
  analogWrite(whiteLED, WVal);
  analogWrite(redLED, RVal);
  analogWrite(greenLED, GVal);
```

```
  analogWrite(blueLED, BVal);
}
```

The Lamp Class

The goal of this chapter is to create a lighting installation that is aware of the users and behaves according to their position in space. The lamps will also be controlled by intentional body gestures performed by the users. You're going to spend the rest of this chapter developing the software that will gather user information from Kinect and translate it into precise orders for each of your lamps, which will be broadcast with an XBee module attached to your computer.

At this moment, you have presumably fabricated a few lamps, each equipped with an XBee wireless module, a battery clip, and an Arduino Nano with specific code on board. It would be very useful for you to create a piece of code in Processing that could define the present state of each lamp and allow you to set each of the lamps' parameters according to the information from the Kinect. This code structure should be implemented as a class.

This is the first class you are implementing from scratch in the book, so it's a good moment to give you a refresher on object-oriented programming (OOP). If this sounds scary, don't panic. You have been using OOP throughout all of the previous chapters and projects; we just haven't talked about it explicitly. OOP is a vast and complex subject, and it would be out of the scope of this book to attempt a thorough introduction, but we're hoping this will be a reasonable starting point for you to get acquainted with the topic so you can understand the following chapters.

Object-Oriented Programming

First of all, object-oriented programming is a programming *paradigm*, or programming style, based on the use of objects for the design of computer programs. There is a long list of available programming paradigms, and they all differ in the abstractions used to represent the elements of a program. This means they are different in the way they *think* about the computation, not in the *type* of computation they perform. You can tackle the same problem using different programming paradigms, and thus many programming languages allow the implementation of code using several of them.

OOP uses *classes* and *objects*, which are instances of classes, to structure the code. Classes are used as blueprints to create objects to represent the elements in the code. In the next pages, you will use classes, fields, constructors, and methods to implement the class `Lamp`, which represents the lamps you have just built and programmed in the previous sections. This class allows you to create instances of your lamp objects, and store and affect their positions and RGB values in a structured way.

Java, the programming language on which Processing is based, is an object-oriented programming language and as such, every time you programmed in Processing, you were using OOP. In Java, the convention is to store each class in a separate file, and in Processing this translates to writing each class in a separate tab (Figure 7-20). This will automatically create a new `.pde` file in your sketch folder with the name of the tab.

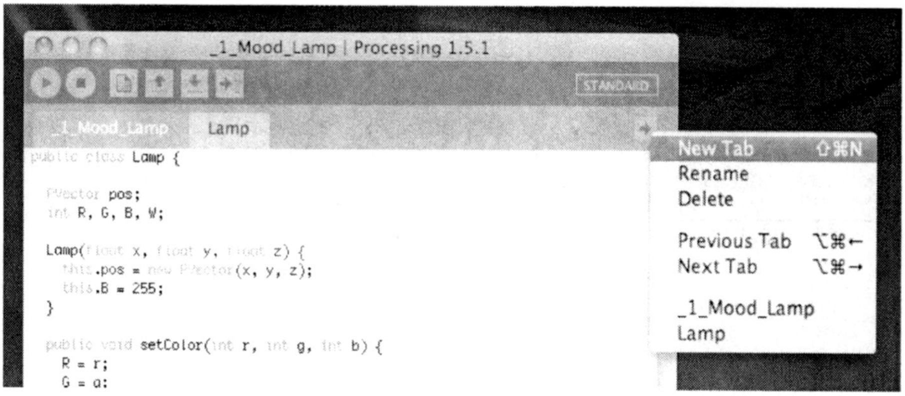

Figure 7-20. Creating a new tab in Processing

Class Declaration

The first line of code for the new class starts with the word "class," followed by the name you have chosen for the class. It is good coding practice to start class names with a capital letter. All the rest of the class code is contained within the curly brackets that you open after the class's name.

```
class Lamp {
   // Your code goes here
}
```

Field Description

After declaring your class, you declare the fields, or properties of the class. In your case, you want to be able to control the position of each lamp object and the brightness of its four LEDs, so you declare one PVector to keep track of the lamp's position and four integers for LED brightness.

```
class Lamp {
PVector pos;
int R, G, B, W;
```

In Java, classes, fields, and methods can be preceded by *access specifiers* defining the visibility of the members. The access specifier is a keyword preceding the declaration of the element. There are three main types of visibility in Java.

- `private`: The field can only be accessed only by the class.
- `protected`: the field can be accessed by the class and its subclasses.
- `public`: The member can be accessed from anywhere.

■ **Note** In Processing, the using of public members is encouraged at all times, so the visibility of every member will be `public` unless otherwise specified. This is the reason why you normally won't be explicitly defining the visibility of a member in your code.

Constructor

The constructor is a subroutine or method that has the same name as the class and doesn't have an explicit return type. The constructor is called at the creation of an object, and it initializes the data members (fields) of the class. There can be multiple constructors for the same class as long as they take different sets of parameters at the creation of the objects.

In your Lamp class, you only need one constructor; it will take three floats as parameters. The constructor initializes the PVector defining the position of the lamp to the x, y, and z parameter values and sets the blue LED to full brightness so you can see that the lamp has been initialized.

```
Lamp(float x, float y, float z) {
  this.pos = new PVector(x, y, z);
  this.B = 255;
}
```

Methods

Methods are functions declared within the class. Public methods can be accessed from other classes, and they can be declared as void if you don't need them to return a value or any kind of data. *Getters* and *setters* are just special ways of calling methods used to change or acquire a specific property of the class. They are sometimes called *accessor methods.*

The first method in your class sets the lamp's RGB parameters. It takes three integers and directly applies them to the R, G, and B fields of your object.

```
public void setColor(int r, int g, int b) {
  R = r;
  G = g;
  B = b;
}
```

You also need the current state of the LEDs in the lamp, so you implement a function that returns an array of three integers containing the RGB values.

```
public int[] getColor() {
  int[] colors = { R, G, B };
  return colors;
}
```

Although you won't be using this function in your code, it could be useful for some other application to be able to change the lamp's position from another class, including your main Processing draw() loop, so implement a setter for your lamp's position PVector.

```
public void setPos(PVector pos) {
  this.pos = pos;
}
```

Implement a draw() method that takes care of the display of the lamp on screen. It's called from the main draw() loop, so you can see your lamps at their right position and colors. Note that this function could be called render(), display(), or any other name that you feel is more descriptive.

```
public void draw() {
  pushMatrix();
  translate(pos.x, pos.y, pos.z);
  pushStyle();
  if (W == 255) {
```

```
      fill(W);
    }
    else {
      fill(R, G, B);
    }
    noStroke();
    box(100, 150, 100);
    popStyle();
    popMatrix();
  }
```

Figure 7-21. The lamp closer to the user will glow at full intensity

The method updateUser() takes the user's position in space and checks if the lamp is closer than a certain threshold distance. If it is, it makes the white LED glow at full brightness, assuming that the user could use a bit of light for reading or finding his way around (Figure 7-21).

```
  public void updateUserPos(PVector userCenter) {
    float dist = pos.dist(userCenter);
    if (dist < 1000) {
      W = 255;
    }
    else {
      W = 0;
    }
    stroke(200);
    line(pos.x, pos.y, pos.z, userCenter.x, userCenter.y, userCenter.z);
  }
```

Finally, drawSelected() is called when the user has successfully selected a lamp by pointing at it, and it will display a gizmo around the lamp so you know the lamp is selected.

```
  public void drawSelected() {
    pushMatrix();
    translate(pos.x, pos.y, pos.z);
    pushStyle();
    noFill();
```

```
    stroke(R, G, B);
    box(150, 225, 150);
    popStyle();
    popMatrix();
  }
```

Remember that all this code should be within the curly braces you opened after your class declaration. After you close the last method, there should be a closing curly brace to close the class.

User Control Sketch

You have implemented a Lamp class that you can use to represent your physical lamps in the code. Now you need to write the necessary routines to allow for gesture-based user control and the sending of serial data from your computer. You're going to use NITE's user and skeleton tracking capabilities for this purpose.

You encountered skeleton tracking in previous projects, but this time you are going to be using three-dimensional skeleton tracking to check where the user is pointing in space. You're also going to implement a simple gesture-recognition routine to trigger the creation of lamps.

First and foremost, you need to tell Processing to import all the libraries that you will be using in the project. These are Processing's OpenGL and Serial libraries, KinectOrbit, and Simple-OpenNI.

```
import processing.opengl.*;
import kinectOrbit.KinectOrbit;
import processing.serial.Serial;
import SimpleOpenNI.SimpleOpenNI;
```

Variable Declaration

You now need to declare all the libraries' objects and create a File object where you store your lamp's positions and colors. This is detailed in the data storage and retrieval section.

```
SimpleOpenNI kinect;
KinectOrbit myOrbit;
Serial myPort;
File lampsFile;
```

The Boolean serial defines if you are initializing and using serial communication. The userID integer keeps track of the current user, and the PVector userCenter contains the current position of the user. You also need to create an ArrayList of Lamp objects so it's easy to add more lamps in runtime.

```
boolean serial = true;
int userID;
PVector userCenter = new PVector();
ArrayList<Lamp> lamps = new ArrayList<Lamp>();
```

Setup() Function

The setup() function sets the frame's size and renderer, initializes the Simple-OpenNI object, and enables the Mirror, depth map, and user capabilities.

```
public void setup() {
  size(800, 600, OPENGL);
  smooth();
```

```
  kinect = new SimpleOpenNI(this);
  kinect.setMirror(true);
  kinect.enableDepth();
  kinect.enableUser(SimpleOpenNI.SKEL_PROFILE_ALL);
```

The KinectOrbit object is initialized now, passing this as a parameter, and you set the gizmo drawing to true. The file lampsFile is also initialized now; it's important that you do it now instead of doing it in the Variable Declaration section, as otherwise the data path won't refer to the data folder in your Processing sketch.

```
  myOrbit = new KinectOrbit(this, 0);
  myOrbit.drawGizmo(true);

  lampsFile = new File(dataPath("lamps.txt"));
```

To finish the setup(), initialize the serial communication if you have set your serial variable to true.

```
  if (serial) {
    String portName = Serial.list()[0];
    myPort = new Serial(this, portName, 9600);
  }
}
```

Draw() Function

The first lines in your draw() function update the data from Kinect and set the background color to black. You then push the KinectOrbit so you can control the camera view with your mouse and call the function drawPointCloud(), which draws the raw point cloud from Kinect on screen. You only display one every six points for speed.

```
public void draw() {
  kinect.update();
  background(0);

  myOrbit.pushOrbit();
  drawPointCloud(6);
```

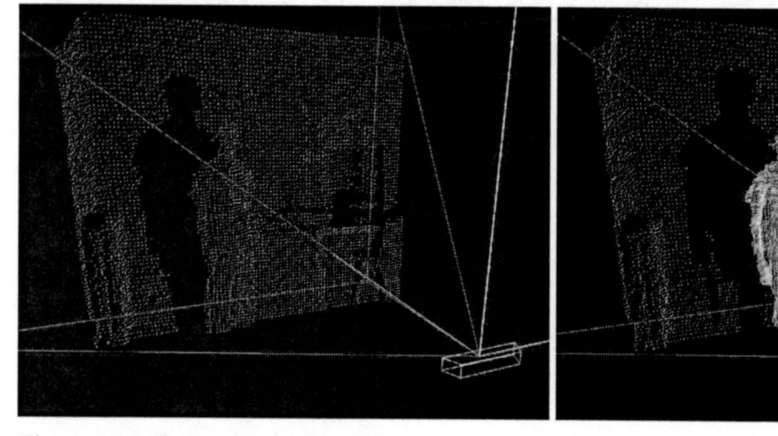

Figure 7-22. User recognized by Kinect

Then you check if there is a user in the Kinect's field of view, and if so, you get his center of mass using the Simple-OpenNI function getNumberOfUsers(), which sets the second parameter (your PVector userCenter) to the center of mass of the user userID (Figure 7-22). You use the center of the user's mass to update each of the lamps in your ArrayList.

```
if (kinect.getNumberOfUsers() != 0) {
  kinect.getCoM(userID, userCenter);
  for (int i = 0; i < lamps.size(); i++) {
    lamps.get(i).updateUserPos(userCenter);
  }
```

Then, if apart from having detected a user, you are also tracking his skeleton, you call the function userControl(), and you draw the skeleton on screen. If you're not tracking a skeleton, you simply print the user's point cloud on screen for reference (Figure 7-23).

```
  if (kinect.isTrackingSkeleton(userID)) {
    userControl(userID);
    drawSkeleton(userID);
  }
  else {
    drawUserPoints(3);
  }
}
```

Figure 7-23. *Use point cloud (left) and user skeleton (right)*

Now you draw each lamp on screen and the Kinect camera frustum, and you pop the KinectOrbit object. If the serial Boolean is true, you call the sendSerial() function.

```
  for (int i = 0; i < lamps.size(); i++) {
    lamps.get(i).draw();
  }
  kinect.drawCamFrustum();
  myOrbit.popOrbit();

  if (serial)  {
    sendSerialData();
  }
}
```

User Control

The function userControl() performs the computations that are most specific to this project. This function gets the user's right arm position from Kinect and uses the joint alignment to check whether the user is pointing at a lamp or not. Then it uses the left hand position to change the color of the selected lamp. This function also contains the implementation of the lamp creation gesture recognition.

First, you create PVectors to store the three-dimensional positions of the body joints you need for your user control. Then you use the Simple-OpenNI function getJointPositionSkeleton() to get the joint's coordinates and set your vectors to these positions.

```
private void userControl(int userId) {
  PVector head = new PVector();
  // Right Arm Vectors
  PVector rHand = new PVector();
  PVector rElbow = new PVector();
  PVector rShoulder = new PVector();
  // Left Arm Vectors
  PVector lHand = new PVector();

  // Head
  kinect.getJointPositionSkeleton(userId, SimpleOpenNI.SKEL_HEAD, head);
  // Right Arm
  kinect.getJointPositionSkeleton(userId, SimpleOpenNI.SKEL_RIGHT_HAND,
  rHand);
  kinect.getJointPositionSkeleton(userId, SimpleOpenNI.SKEL_RIGHT_ELBOW,
  rElbow);
  kinect.getJointPositionSkeleton(userId,
  SimpleOpenNI.SKEL_RIGHT_SHOULDER, rShoulder);
  // Left Arm
  kinect.getJointPositionSkeleton(userId, SimpleOpenNI.SKEL_LEFT_HAND,
  lHand);
```

Lamp Control

The way you check if the user is pointing at a lamp is by checking that his right arm is stretched; if it is, you check where in space it is pointing. For this, you need the vectors defining the user's forearm and arm; these are easily found by subtracting the vector defining the shoulder position from the one defining the elbow position, and doing the same operation with the elbow and the hand PVectors (Figure 7-24).

```
  PVector rForearm = PVector.sub(rShoulder, rElbow);
  PVector rArm = PVector.sub(rElbow, rHand);
```

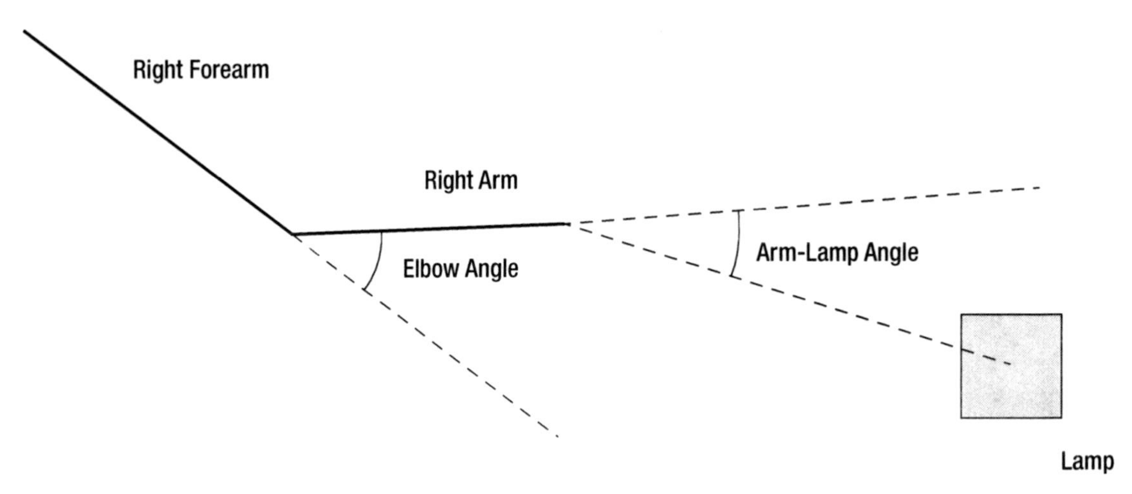

Figure 7-24. Right arm angles and lamp

Once you have these vectors, you can check whether the user's right arm is stretched. You do this by testing if the angle between the two PVectors (the complementary of the elbow angle) is lesser than the threshold you have set to PI/8, or 22.5 degrees. If the right arm is indeed stretched, you test the angle between the right arm's PVector and the angle defining the direction from the right hand to the lamp's position. The smaller the arm-lamp angle is, the more directly the right hand is pointing at the lamp. You have set this threshold to PI/4, or 45 degrees, so it's easier for the user to select a lamp.

If you are under the threshold, you use the `Lamp` object's method `setColors()`, passing as a parameter the halved x, y, and z distances of the left hand to the user's head. This way the user can change the color parameters of the lamp using unique body gestures (Figure 7-25).

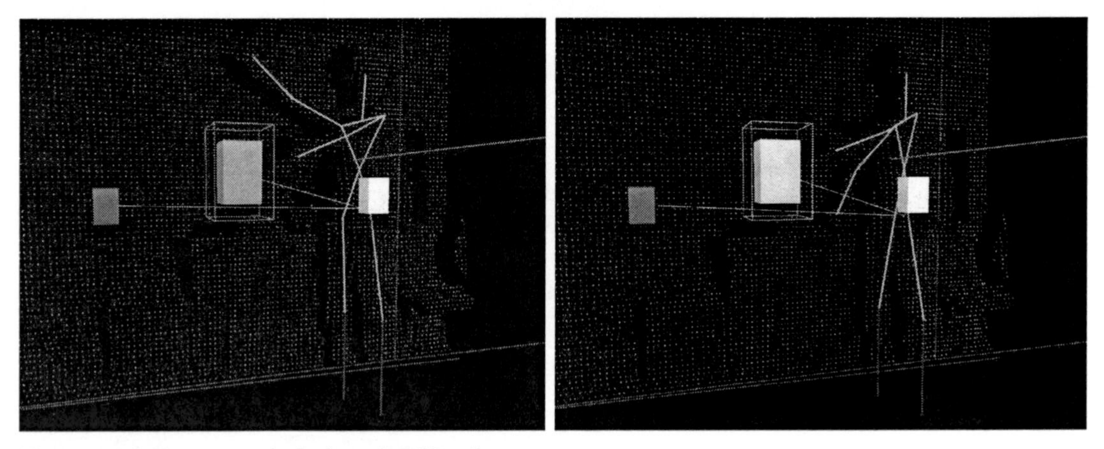

Figure 7-25. User control of a lamp's RGB values

```
if (PVector.angleBetween(rForearm, rArm) < PI / 8f) {
for (int i = 0; i < lamps.size(); i++) {
      PVector handToLamp = PVector.sub(rHand, lamps.get(i).pos);
      if (PVector.angleBetween(rArm, handToLamp) < PI / 4) {
```

```
PVector colors = PVector.sub(head, lHand);
        lamps.get(i).setColor((int) colors.x / 2,
        (int) colors.y / 2, (int) colors.z / 2);
        lamps.get(i).drawSelected();
      }
    }
  }
```

Lamp Creation

Until now, we have been talking about lamps that you haven't created yet. You have a series of physical lamps that you want to be able to control from your code, and you have implemented a way to control virtual lamps in your program. Now, if you can make the coordinates of your virtual lamps coincident to the physical lamps, you'll be getting somewhere.

The way you do this is by implementing gesture recognition that allows the user to tell the program where the lamps are placed physically. You basically place the lamps in the space and then stand by each of the lamps and tell the computer, "Hey, here is a lamp!"

This gesture should be something the user can do without too much effort but one that is not common enough to trigger a random series of lamps every minute. The gesture we came up with is touching your head with two hands, which is a reasonably good balance between easiness and rareness; it also has the advantage of being extremely easy to detect. You only need to check the distances of the right and left hands to the head; if both are smaller than 200mm you create a lamp positioned at the user's center of mass. This is not their exact location, but it's a reasonably good approximation.

There is one more test you need to perform. If you don't explicitly avoid it, the program will create lamps at every loop the whole time your hands are on your head. To avoid this, set a proximity limit so if the lamp about to be created stands closer than 200mm to any other lamp, it won't be created. This way, even if you keep your hands on your head for a minute, only one lamp will be created until you move away by the specified distance (Figure 7-26).

```
if (head.dist(rHand) < 200 && head.dist(lHand) < 200) {
    boolean tooClose = false;
    for (int i = 0; i < lamps.size(); i++) {
      if (userCenter.dist(lamps.get(i).pos) < 200) {
        tooClose = true;
      }
    }
    if (!tooClose) {
      Lamp lampTemp = new Lamp(userCenter.x, userCenter.y,
      userCenter.z);
      lamps.add(lampTemp);
    }
  }
}
```

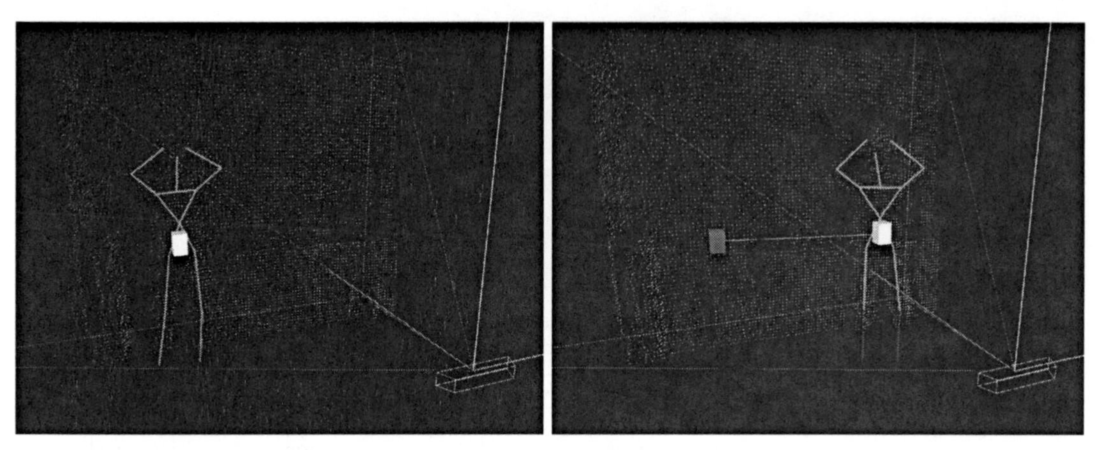

Figure 7-26. Lamp generation by body gestures

Data Storage and Retrieval

You have installed your lamps all over your living room, and you have told your program where each of them is located by using your lamp-creation gesture. But you don't want to have to do this every time you restart your sketch. It would be great to be able to save the current lamp layout together with their individual states to an external file that can be retrieved in future sessions. This is the purpose of the functions `saveLamps()` and `loadLamps()`.

You use the previously defined `lampsFile` file to store the information, employing the Processing function `saveStrings()` for this purpose. This function allows you to save an array of strings into a file with just one line of code. The first parameter that the function takes is the file you want to save your data to and the second one is the array of strings.

In your case, you need first to transform each of your lamps' position coordinates and RGB values to strings, and then store the information for each lamp in one line of code, separating the data with blank spaces so you can find them later.

```
void saveLamps() {
  String[] lines = new String[lamps.size()];
  for (int i = 0; i < lamps.size(); i++) {
    lines[i] = String.valueOf(lamps.get(i).pos.x) + " "
      + String.valueOf(lamps.get(i).pos.y) + " "
      + String.valueOf(lamps.get(i).pos.z) + " "
      + Integer.toString(lamps.get(i).getColor()[0]) + " "
      + Integer.toString(lamps.get(i).getColor()[1]) + " "
      + Integer.toString(lamps.get(i).getColor()[2]);
  }
  saveStrings(lampsFile, lines);
}
```

The function `loadLamps()` performs the inverse task to the previous one. You use the Processing function `loadStrings()` to save each line in the file to a string in the array of strings `stringArray`, and then you split each line at the blank spaces and convert each substring into floats defining the position of the lamp and three integers defining its color. You then create the necessary lamps and set their color values to the ones loaded from the file.

```
void loadLamps() {
  String lines[] = loadStrings(lampsFile);
  for (int i = 0; i < lines.length; i++) {
    String[] coordinates = lines[i].split(" ");
    Lamp lampTemp = new Lamp(Float.valueOf(coordinates[0]),
    Float.valueOf(coordinates[1]),
    Float.valueOf(coordinates[2]));
    lampTemp.setColor(Integer.valueOf(coordinates[3]),
    Integer.valueOf(coordinates[4]),
    Integer.valueOf(coordinates[5]));
    lamps.add(lampTemp);
  }
}
```

These two functions are called when you press the s and l (the letter l, not number 1) keys on your keyboard. The Processing callback function keyPressed() takes care of invoking them when a new keyboard event is triggered.

```
public void keyPressed() {
  switch (key) {
  case 's':
    saveLamps();
    break;
  case 'l':
    loadLamps();
    break;
  }
}
```

Serial Communication

The serial function performs the task of sending the necessary data to coordinate the physical lamps to the ones you are seeing on your screen. For each lamp you send the event trigger 'S', followed by the number of the lamp the message is intended for, and then the W, R, G, and B values.

```
void sendSerialData() {
    for (int i = 0; i < lamps.size(); i++) {
    myPort.write('S');
    myPort.write(i);
    myPort.write(lamps.get(i).W);
    myPort.write(lamps.get(i).R);
    myPort.write(lamps.get(i).G);
    myPort.write(lamps.get(i).B);
  }
}
```

Display Functions

The following functions are used to print the Kinect point clouds and user skeleton on screen. You saw how to display the Kinect point cloud in Chapter 3.

```
void drawPointCloud(int steps) {
  int[] depthMap = kinect.depthMap();
```

```
  int index;
  PVector realWorldPoint;
  stroke(255);
  for (int y = 0; y < kinect.depthHeight(); y += steps) {
    for (int x = 0; x < kinect.depthWidth(); x += steps) {
      index = x + y * kinect.depthWidth();
      if (depthMap[index] > 0) {
        realWorldPoint = kinect.depthMapRealWorld()[index];
        point(realWorldPoint.x, realWorldPoint.y, realWorldPoint.z);
      }
    }
  }
}
```

The drawing of the user points builds on the same principles as the previous function, but you sort the depth map points according to the color of the equivalent point in the user map. The user map is an array of pixels with value 0 if the Kinect determines they are background pixels, and a higher value, starting at 1, for the pixels defining the users.

This function allows you to print the user's pixels in a different color, so you can observe on screen what Kinect is considering to be your user.

```
void drawUserPoints(int steps) {
  int[] userMap = kinect.getUsersPixels(SimpleOpenNI.USERS_ALL);
  // draw the 3d point depth map
  PVector[] realWorldPoint = new PVector[kinect.depthHeight() * kinect.depthWidth()];
  int index;
  pushStyle();
  stroke(255);
  for (int y = 0; y < kinect.depthHeight(); y += steps) {
    for (int x = 0; x < kinect.depthWidth(); x += steps) {
      index = x + y * kinect.depthWidth();
      realWorldPoint[index] = kinect.depthMapRealWorld()[index].get();
      if (userMap[index] != 0) {
        strokeWeight(2);
        stroke(0, 255, 0);
        point(realWorldPoint[index].x, realWorldPoint[index].y,
        realWorldPoint[index].z);
      }
    }
  }
  popStyle();
}
```

The next two functions are from Simple-OpenNI's 3d skeleton tracking example. They both use functions specific to Simple-OpenNI to draw the user's limbs on screen.

```
void drawSkeleton(int userId) {
  strokeWeight(3);
  // to get the 3d joint data
  drawLimb(userId, SimpleOpenNI.SKEL_HEAD, SimpleOpenNI.SKEL_NECK);
  drawLimb(userId, SimpleOpenNI.SKEL_NECK,
  SimpleOpenNI.SKEL_LEFT_SHOULDER);
  drawLimb(userId, SimpleOpenNI.SKEL_LEFT_SHOULDER,
  SimpleOpenNI.SKEL_LEFT_ELBOW);
  drawLimb(userId, SimpleOpenNI.SKEL_LEFT_ELBOW,
```

```
  SimpleOpenNI.SKEL_LEFT_HAND);

  drawLimb(userId, SimpleOpenNI.SKEL_NECK,
  SimpleOpenNI.SKEL_RIGHT_SHOULDER);
  drawLimb(userId, SimpleOpenNI.SKEL_RIGHT_SHOULDER,
  SimpleOpenNI.SKEL_RIGHT_ELBOW);
  drawLimb(userId, SimpleOpenNI.SKEL_RIGHT_ELBOW,
  SimpleOpenNI.SKEL_RIGHT_HAND);

  drawLimb(userId, SimpleOpenNI.SKEL_LEFT_SHOULDER,
  SimpleOpenNI.SKEL_TORSO);
  drawLimb(userId, SimpleOpenNI.SKEL_RIGHT_SHOULDER,
  SimpleOpenNI.SKEL_TORSO);

  drawLimb(userId, SimpleOpenNI.SKEL_TORSO, SimpleOpenNI.SKEL_LEFT_HIP);
  drawLimb(userId, SimpleOpenNI.SKEL_LEFT_HIP,
  SimpleOpenNI.SKEL_LEFT_KNEE);
  drawLimb(userId, SimpleOpenNI.SKEL_LEFT_KNEE,
  SimpleOpenNI.SKEL_LEFT_FOOT);

  drawLimb(userId, SimpleOpenNI.SKEL_TORSO, SimpleOpenNI.SKEL_RIGHT_HIP);
  drawLimb(userId, SimpleOpenNI.SKEL_RIGHT_HIP,
  SimpleOpenNI.SKEL_RIGHT_KNEE);
  drawLimb(userId, SimpleOpenNI.SKEL_RIGHT_KNEE,
  SimpleOpenNI.SKEL_RIGHT_FOOT);

  strokeWeight(1);
}

void drawLimb(int userId, int jointType1, int jointType2) {
  PVector jointPos1 = new PVector();
  PVector jointPos2 = new PVector();
  float confidence;

  // draw the joint position
  confidence = kinect.getJointPositionSkeleton(userId, jointType1,
  jointPos1);
  confidence = kinect.getJointPositionSkeleton(userId, jointType2,
  jointPos2);
  stroke(255, 0, 0, confidence * 200 + 55);
  line(jointPos1.x, jointPos1.y, jointPos1.z, jointPos2.x, jointPos2.y,
  jointPos2.z);
}
```

Simple-OpenNI Callbacks

The Simple-OpenNI callbacks are the same functions you used in the previous chapter. The only point to notice is that you update the `userID` integer to the last user you detected in the `onNewUser()` function.

Speaking of classes and member visibility, you should add the keyword "`public`" before every callback function. This is unnecessary in Processing, but if you are using another IDE, such as Eclipse, you will need to set all the callbacks as public or Simple-OpenNI won't invoke them.

```
public void onNewUser(int userId) {
  println("onNewUser - userId: " + userId);
  println("  start pose detection");
  kinect.startPoseDetection("Psi", userId);
  userID = userId;
}

public void onLostUser(int userId) {
  println("onLostUser - userId: " + userId);
}

public void onStartCalibration(int userId) {
  println("onStartCalibration - userId: " + userId);
}

public void onEndCalibration(int userId, boolean successfull) {
  println("onEndCalibration - userId: " + userId + ", successful: " + successfull);
  if (successfull) {
    println("  User calibrated !!!");
    kinect.startTrackingSkeleton(userId);
  }
  else {
    println("  Failed to calibrate user !!!");
    println("  Start pose detection");
    kinect.startPoseDetection("Psi", userId);
  }
}

public void onStartPose(String pose, int userId) {
  println("onStartdPose - userId: " + userId + ", pose: " + pose);
  println(" stop pose detection");
  kinect.stopPoseDetection(userId);
  kinect.requestCalibrationSkeleton(userId, true);
}

public void onEndPose(String pose, int userId) {
  println("onEndPose - userId: " + userId + ", pose: " + pose);
}
```

Summary

You were introduced to a broad range of topics in this chapter. You learned how to implement three-dimensional skeleton tracking and use it to perform custom gesture recognition. You also had a refresher on object-oriented programming, and you learned how to write a new class from scratch.

On the hardware side of things, you built Arduino-based autonomous devices, each with its own power source and wireless module, which permitted the communication of the units with your computer. You also learned about the RGB color space and that you can reproduce the whole color spectrum with only three RGB LEDs (Figure 7-27). You also learned how to calculate the appropriate resistors for the LEDs. In the end, you were able to control a whole lighting system with body gestures and make it aware of your position in space so it could respond by lighting the closest lamps to you with a higher intensity.

Figure 7-3. RGB lamp color change

You can think of this chapter as a general introduction to building any sort of application for intelligent houses. Imagine applying this sort of gesture control to any of your home appliances. The introduction of the XBee wireless module adds a new set of possibilities for using Arduino remotely, and you will explore these possibilities in future chapters of this book.

CHAPTER 8

Kinect-Driven Drawing Robot

by Przemek Jaworski

In this chapter, you will be using some of the techniques that you learned in previous chapters and exploring them further with a different goal in mind. You will be using servos to build a robotic arm that is able to draw algorithmic patterns on paper (Figures 8-1 and 8-2).

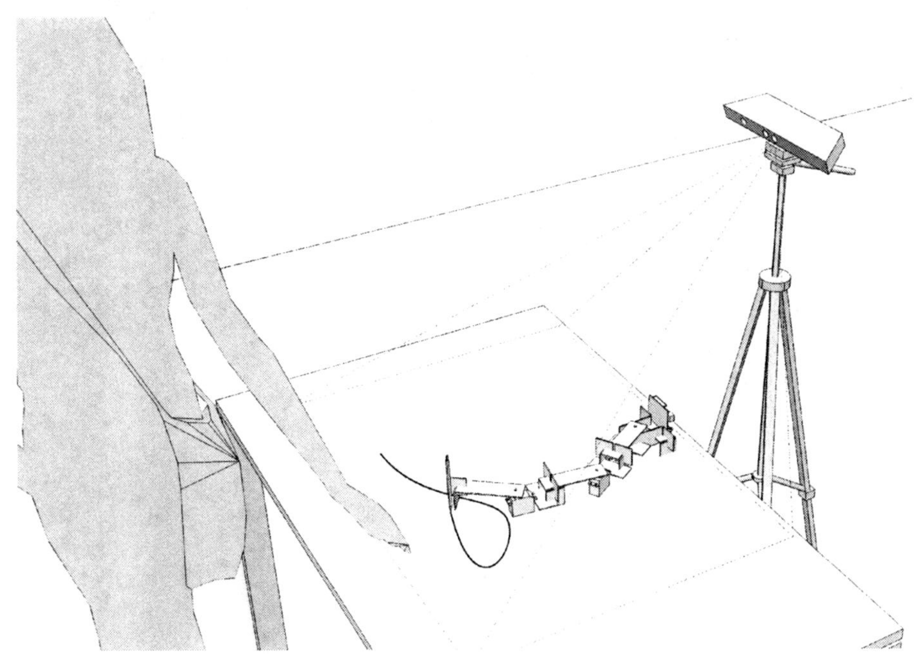

Figure 8-1. Visualization of the installation

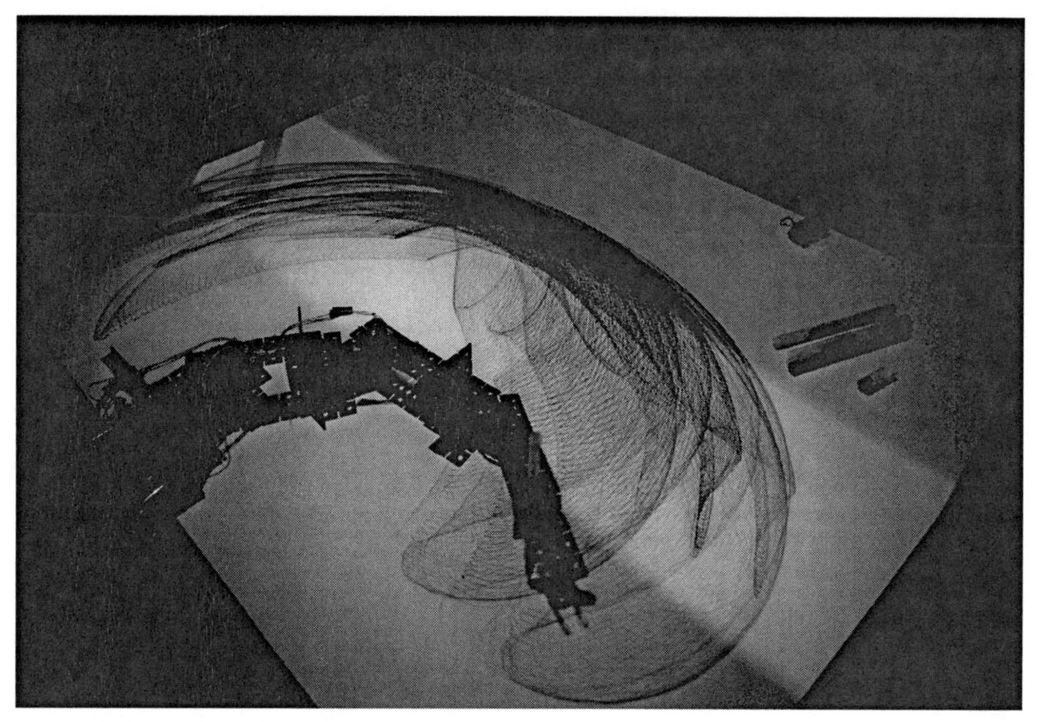

Figure 8-2. Drawing robot in action

Why an arm and not a two-axis pen plotter? Actually, an arm-like robot is simpler to construct and much easier to code. As in Chapter 6, you're going to use servo motors, which are less complicated than steppers (no need for endstops and other sensors). In the course of this chapter, you will use Firmata, a library that allows you to control your Arduino board from Processing without the need to write any Arduino code.

Your robot will be controlled by a Kinect-based tangible table, through which you will be able to issue instructions to the robot by moving your hands. It will be as simple as pointing your finger to an area and your robot will follow, drawing either your finger strokes or some sort of pre-defined algorithmic pattern. Table 8-1 lists the required parts and Figure 8-3 shows them.

Table 8-1. The Parts List

Part	Description	Price
Screws, medium	M4 screws, 25 mm length, and nuts	$5
Screws, small	M3 screws, 20mm length, and nuts	$3
K'lik construction system	Choose the Starter Pack with a mix of various sized plates	$12
Power Supply	Optional 5V, could be hacked ATX one	$34.99
Three servo motors	Hitec HS-311 or similar	$7.99 each

Part	Description	Price
Extension lead cables	Three for each servo	$1
Paper clips	To attach the device to the table	$5 for pack
Two right angle metal brackets	To mount the robot	$1

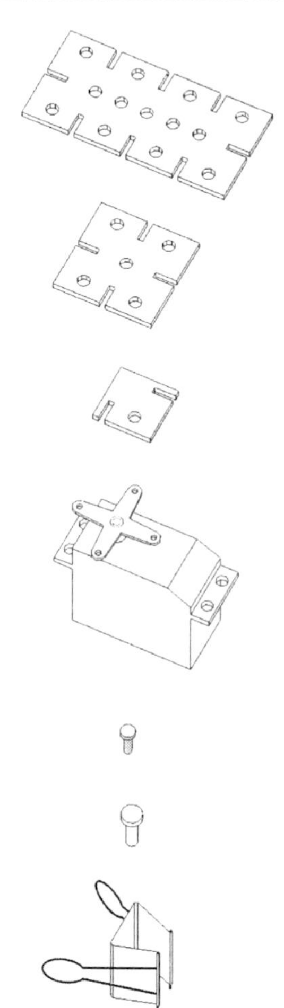

Figure 8-3. *Some parts shown graphically*

The parts taken from a slotted-plate construction system are 13 large plastic plates, 14 mid-sized plates, and 14 small joints (to make the structure more rigid). Apart from three servos (standard 0-180 degree servos, similar to those used in earlier exercises), you'll need six small screws to assemble the arms to the servo horns (they're usually supplied with the servos) and 14 short M4 screws (16 or 20mm length). Paper clips will be used to fasten the pen to the robot and the base to the table.

Building the Robot

Before you start, a few words must be said about the construction. I used a slotted-plate construction kit (K'lik - available from web sites like `http://www.mindsetsonline.co.uk`). You can use any similar system consisting of perforated plates that can be connected at straight angles or by screws. The robot itself is actually quite simple. It consists of three moveable parts: a base that holds Arduino and one servomotor, and three parts constituting the arms.

The construction doesn't really have to look exactly as shown in Figure 8-4, as there are many ways of assembling a similar robot. Parts can be designed differently, even 3D printed or CNC machined. I leave this to your creativity. Just note that I'll be describing to a slotted-plate construction system. Optionally, you could make all required parts out of plastic or Perspex.

The most important thing is to build a steady base with one servo (it will move the rest of the mechanism, so it has to be quite sturdy) and make sure it can be securely attached to the table. The other three parts, shown in Figure 8-4, should be made as light as possible; this will make it easier and faster to move.

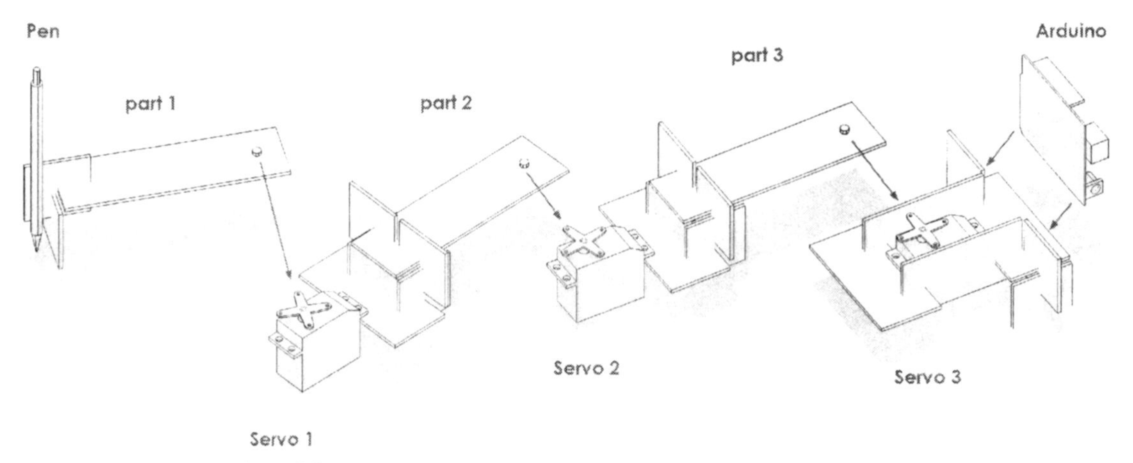

Figure 8-4. Construction of the arm

Perhaps it's best to start with easiest elements, so let's focus on the arm at the end of the robot (Part 1).

Part 1

Part 1 is the part that holds the pen, so it's the lightest part of all. The assembly drawing in Figure 8-5 shows the elements and their contact points for Part 1. Use two M4 screws to fasten two plates together and two small M3 screws to attach it to servo horn. The bracket needs to be attached to the servo via the screw that comes with it. Figure 8-6 shows the finished part.

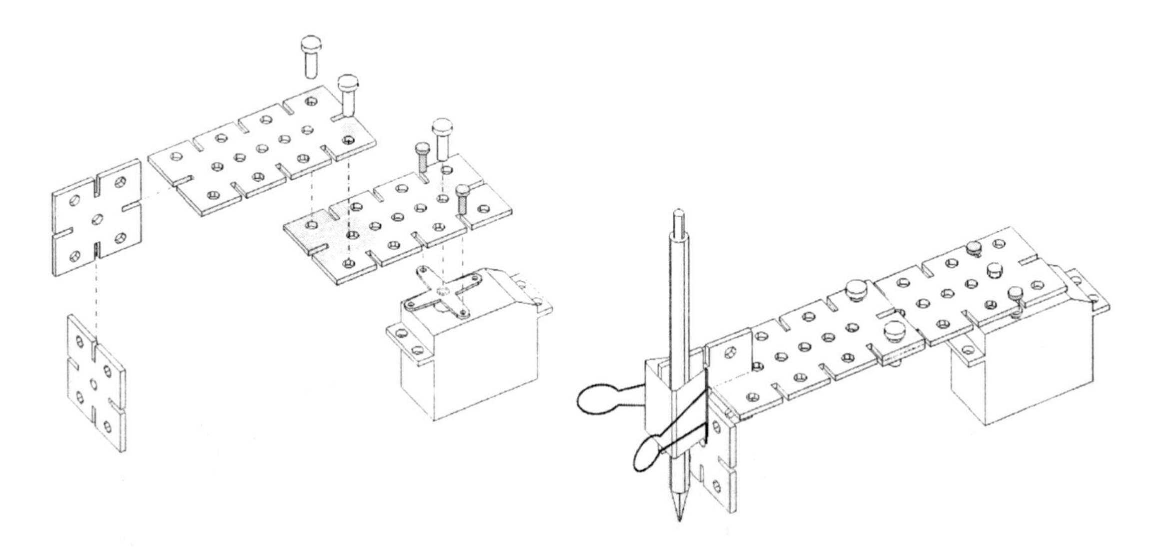

***Figure 8-5.** Part 1 assembly*

***Figure 8-6.** Part 1 finished*

Part 2

Part 2 is a bit more complex as it includes an assembly of plates that hold the servo that connects Part 2 to Part 1. The assembly should look like Figure 8-7.

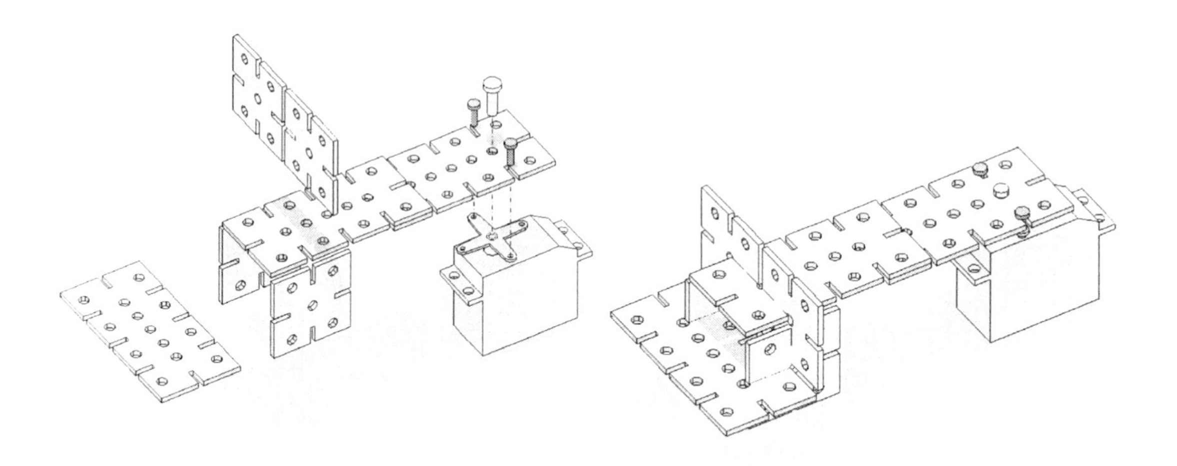

Figure 8-7. Part 2 assembly

Note that the top part is a set of two mid-sized plates, and not one large plate. They both stiffen the end so the piece can hold the servo attached to the front, as shown in Figure 8-8. The servo is attached with four standard M3 screws and one central screw.

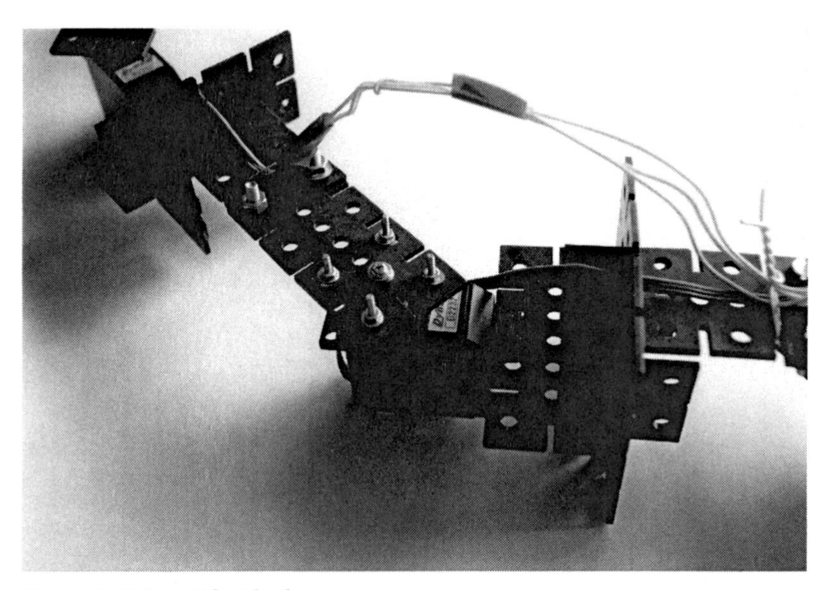

Figure 8-8. Part 2 finished

Part 3

Part 3 is the one closest to the base. Its construction is very similar to Part 2. The only difference is that it contains one more large plate at the bottom, additionally stiffening the structure (Figure 8-9). This plate is fitted here because this part is positioned higher above the table than other parts, so there is enough clearance to do so.

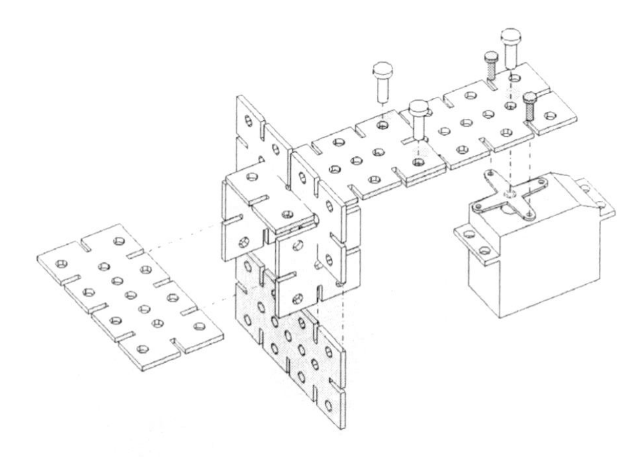

***Figure 8-9.** Part 3 assembly*

To connect the servo to the previous part, note that the spacing between the holes of the construction plates and those of the servo screws are different. You can drill new holes in the construction set or you can attach the servo to the arm using rigid wire (tightening it with pliers and rotating the ends spirally); see Figure 8-10. If you laser-cut your own parts, the holes will align, so you can simply use screws.

***Figure 8-10.** Part 3 finished*

The Base

The base consists of four large plates and four small plates holding the servo in the middle. Additionally, there are two sets of three small plates joined in a corner shape to increase the rigidity of the part. Each corner shape is positioned at one side of the base (Figure 8-11).

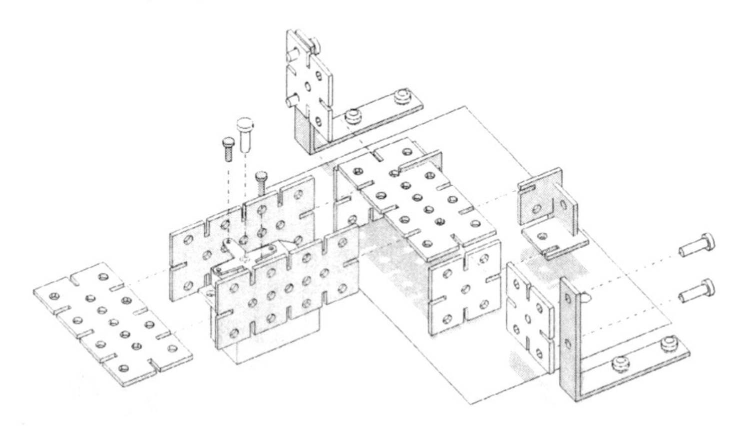

Figure 8-11. *Base assembly*

The Arduino board can be attached by screwing it to the plastic parts using the mounting holes in the PCB and one M4 screw and nut (Figure 8-12).

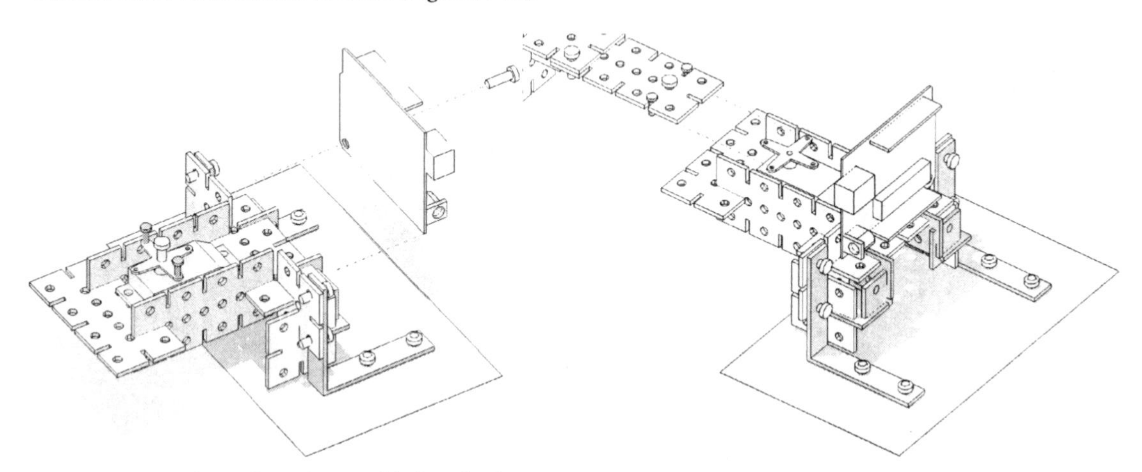

Figure 8-12. *Arduino board assembled to the base*

Finally, attach the base to a small pad, which you then clip to the table (Figure 8-13). This way the robot stays in place, but you can move it easily if needed.

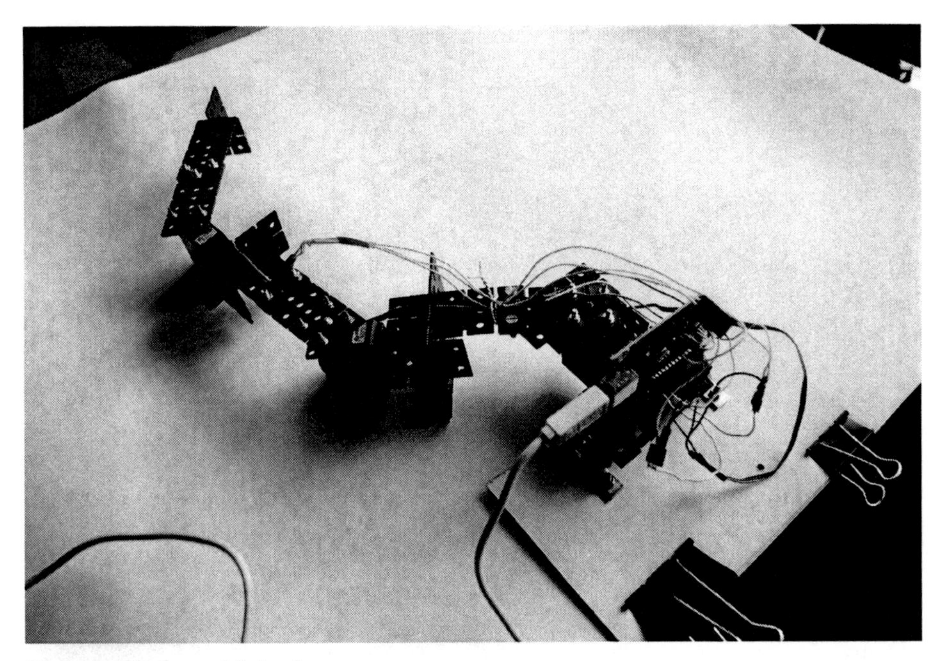

Figure 8-13. Assembled robot

Building the Circuit

The wiring of this project is quite simple as you are only using three servos. Just connect each servo's GND connection with GND on Arduino, +5V wire to +5V on Arduino, and the signal wires to pins 9, 10, and 11 (see Table 8-2 and Figure 8-14). The positive voltage wire (+5V) is usually a red one, ground is black, and the signal wire is orange or yellow.

Table 8-2. Servo Connections

Servo 1 Ground	Servo 1 +5V	Servo 1 Signal	Servo 2 Ground	Servo 2 +5V	Servo 2 Signal	Servo 3 Ground	Servo 3 +5V	Servo 3 Signal
Arduino GROUND	Arduino 5V	Arduino Pin 9	Arduino GROUND	Arduino 5V	Arduino Pin 10	Arduino GROUND	Arduino 5V	Arduino Pin 11

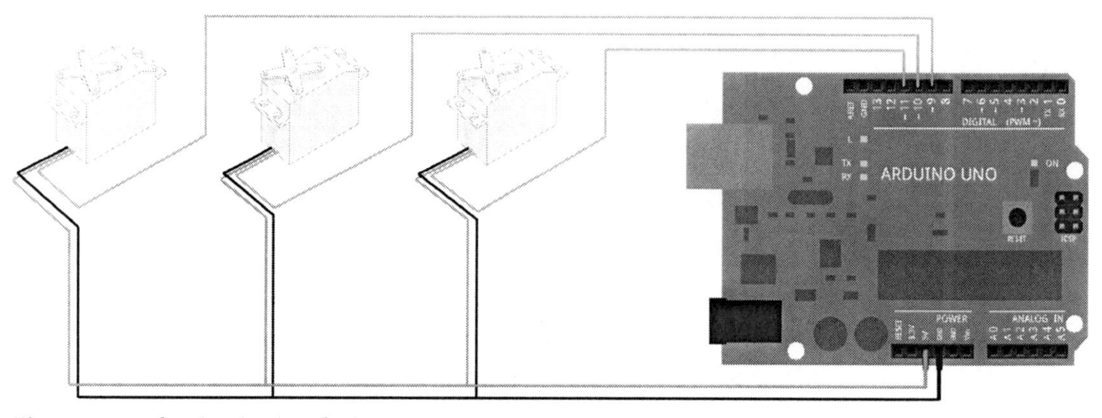

Figure 8-14. *The circuit (signal wires connect to pins 9, 10, and 11 at the top.)*

If you have a +5V power supply that can provide more than 0.5A, simply connect it to Arduino using the built-in socket. This is only necessary if your computer/laptop USB port is not able to meet the demand. In some newer computers, USB ports are powerful enough to drive three small servos without a problem.

Testing the Circuit

Next, test the servos. In previous projects, you learned how to drive your servos using Arduino code and how to control their positions using serial communication from Processing. In this project, you will take a different approach: you will utilize a Processing library to manipulate the servos directly from Processing.

Firmata and the Arduino Library for Processing

The Arduino library is one of Processing's core libraries so you don't need to install it. This library communicates to a firmware installed in your Arduino board called Firmata, eliminating the need to write any separate Arduino code to drive your servos.

As opposed to previous examples, everything is in one piece of code in this project. This approach is very advantageous for some uses but is not applicable to freestanding Arduino projects. Arduino can't be used without a computer if you are using this library, so it always needs to be connected to the computer via USB cable or other wireless means. If the library doesn't exist by default on your Arduino IDE, you can download it from `http://www.arduino.cc/playground/Interfacing/Processing`.

Before you start testing the circuit, you must upload the Firmata code to Arduino so it's ready to listen to the instructions via serial port. Open Arduino environment and navigate to File ➤ Examples ➤ Firmata ➤ Servo Firmata. Make sure you have the proper board and COM port selected, and press the Upload button. After the successful upload, you can write your first servo navigation program.

Servo Test

Open the Processing environment and type the following code:

```
import processing.serial.*;
import cc.arduino.*;
```

```
Arduino arduino;
int servoPin = 11; // Control pin for servo motor
void setup(){
  size (180, 100);
  arduino = new Arduino(this, Arduino.list()[0]);
  arduino.pinMode(servoPin, Arduino.OUTPUT);
}

void draw(){
  // the servo moves to the horizontal location of the mouse
  arduino.analogWrite(servoPin, mouseX);
}
```

Note that you're using pin 11 for connecting the servo, which is a PWM pin. You're also not using any COM ports in this program. Instead you're asking the computer for a list of available Arduinos (there might be more than one) and specifying which one you're using. That's what `Arduino.list()[0]` means. However, sometimes it detects other devices connected to COM ports as Arduinos, so depending on your setup, you might have to use `Arduino.list()[1]` instead.

Once you run this simple Processing application, your servo should rotate as you move your mouse within the app's window. The servo accepts angles from 0 to 180; hence the 180-pixel width used.

Upon successful testing of this simple setup, you can move on to construct the kinetic part of the installation. The difference between the previous example and your final piece is that you're going to be using three servos connected to PWM pins 9, 10, and 11. (Depending on your Arduino version, you might have more PWM pins. It's easy to recognize them by a small white PWM mark). All GND wires should go to GND on Arduino and all power wires to +5V.

Robot Simulation

The main routine of the installation is the following: you're going to scan a tangible table area, detect a hand (or a pointer) over the table, filter the point cloud, and extract the X and Y coordinates of where the robot should move. These X and Y values will be translated into servo angles in order to make the robot move to the same location. Let's have a look at the geometry of your robot.

You have three modules, each driven by separate servo and each able to rotate between 0 and 180 degrees. Let's assume the length of each module is exactly 150mm. This means you must translate these X and Y coordinates into angular values sent to each servo. This might look complicated, but fortunately you can use some simple geometric dependencies to find and use the angles.

Look at the diagram in Figure 8-15. You can see that each X and Y coordinate can be translated into polar coordinates: length L and rotation angle. This is true for all points lying within maximum reach of the robot and within a 0-180 angle in relation to upper horizontal edge of your working area.

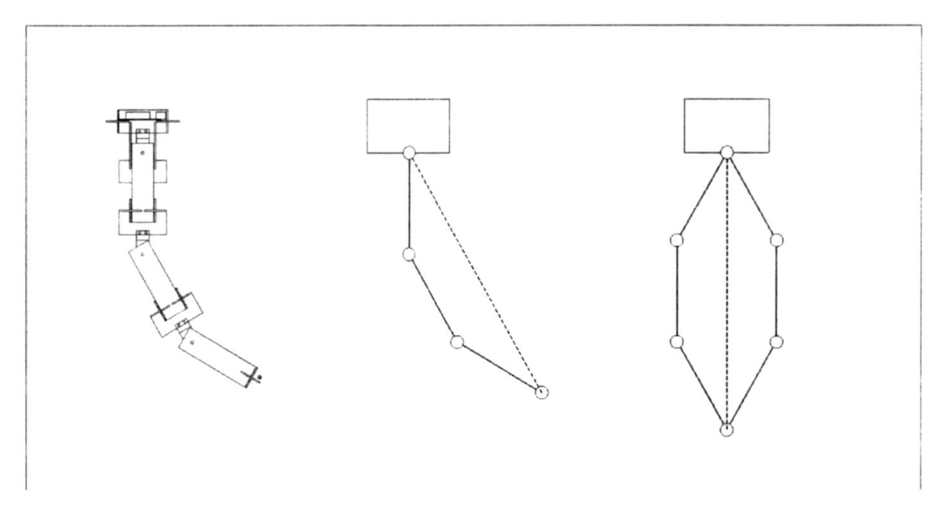

Figure 8-15. Robot angles

This translation, however, poses a small problem. As you can see in the right drawing in Figure 8-15, sometimes (or most of the time) there are two ways of reaching a given position. The tip of the pen will reach XY point in both cases, so how do you decide which one to use?

The truth is, that once you choose one side, you can't reach all the points in your working area simply because your robot is hitting the limits of its arm's servo rotation. Each servo can only rotate from -90 to 90 degrees (or 0 to 180, depending on reference angle), so the pen's closer position to the base will be reached by the robot bending into a C shape.

For the purpose of this exercise, you're going to simplify the problem. You'll assume that your robot is always using only one side. To do that, you're going to rotate the base by 45 degrees, as shown in Figure 8-16.

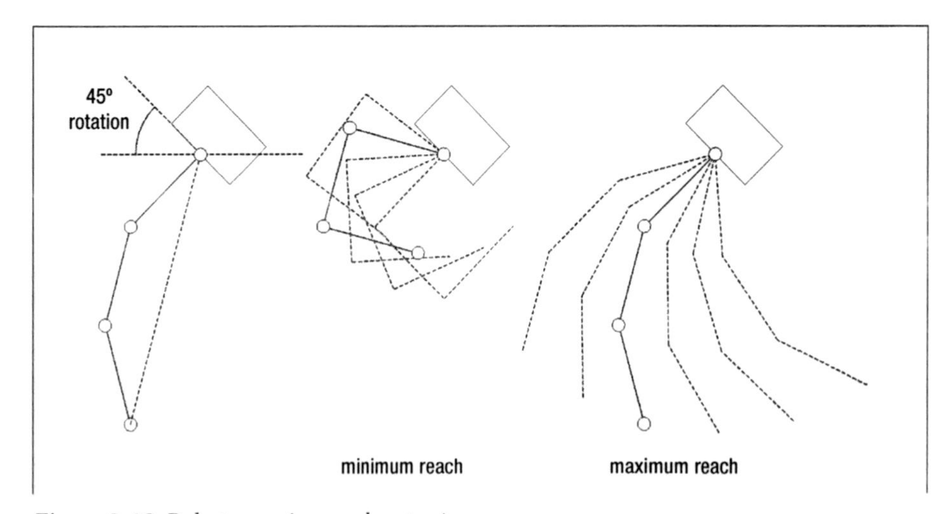

Figure 8-16. Robot span (example setup)

This way, the working area resembles a slice of the circle with its minimum and maximum reach arcs cut off at the top and bottom (Figure 8-17). Of course, you could use a more intelligent algorithm allowing use of more space, but to keep things uncomplicated, just stick to this setup.

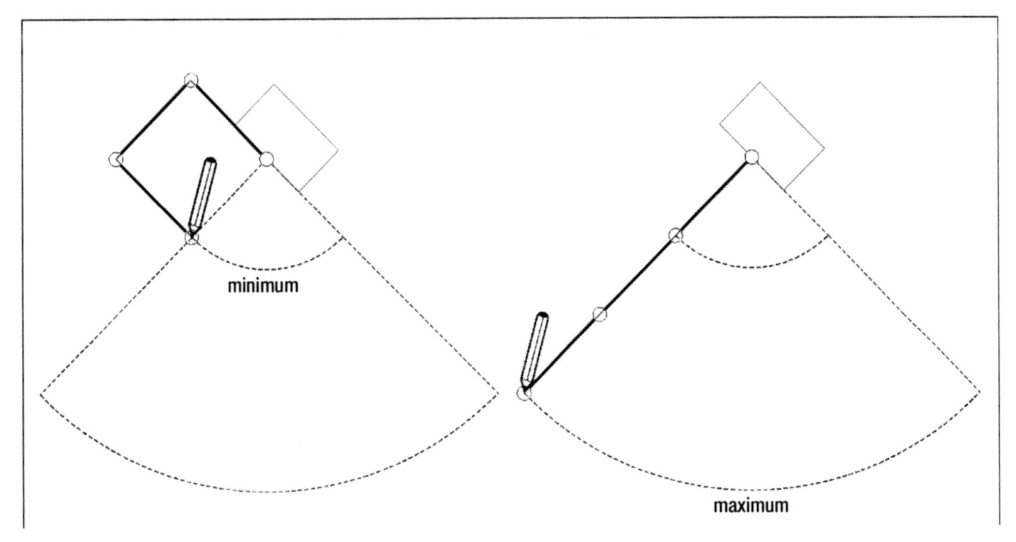

Figure 8-17. *Robot's working area*

Angle Measuring Program

Finally, you must tackle the translation from Cartesian coordinates (XY) to polar coordinates (angle/distance). Knowing the X and Y and your origin of coordinates (the pivotal point of the first segment's servo), you can easily calculate the distance to the pen's desired X and Y position using Processing's `dist()` function. The angle doesn't pose a big challenge because it's usually derivable from the arcsine. The functions return an angle based on the proportion of the sides of a triangle formed by the XY point, the origin (0,0) point, and its projections on the axis of the coordinate system. In this case, your angle will be arcsine(Y coordinate/total length L), assuming the base is at the (0,0) point.

To make more sense of it, let's test the angle-measuring program.

```
void setup(){
size (800, 600);
}

void draw(){
  background(255);
```

First, center your X coordinate in the middle of the screen.

```
  translate(width/2, 0);
```

Then, capture the coordinates of your pen tip (mouse).

```
  float penX = mouseX - width/2;
  float penY = mouseY;

  ellipse(penX, penY, 20, 20);//draw pen tip
```

```
  ellipse(0, 0, 20, 20);

  line(0, 0, penX, penY);
  float len = dist(0, 0, penX, penY); //let's measure the length of your line
```

The asin() function returns an angle value in a range of 0 to PI/2, in radians. The following line makes sure that the angle is greater than PI/2 (90 degrees) when penX is negative:

```
  float angle = asin(penY/len);
  if (penX < 0) { angle = PI - angle; }
```

Then output the angle value converted from radians to degrees.

```
  println("angle = " + degrees(angle));
  println("length = " + len);//print out the length
```

And finally, draw your angle as an arc.

```
  arc(0, 0, 200, 200, 0, angle);
}
```

This demonstrates how to extract any angle from the proportions of a triangle defined by three points. But the goal is to find not one but three angles, one for each servo; therefore you need to go a bit further. If you look at Figure 8-18, you can see how to perceive the virtual model of the robot. This should make it much easier to understand the math behind it.

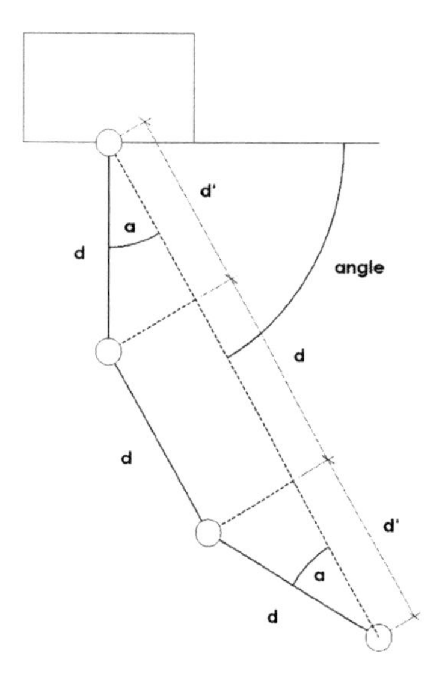

Figure 8-18. Robot's angles (assuming no 45 degree rotation has been applied yet)

Robot Simulation Program

Let's imagine that the length of your reach (described previously as L) can be split into three components, being perpendicular projections of each sub-arm onto a straight line. They're called d', d, and d' again (the first and last ones are the same). Using this method, you can codify each polar coordinate (angle and length) as an angle and two different lengths (d' and d). But wait a minute; don't you need the angles, not the lengths, to drive the servos?

That's right. Therefore, having d' and d, you can use their proportion, and the `asin()`function again, to determine it!

So, the length of your reach is L.

L = dist(0, 0, penX, penY)

But also, looking closely at Figure 8-18, you can determine that

L = d' + d + d' = 2 * d' + d

D is given (150mm in your robot), so

d' = (L – 150mm) / 2

Therefore your angle (a) is

a = acos(d'/d) = acos(d'/150)

Voila! So now you have all you need: the polar coordinates of the XY point (angle and length) and the internal angles of the trapezoidal shape formed by your arms. Using that, you can deduct the following:

- Angle of first servo: main angle + a
- Angle of second servo: 90 – a
- Angle of third servo: 90 - a

Note that these angles are relative to the position of each servo.

You are going to build this sketch from the Angle Measuring sketch (described before), so add the following code at the very end of `draw()` function, just before the closing curly bracket.

First, constrain the length to the physical size of the robot's parts.

```
if (len > 450) { len = 450; }
if (len < 150) { len = 150; }
```

Then, use what you have just learned to calculate your three servo angles.

```
float dprime =  (len - 150) / 2.0;
float a = acos(dprime / 150);
float angle1 = angle + a;
float angle2 = -a;
float angle3 = -a;
```

And finally, use the angles to draw the robot on screen.

```
rotate(angle1);
line(0, 0, 150, 0);
translate(150, 0);

rotate(angle2);
line(0, 0, 150, 0);
translate(150, 0);

rotate(angle3);
line(0, 0, 150, 0);
translate(150, 0);
```

After running it, you should see the image shown in Figure 8-19.

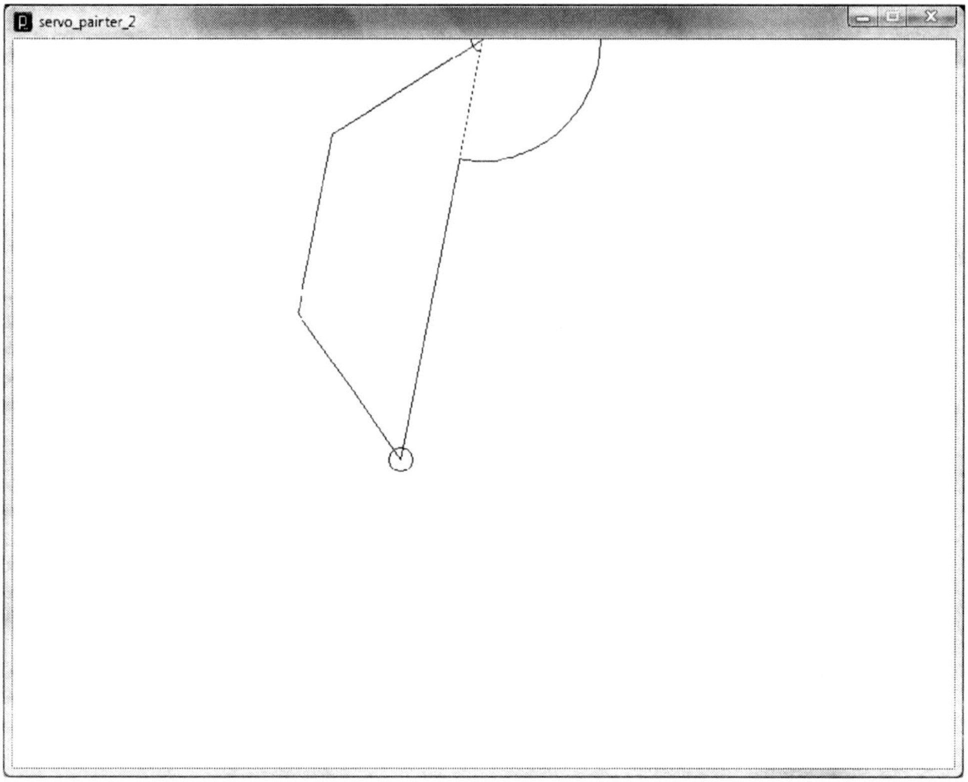

Figure 8-19. *Screenshot from the program*

The full code for this exercise, declaring the angles as global variables, is as follows:

```
float angle1 = 0;
float angle2 = 0;
float angle3 = 0;

//servo robot testing program
void setup(){
size (800, 600);
}

void draw(){
  background(255);
  translate(width/2, 0); //you need to center your X coordinates to the middle of the screen
  float penX = mouseX - width/2; //you're capturing coordinates of your pen tip (mouse)
  float penY = mouseY;
  ellipse(penX, penY, 20,20);
  ellipse(0,0, 20,20);
```

```
  line(0,0, penX, penY);
  float len = dist(0,0, penX, penY); //let's measure the length of your line
  float angle = asin(penY/len); //asin returns angle value in range of 0 to PI/2, in radians
  if (penX<0) { angle = PI - angle; }  // this line makes sure angle is greater than PI/2 (90
// deg) when penX is negative
  println("angle = " + degrees(angle)); //you're outputting angle value converted from radians
// to degrees
  println("length = " + len);//print out the length
  arc(0,0,200,200, 0, angle); //let's draw your angle as an arc
  if (len>450) len = 450;
  if (len<150) len = 150;

  float dprime = (len - 150)/2.0;
  float a = acos(dprime/150);
  angle1 = angle + a;
  angle2 =  - a;
  angle3 =  - a;

  rotate(angle1);
  line(0,0,150,0);
  translate(150,0);

  rotate(angle2);
  line(0,0,150,0);
  translate(150,0);

  rotate(angle3);
  line(0,0,150,0);
  translate(150,0);

}
```

The simulation shows the position of the robot on screen. Notice that the len variable is limited to a range of 150 to 450 in order to be physically reachable. This just makes sure that you're working in realistic working area.

This concludes the virtual testing phase. Let's get back to the wires and the Arduino.

Driving the Physical Robot

You should have your robot built. The Firmata code should be uploaded to the Arduino board and the three servos should be connected to pins 9, 10 and 11. Using the servo-testing program, make sure that servo1pin is the one connected to Part 1, servo2pin to Part 2, and servo3pin to Part 3 and the base.

Add the Serial and Arduino libraries to your previous code, and use them to send the angles to the servos through the Arduino board.

```
import processing.serial.*;
import cc.arduino.*;
Arduino arduino;

int servo1pin = 11;
int servo2pin = 10;
int servo3pin = 9;
```

```
float angle1 = 0; //declaration of all 3 angles for servos
float angle2 = 0;
float angle3 = 0;
void setup(){
size (800, 600);
arduino = new Arduino(this, Arduino.list()[0]);
arduino.pinMode(servo1pin, Arduino.OUTPUT);
arduino.pinMode(servo2pin, Arduino.OUTPUT);
arduino.pinMode(servo3pin, Arduino.OUTPUT);
}

void draw(){
  background(255);
  translate(width/2, 0);

  float penX = mouseX-width/2;
  float penY = mouseY;

  ellipse(penX, penY, 20,20);
  ellipse(0, 0, 20, 20);

  line(0, 0, penX, penY);
  float len = dist(0, 0, penX, penY); //let's measure the length of your line
  float angle = asin(penY/len);
  if (penX < 0) { angle = PI - angle; }

  println("angle = " + degrees(angle)); //output angle in degrees
  println("length = " + len);//print out the length
  arc(0,0,200,200, 0, angle); //let's draw your angle as an arc
  if (len > 450) { len = 450; }
  if (len < 150) { len = 150; }

  float dprim = (len - 150)/2.0;
  float a = acos(dprim/150);
  angle3 = angle + a;
  angle2 = -a;
  angle1 = -a;

  rotate(angle3);
  line(0, 0, 150, 0);
  translate(150, 0);

  rotate(angle2);
  line(0, 0, 150, 0);
  translate(150, 0);

  rotate(angle1);
  line(0, 0, 150, 0);
  translate(150, 0);
```

Use the Arduino library to send the values of the angles to the Arduino board.

```
  arduino.analogWrite(servo1pin, 90-round(degrees(angle1))); // move servo 1
  arduino.analogWrite(servo2pin, 90-round(degrees(angle2))); // move servo 2
  arduino.analogWrite(servo3pin, round(degrees(angle3))); // move servo 3
}
```

These last three lines are the instructions to move the servos. Due to some differences between graphical interpretation on screen and the way the servos operate, you must add some modifications. First, you must convert angles from radians to degrees, but you also must change the values by making them negative (flipping direction) and adding a half-rotation. This is because the middle of the angular scope of rotation is not 0 (then the scope would be from -90 to 90 degrees), but 90 (as it goes from 0 to 180). The exception from this rule is your base servo, which takes the same angle as in the graphic representation.

After connecting the Arduino and launching this program, you'll see all the servos straighten up to one line. Then they start following the on-screen geometry. Try to use gentle movements; otherwise the servos will move abruptly and could unfasten the assembly if it's slightly weak. Generally, it's a good idea to calibrate the servos first by dismantling the sub-arms and connecting them again once the servos get a signal from Arduino (without the program running). They are usually set to 90 degrees automatically, so once you secure the robot in straight position, they retain this as starting point.

After some experimenting, depending on your mechanical setup, you might want to fine-tune the way the robot applies angles to joints. It's best to do this by using multipliers on angles (shown in bold for emphasis).

```
  arduino.analogWrite(servo1pin, 90-round(degrees(angle1*1.25))); // move servo 1
  arduino.analogWrite(servo2pin, 90-round(degrees(angle2*1.35))); // move servo 2
  arduino.analogWrite(servo3pin, round(degrees(angle3))); // move servo 3
```

It's a matter of trial and error, and some common sense, to find the best values. The goal is to match the orientation of the robot with what you see on screen.

After verifying that everything works, use a paper clip to attach the pen to the furthest sub-arm. Now you can use your mouse to express your creativity on paper!

Kinect: Tangible Table Interface

To make the installation complete, you need to add the sensing part. It will allow you to use gestures to control the drawing or, you could also say, make your robot interpret your posture and draw something on its own.

You then need to add another layer of computational mechanisms, the one that interprets gestures and translates them into a set of XY coordinates to pass to the robot. In this way, you connect the two sides together: the Kinect scans your hand or finger, extracts XY coordinates from it, and passes it on to the robot simulation Processing sketch, which converts the data into a series of servo rotations.

The idea of extracting gestures or coordinates from your hands and fingers is not that simple, though, because the orientation of the point cloud is a bit tricky in relation to the table. You could, of course, hang the Kinect from the ceiling, make sure it's perpendicular to the table, and align it with your (0,0) point, which is the base of the robot. In practice, however, this could be hard to realize; the orthogonal aligning, precision, and installation of the sensor are quite complicated tasks (and drilling holes in the ceiling might not be the best idea).

You're going to pursue a slightly different path. You'll assume that the Kinect is somewhere in space, in any orientation, and you will try to fine-tune it with the scene inside the program. In practice, this is a much more flexible approach and less prone to errors than other techniques.

To make things easy, use a tripod to mount the Kinect and make it point to the middle of your working area. Make sure the Kinect is in a horizontal position, only tilting towards the table (not tilting left or right), as shown in Figure 8-20.

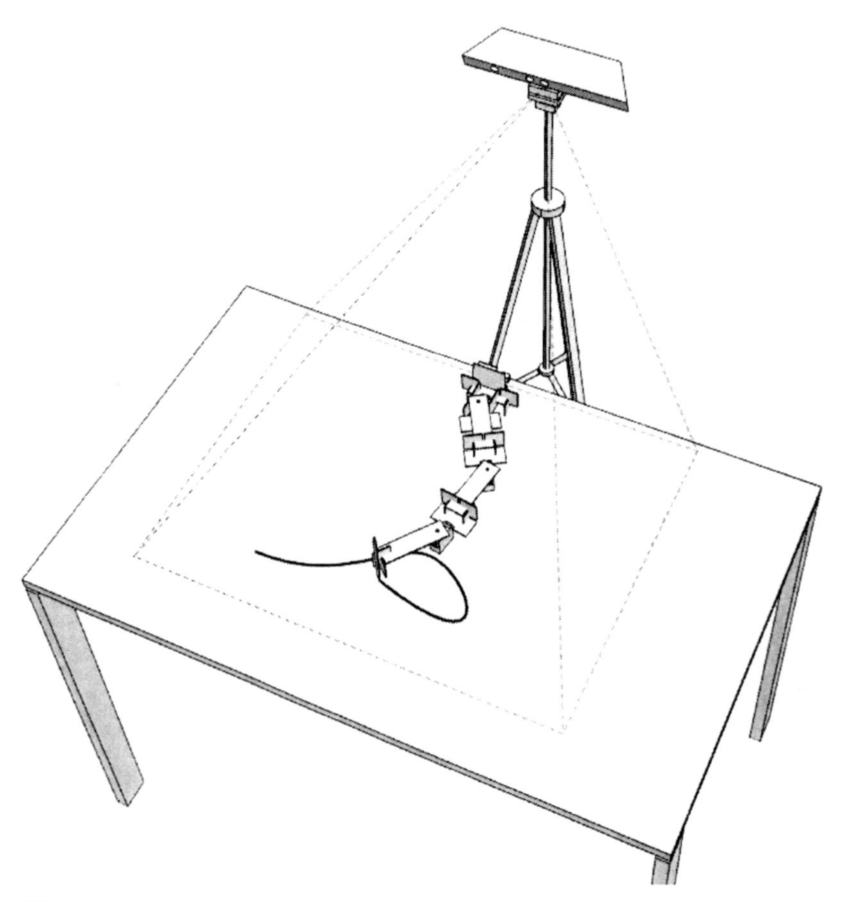

Figure 8-20. *Tangible table setup; scanning volume is shown with dotted lines.*

After positioning the Kinect on the tripod, open the DepthMap3d example from the Simple-OpenNI examples and look at the point cloud. You will see that the surface of the table is at an angle with the Kinect's coordinate system, making it difficult to extract coordinates from your position in space (Figure 8-21). With the example running, you can rotate the point cloud to see that, when viewed at a certain angle (from the side), the table is indeed angled. Moreover, the coordinate system to which these points refer is related to the Kinect itself; the (0,0,0) point is the sensor's camera. The task is to make the point cloud coexist in the same coordinate system as your robot.

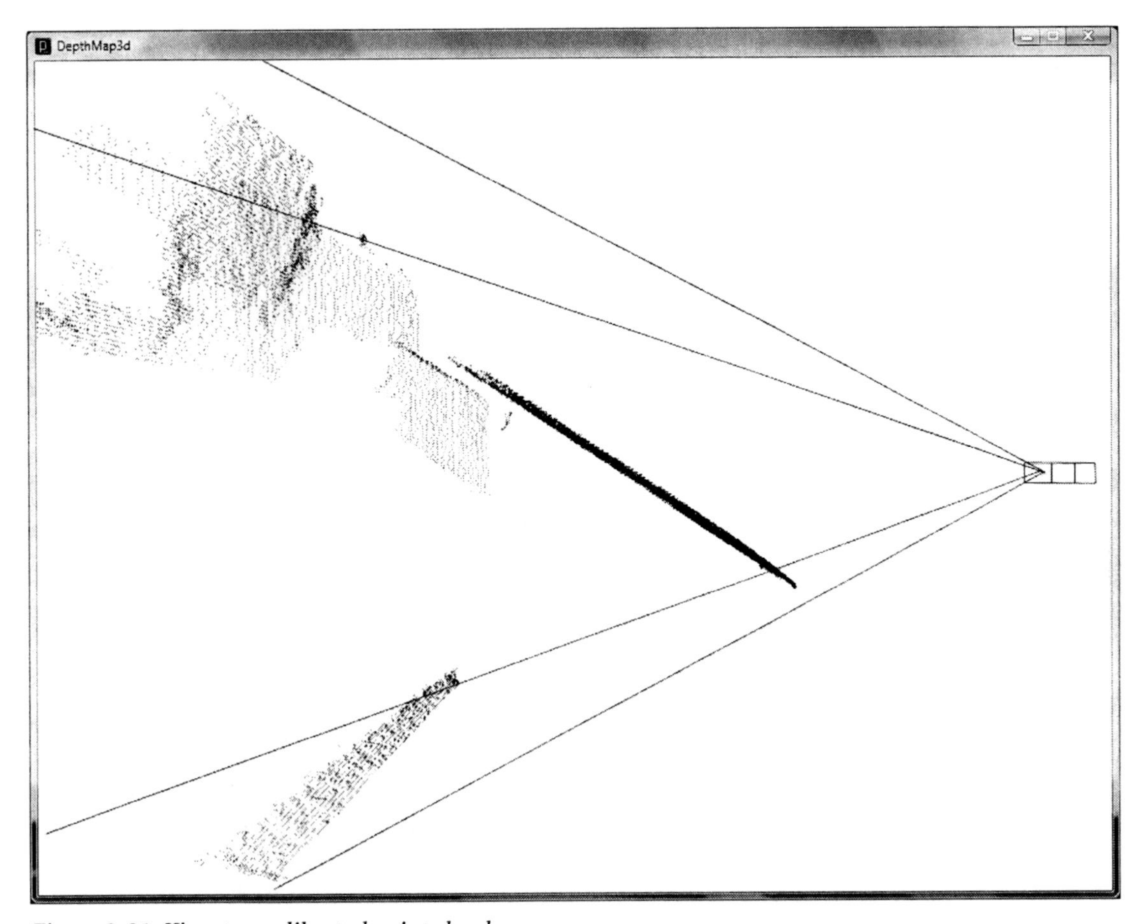

Figure 8-21. Kinect uncalibrated point cloud

Calibrating the Point Cloud

You need to align the table surface to your coordinate system by transforming the Kinect point cloud coordinates. What does this mean?

You must virtually rotate the table (tilt it) so it stays perpendicular to the Kinect axis (therefore being in the XY plane; see Figure 8-22). Also, you must shift it a bit so the origin (0,0) point matches your robot's origin of coordinates.

Once the point cloud has been aligned, the z-coordinate (or depth) of each point allows you to filter the point cloud and extract only the relevant data. In your case, this means the finger position, whenever it is close enough to the table.

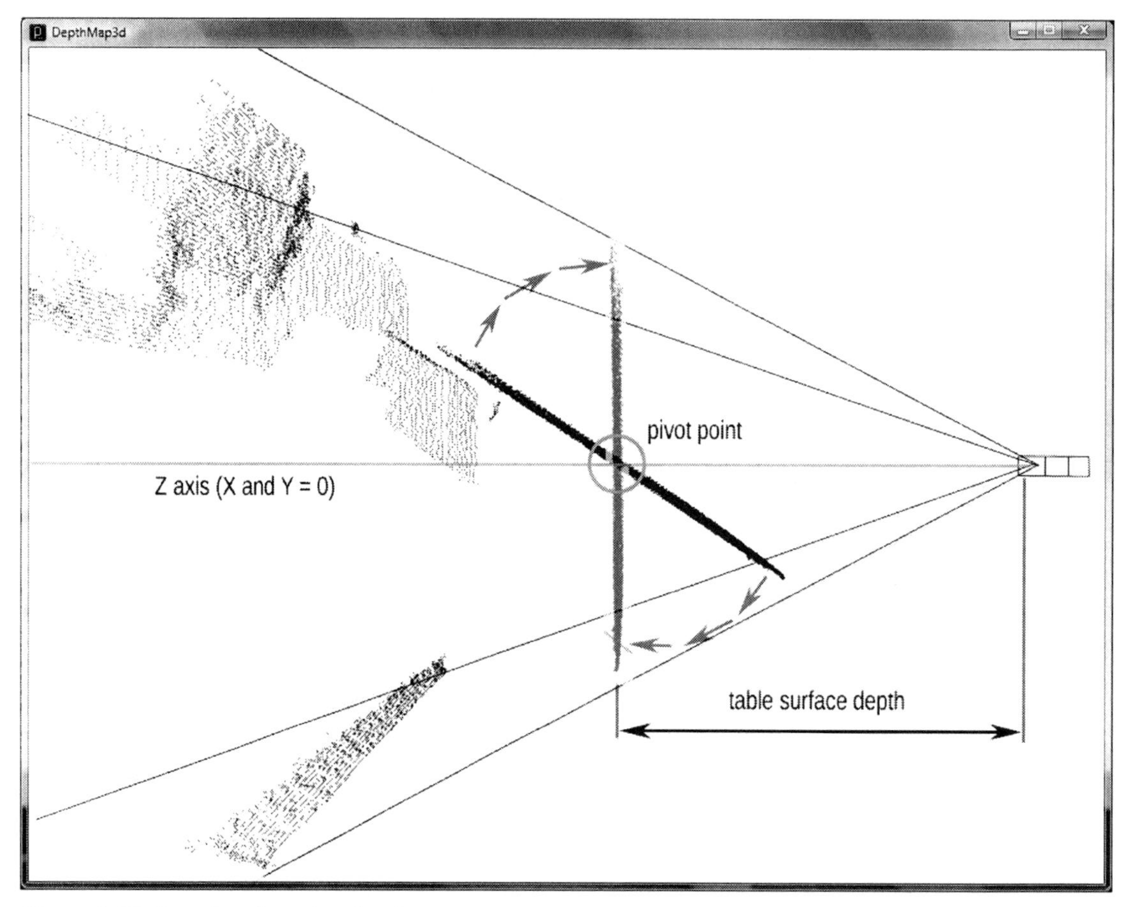

Figure 8-22. Rotating the point cloud

This might seem a little complicated at first, but the effort is worth it as you can now fine-tune the calibration with program variables, which is not possible in an orthogonal, mechanically fixed setup (or it requires moving the sensor).

Figure 8-22 illustrates the strategy: after rotating the table (point cloud) to the perpendicular position, you get the Z axis and mark all points that have X and Y = 0. What's most important, though, is that you can now *filter* the point cloud and cut out unnecessary bits like the background.

How do you do that? You won't use conventional Processing rotations with matrices such as `rotateX()`. Rotations and translations with standard 3D matrices keep the coordinates of each point unchanged. This means that even though the point (or object) appears in completely different place, it still has the same X, Y, and Z coordinate. It's more like moving coordinate system around, together with all its contents.

Rotation Equations

Back in the old days when there was no hardware graphics acceleration or OpenGL, there was no way of using `glTranslate` and `glRotate` to draw scenes in 3D. Coders working on 3D interactive games had to create mathematical descriptions of their 3D worlds by themselves. Even in 2D games, there was a need to rotate things on screen.

Well, to keep things simple, let's have a look at equations they were using. If you get a point on plane (X, Y) and want to rotate it, you have to calculate it like this:

angle = your rotation

newX = X * cos(angle) – Y * sin(angle)

newY = X * sin (angle) + Y * cos(angle)

Yes, that's it! If you want to understand these equations, don't Google them, or you risk being scared to death by the hundreds of scientific papers about 3D rotations of geometric entities! The best way to find an explanation is to search for old game-writing tutorials or simple computer graphics books from the 90s (such as those by Bowyer and Woodwark or Foley and Van Damme).

How does it work? Imagine you have a point on plane with coordinates (1, 0). If you want to rotate it, you multiply it with the previous operation and you get

newX = 1 * cos(angle)

newY = 1 * sin(angle)

This is, in fact, an equation for a circle, where the angle is between 0 and 360 degrees. Using sin, cos, and the previous operations, you can rotate any two coordinates around a (0,0) point, getting in return true numerical values of its coordinates (which is what you need!). So, after this lengthy explanation, let's put these concepts to use.

Rotating the Point Cloud

Let's start coding another example from scratch. The following code is derived from standard DepthMap3D code found in Simple-OpenNI examples. It displays the point cloud and allows you to rotate it. Then, after rotating the view so you look at the table from the side, you virtually rotate the point cloud until the table stands in a perpendicular position to the Kinect. You then hardcode the rotation angle into your code so the program is calibrated.

Import all the necessary libraries and initialize the Simple-OpenNI object. Then declare the variables that control the perspective.

```
import processing.opengl.*;
import SimpleOpenNI.*;
SimpleOpenNI kinect;

float        zoomF = 0.3f;
float        rotX = radians(180);
float        rotY = radians(0);
```

The float `tableRotation` stores your rotation angle. This is the value that needs to be assigned a default value for the calibration to be permanent.

```
float        tableRotation = 0;
```

The `setup()` function includes all the necessary functions to set the sketch size, initialize the Simple-OpenNI object, and set the initial perspective settings.

```
void setup()
{
  size(1024, 768, OPENGL);
```

```
  kinect = new SimpleOpenNI(this);
  kinect.setMirror(false);      // disable mirror
  kinect.enableDepth(); // enable depthMap generation
  perspective(95, float(width)/float(height), 10, 150000);
}
```

Within the draw() function, update the Kinect data and set the perspective settings using the functions rotateX(), rotateY(), and scale().

```
void draw()
{
  kinect.update();    // update the cam
  background(255);
  translate(width/2, height/2, 0);
  rotateX(rotX);
  rotateY(rotY);
  scale(zoomF);
```

Then declare and initialize the necessary variables for the drawing of the point cloud, as you have done in other projects.

```
  int[]   depthMap = kinect.depthMap(); // tip - this line will throw an error if your Kinect
// is not connected
  int     steps   = 3;  // to speed up the drawing, draw every third point
  int     index;
  PVector realWorldPoint;
```

Set the rotation center of the scene for visual purposes to 1000 in front of the camera, which is an approximation of the distance from the Kinect to the center of the table.

```
  translate(0, 0, -1000);
  stroke(0);

  PVector[] realWorldMap = kinect.depthMapRealWorld();
  PVector newPoint = new PVector();
```

To make things clearer, mark a point on Kinect's z-axis (so the X and Y are equal to 0). This code uses the realWorldMap array, which is an array of 3D coordinates of each screen point. You're simply choosing the one that is in the middle of the screen (hence depthWidth/2 and depthHeight/2). As you are using coordinates of the Kinect, where (0,0,0) is the sensor itself, you are sampling the depth there and placing the red cube using the function drawBox() (which you implement later).

```
  index = kinect.depthWidth()/2 + kinect.depthHeight()/2 * kinect.depthWidth();
  float pivotDepth = realWorldMap[index].z;
  fill(255,0,0);
  drawBox(0, 0, pivotDepth, 50);
```

When you finish the program and run it, you'll discover that the point on the z-axis hits the table and lands somewhere in the middle of it. If it doesn't, move your Kinect and tripod so it hits the middle of the table. Feel free to adjust the tilt and the height of the sensor but don't rotate it around the tripod; keep it perpendicular to the table.

The red cube is a special point. It's used to determine the axis of your rotation. To make it clearer, and accessible in the rest of the code, the local variable pivotDepth is assigned this special depth value.

You are now ready to virtually tilt the table and display the tilted points. But which pair of coordinates should you use for rotation? You have three: X, Y, and Z. The answer is Y and Z. The X coordinate stays the same (right side of the table, which is positive X, stays on the right; the left side, negative X, on the left).

```
  for(int y=0; y < kinect.depthHeight(); y+=steps)
  {
    for(int x=0; x < kinect.depthWidth(); x+=steps)
    {
      index = x + y * kinect.depthWidth();
      if(depthMap[index] > 0)
      {
        realWorldPoint = realWorldMap[index];
        realWorldPoint.z -= pivotDepth;

        float ss = sin(tableRotation);
        float cs = cos(tableRotation);

        newPoint.x = realWorldPoint.x;
        newPoint.y = realWorldPoint.y*cs - realWorldPoint.z*ss;
        newPoint.z = realWorldPoint.y*ss + realWorldPoint.z*cs + pivotDepth;
        point(newPoint.x, newPoint.y, newPoint.z);
      }
    }
  }
```

Now, display the value of your `tableRotation`. It tells you the value of the tilt, as you will be rotating it by eye. After you find the proper alignment, it must be written down and assigned to the variable at the beginning of the program (so the table stays rotated all times).

```
  println("tableRot = " + tableRotation);
  kinect.drawCamFrustum();    // draw the kinect cam
}
```

Within the `KeyPressed()` callback function, add code that reacts to the pressing of keys 1 and 2 by changing the values of `tableRotation` in small increments/decrements. Also, listen for input from the arrow keys to change the point of view.

```
void keyPressed()
{
  switch(key)
  {
  case '1':
    tableRotation -= 0.05;
    break;
  case '2':
    tableRotation += 0.05;
    break;
  }

  switch(keyCode)
  {
  case LEFT:
    rotY += 0.1f;
    break;
  case RIGHT:
    // zoom out
    rotY -= 0.1f;
    break;
  case UP:
```

```
      if(keyEvent.isShiftDown())
      {
        zoomF += 0.02f;
      }
      else
      {
        rotX += 0.1f;
      }
      break;
    case DOWN:
      if(keyEvent.isShiftDown())
      {
        zoomF -= 0.02f;
        if(zoomF < 0.01)
          zoomF = 0.01;
      }
      else
      {
        rotX -= 0.1f;
      }
      break;
  }
}
```

Finally, add a small, useful function that draws the box. Create another tab (call it functions, for example), and paste the following:

```
void drawBox(float x, float y, float z, float size)
{
  pushMatrix();
    translate(x, y, z);
    box(size);
  popMatrix();
}
```

Run the code and use the arrow keys (LEFT/RIGHT) to view the cloud from the side (meaning you rotate it graphically, without changing anything with coordinates). The table surface should appear as a thin line of tilted points. Then, using keys 1 and 2, you should be able to tilt the table (proper rotation, changing Y and Z coordinates) and find its best position. Aim to make it close to perpendicular to the Kinect axis, as per Figure 8-23.

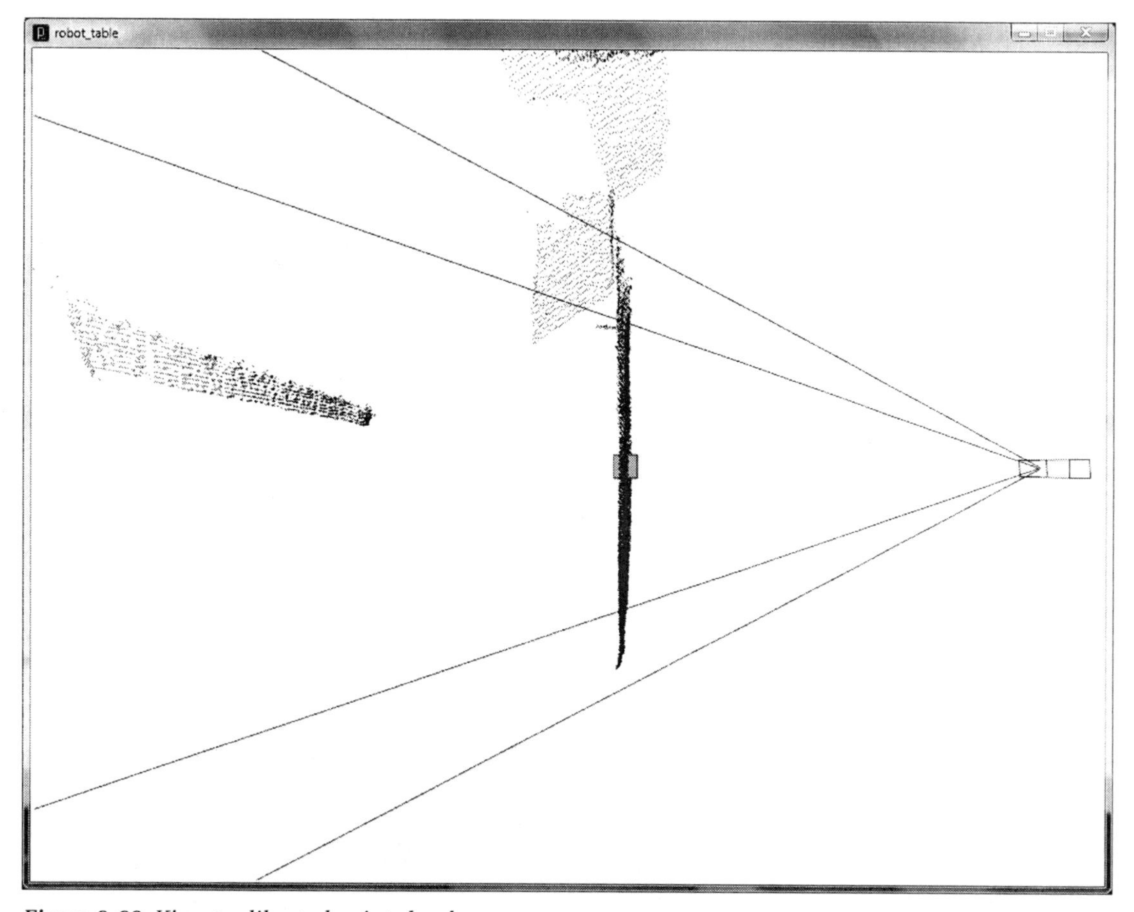

Figure 8-23. Kinect calibrated point cloud

And there you go! Your table is virtually rotated now, so it seems like the Kinect is looking straight at it. Using the output from Processing, hardcode the value of this rotation into the program. Simply change `tableRotation` at the beginning of the code to whatever the readout is on your console (mine was –1.0, but yours might be different). If you run the program again, you'll see the table straight from the top.

You also need to hardcode the value of `pivotDepth` so when you move your hand on the table, the rotation pivot won't change. Substitute this line in your code:

```
float pivotDepth = realWorldMap[index].z;
```

By this one:

```
float pivotDepth = 875;
```

In our case, the centre of our tangible table was 875mm away from the Kinect, but you will need to extract this value from your table rotation code.

Point Cloud Filtering

As you look at the scene from the side, try to manipulate the objects on the table by looking at the screen. What's visible? You can probably see some objects, your hands, and some background (like the floor). You should also see the table surface. The thing is, most of this is not necessary for your tracking. The only object you're interested in is the hand/pointer (anything that can be clearly distinguished from the rest of the point cloud).

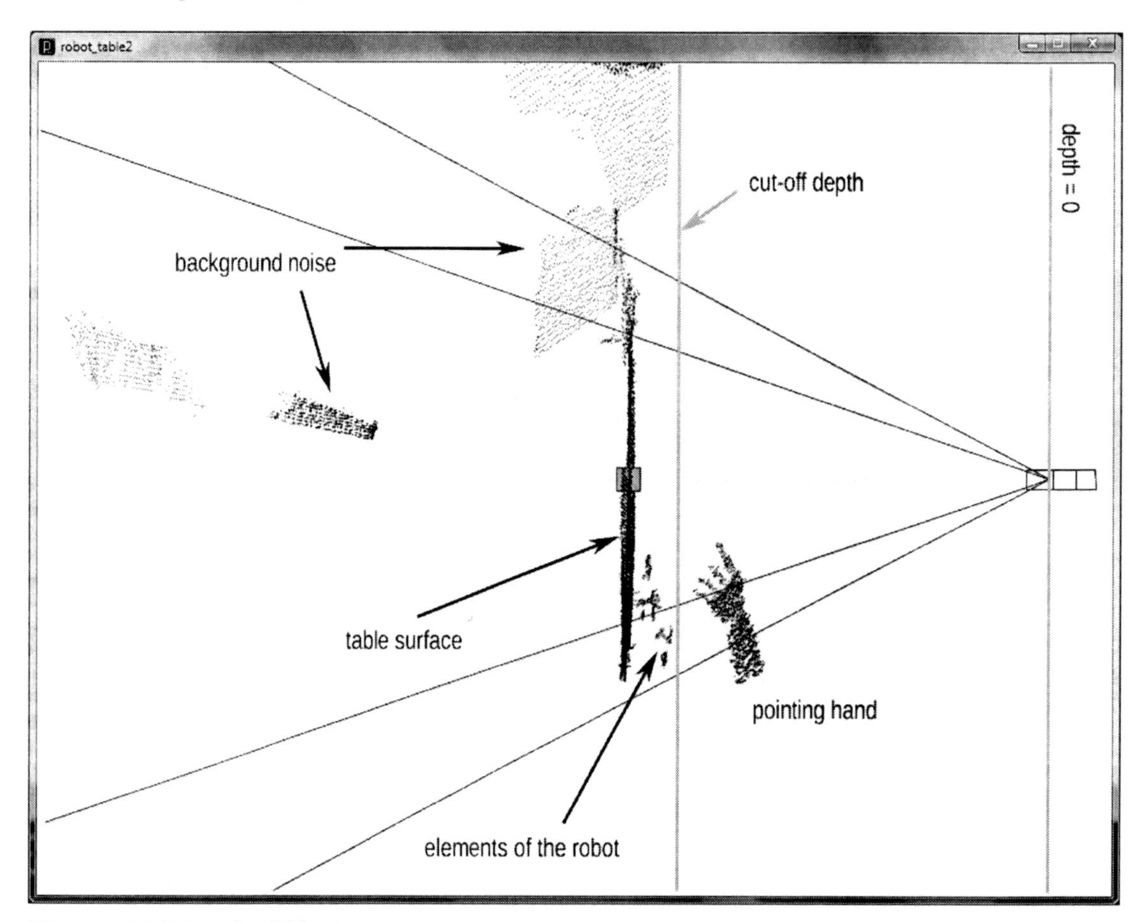

Figure 8-24. *Point cloud filtering*

Take a look at Figure 8-24. It clearly shows how the point cloud is structured and which depths you can ignore. By simply adding an `if()` statement around the point drawing function, you can get rid of most of the unnecessary data.

First, declare the `zCutOffDepth` global variable at the top of your sketch.

```
float        zCutOffDepth = 1200;
```

Then, add a conditional wrapper around the point drawing function.

```
if (newPoint.z < zCutOffDepth) {
  point(newPoint.x, newPoint.y, newPoint.z);
}
```

To know the value in runtime, add following line at the very end of the `draw()` function (before the closing curly brace):

```
println("cutOffDepth = " + zCutOffDepth);
```

This reads the value and hardcodes it for later. To be able to change this cut-off depth at any time, add two more cases into the `keyPressed()` function, inside the `switch(key)` statement.

```
case '3':
  zCutOffDepth -= 5;
  break;
case '4':
  zCutOffDepth += 5;
  break;
```

Run the code with these additions. Viewing the point cloud sideways, press keys 3 and 4 until you get rid of all the points at the bottom of the table. In fact, you're also going to cut off the robot itself, leaving only some space about 100mm above the table. After performing this, you should see nothing—just a pure white screen. But as soon as you hover your hand above the table, it should appear as black points.

As before, once the calibration is done, you should hardcode the value of `zCutOffDepth` at the beginning of the program. In my setup, it was something around 700mm, but yours might be different, depending on how close the Kinect is to the table.

Finding the Finger Position

Now let's get back to the main idea: extracting the X and Y coordinates of the pointer and passing them on to the robot-driving mechanism (which already knows how to translate Cartesian coordinates to three separate servo angles).

To get the X and Y coordinates, you will use a trick to find a special point in the entire point cloud: your finger tip. In a typical case, it's the *lowest* point above the table, assuming you're considering the hand only. Usually you hold your hand in such a way that the finger is the object closest to the surface (or its tip, actually). Therefore, your task is simply to find the lowest point in scanned area. Mathematically speaking, this is the point with highest Z-coordinate.

To make things clearer and easier to understand, I will first describe the changes necessary to complete this task and then I will include the entire code of the `draw()` function so you can see the changes in context.

Previously, you filtered the point cloud using the depth value and removed points that were too far. Now the time has come remove all the points that are outside of your working area (which is a rectangle).

To filter the cloud a bit more, you must draw your reference area first. I assume this is 800 x 600 mm (it would be useful to place a sheet of paper with these dimensions on the table). The Kinect should be pointing to the middle of it, which will be your (0,0) point. Therefore, to draw it, you need these four lines right after the point drawing loop (be careful to type the coordinates properly).

```
line(-400, -300, pivotDepth,  400, -300, pivotDepth);
line(-400,  300, pivotDepth,  400,  300, pivotDepth);
line( 400, -300, pivotDepth,  400,  300, pivotDepth);
line(-400, -300, pivotDepth, -400,  300, pivotDepth);
```

You are using `pivotDepth` as a z-coordinate so the lines are drawn on the table. After running the program, you should see a rectangle at table depth.

To clean things up a bit more, let's filter out points that are *outside* of this area. So, in the condition that currently resides right before the point-drawing command

```
if (newPoint.z < zCutOffDepth) {
  point(newPoint.x, newPoint.y, newPoint.z);
}
```

change it to

```
if ((newPoint.z < zCutOffDepth) &&
    (newPoint.x < 400) &&
    (newPoint.x > -400) &&
    (newPoint.y < 300) &&
    (newPoint.y > -300)) {
        point(newPoint.x, newPoint.y, newPoint.z);
      }
```

This will get rid of all the points outside of the table boundary. Run the program; you should see a rectangle with red box in it (in the middle of the table). If you hover your hand above the work area, you should be able to see it on screen (and nothing else; once the zCutOffDepth variable is set properly, it should cut out unnecessary noise).

Since the cloud is filtered, you're left with just the space directly above the table. If there are no objects in this space, you receive zero points from the scan. Once you put something there (your hand or a stick), it should appear. Knowing this, you can safely add more filtering conditions, the ones you require to find the lowest point above the table.

Let's declare a special variable, something that will store your target as a point, so all three coordinates are known. This should be at the beginning of the code, so it's defined as a global coordinate.

```
PVector pointer = new PVector();
```

Then, for each loop, you have to reset the maximum depth value found (the depth of the furthest point in Z direction you've found each time). This is reset just before your main scanning loop, right after the drawBox(0,0, pivotDepth, 50); line.

```
float maxDepth = 0;
```

To find the point, go to the core of the scanning loop and add a conditional statement just after the instruction that draws each point. This statement checks if the point you're drawing is the furthest one (closest to the table).

```
if (newPoint.z > maxDepth)
```

Then it stores its value for later, along with the depth value encountered. In other words, it stores two things: the maximum depth found during scanning (the point closest to the table), and the x, y, and z coordinates of this point.

To summarize, these steps are highlighted in bold in the following block of code, which is the main scanning loop in its entirety:

```
float maxDepth = 0;
for(int y=0;y < context.depthHeight();y+=steps)
{
  for(int x=0;x < context.depthWidth();x+=steps)
  {
    index = x + y * context.depthWidth();
    if(depthMap[index] > 0)
    {
      realWorldPoint = realWorldMap[index];
```

```
      realWorldPoint.z -= pivotDepth; //subtract depth from z-coordinate

      float ss = sin(tableRotation);
      float cs = cos(tableRotation);

      newPoint.x = realWorldPoint.x; //x doesn't change
      newPoint.y = realWorldPoint.y*cs - realWorldPoint.z*ss;//rotate Y
      newPoint.z = realWorldPoint.y*ss + realWorldPoint.z*cs;//rotate Z
      newPoint.z += pivotDepth; //add depth back again
    if ((newPoint.z < zCutOffDepth) &&
        (newPoint.x < 400) &&
        (newPoint.x > -400) &&
        (newPoint.y < 300) &&
        (newPoint.y > -300))
        {
          point(newPoint.x,newPoint.y,newPoint.z);  //draw point!
          if (newPoint.z>maxDepth) //store deepest point found
          {
            maxDepth = newPoint.z;
            pointer.x = newPoint.x;
            pointer.y = newPoint.y;
            pointer.z = newPoint.z;
          }
        }
      }

    }
  }
}
```

Finally, after the loop has been executed, you can draw the deepest point as a yellow box by adding the following code right after the loop:

```
fill(255,255,0);
drawBox(pointer.x,pointer.y,pointer.z,50);
```

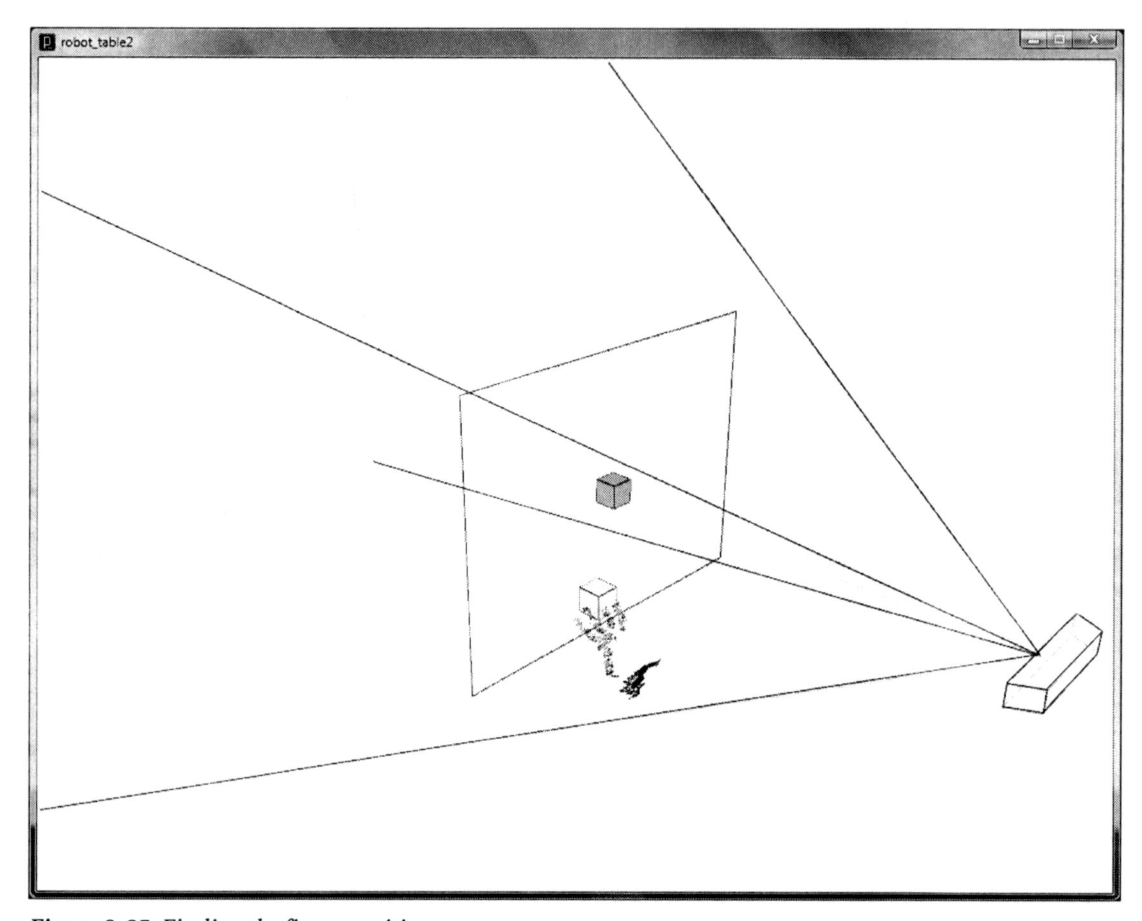

Figure 8-25. Finding the finger position

The yellow box now shows the tip of one of the fingers on the pointing hand. Because all this information is stored in the pointer variable, you can easily retrieve it and direct the robot with it.

You should see something resembling Figure 8-25. The yellow box might be a little bit shaky, but it should always stick to the lowest (closest to the table) part of the point cloud. If you see parts of the table or parts of the robot, it means your zCutOffDepth is miscalibrated. If so, return to the calibration part of the "Point Cloud Filtering" section and fine-tune this value until you get clean scan of the space above the work area.

Virtual Robot Model

Before you connect the robot to this piece of code, you need to make sure it works. Let's start with a virtual version of it. As before, you determine the angle based on XY coordinates; however, you need to align both coordinate systems together. The point cloud's coordinate system has its center where the red box is and you need it aligned with the robot's main pivot point. Knowing that the robot's center point needs to be reconfigurable, let's declare it as two variables. At the end of the draw() loop, add the following code:

```
float robotX = 0;
float robotY = -300;

pushMatrix();
  translate(0,0, pivotDepth);
  ellipse(robotX, robotY, 30,30);
  ellipse(pointer.x, pointer.y, 30,30);
  line(robotX, robotY, pointer.x, pointer.y);
popMatrix();
```

You use pushMatrix() and popMatrix() to make sure you're drawing in the table's depth. By using translate(0,0, pivotDepth) you're pushing your drawing plane; from now on, your Z depth is equal zero at table's surface. This makes the task of drawing on it simpler as you can use two-dimensional commands like ellipse().

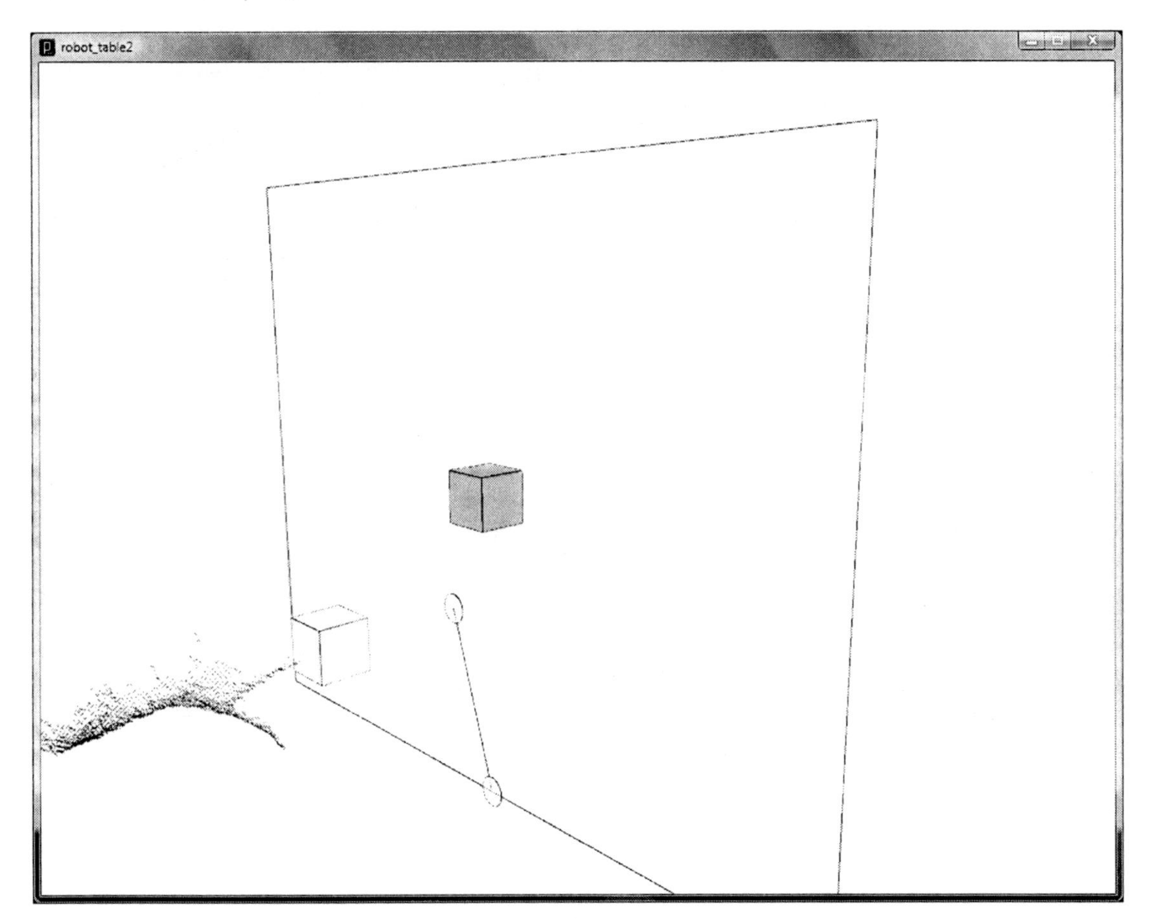

Figure 8-26. *The finger position projected on the table*

What you see in Figure 8-26 is a projection of the tip of the pointer (made by projecting it on the table) connected with the virtual robot's center. This is exactly what you were doing at the beginning, so now you can easily copy and paste the previous code and use it to detect the angles. You should be able to run your code and see something similar.

To draw entire virtual robot with its sub-arms, the last block of code in the draw() function can be modified to look like the following:

```
//===========================
//you're drawing the robot now
//===========================
float robotX = 0;
float robotY = -300;

pushMatrix();
    translate(0, 0, pivotDepth);
    ellipse(robotX, robotY, 30, 30);
    ellipse(pointer.x, pointer.y, 30, 30);
    line(robotX, robotY, pointer.x, pointer.y);

    float penX = pointer.x; //you're capturing coordinates of your pen tip (mouse)
    float penY = pointer.y;
    float len = dist(robotX, robotY, penX, penY); //let's measure the length of your line
    float angle = asin((penY-robotY)/len); //asin returns angle value in range of 0 to PI/2,
// in radians
    if (penX<0) { angle = PI - angle; } // this line makes sure angle is greater than PI/2
// (90 deg) when penX is negative
    println("angle = " + degrees(angle)); //you're outputting angle value converted from
// radians to degrees
    println("length = " + len);//print out the length
    arc(robotX, robotY, 200, 200, 0, angle); //let's draw your angle as an arc
    if (len > 450) { len = 450; }
    if (len < 150) { len = 150; }

    float dprime = (len - 150) / 2.0;
    float a = acos(dprime / 150);
    float angle3 = angle + a;
    float angle2 = - a;
    float angle1 = - a;

    translate( robotX, robotY );
    rotate(angle3);
    line(0, 0, 150, 0);
    translate(150, 0);

    rotate(angle2);
    line(0, 0, 150, 0);
    translate(150, 0);

    rotate(angle1);
    line(0, 0, 150, 0);
    translate(150, 0);

popMatrix();
```

Your robot's center pivot point is not 0,0 anymore. It's been replaced by `robotX` and `robotY`. This is good because you can change it as you recalibrate/retune the robot.

This is quite close to you want to achieve. You already have the following:

- XY coordinates of the finger (pointer)
- All three angles for the servos
- Both coordinate systems tuned together

By this point, you should see the virtual model of the robot in your work area, as shown in Figure 8-27.

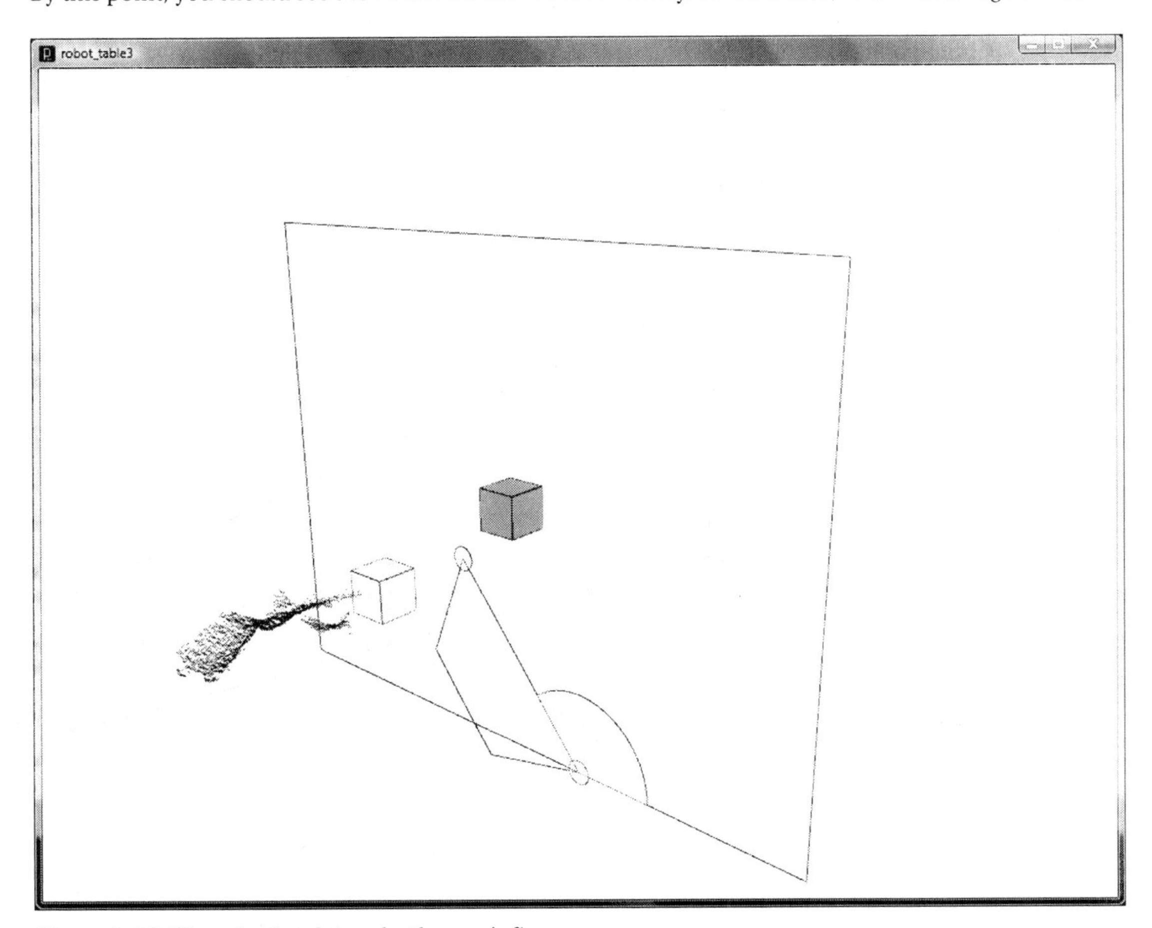

***Figure 8-27.** Virtual robot driven by the user's finger*

Polishing the Input

Is this everything you need to drive the robot? Well, yes, but at this stage there is small danger arising. Looking at the behavior of the virtual instance of the robot, there's something quite quirky about it. It's shaky and performs unstable movements.

Due to the filtering of the point cloud and the way the lowest point above the table is captured, in some cases you can observe very rapid movements of the arm. For example, when you point your finger to the middle of the working area, the robot will move there. But when you retract your hand from it, it will follow it at all costs—and will then suddenly get repositioned somewhere near the boundary (following the path of your hand). Sometimes the robot is somewhere on the table, and once you insert your hand into the scanning area, it will move towards it very quickly.

This is okay in your virtual world. In the context of a physical installation, it could be dangerous. It might damage the robot, which is simply not able to move 50cm in 100 milliseconds. To avoid this, you need to add the following functionalities:

- Movement smoothing
- Further point cloud filtering, cutting off all points above certain height

The first task is to remember all pointer positions and average the last 20 of them. The second is quite easy: just add another condition to the point cloud filtering statement.

Let's start with second one. Replace the `if()` statement in main scanning loop with this code:

```
if ((newPoint.z > zCutOffDepth - 50) &&
    (newPoint.z < zCutOffDepth) &&
    (newPoint.x < 400) &&
    (newPoint.x > -400) &&
    (newPoint.y < 300) &&
    (newPoint.y > -300)) {
```

As you can see, you just added one more statement at the beginning. This, in conjunction with second statement, only allows a thin slice of points to pass the filter. Its thickness is exactly 50mm. In practice, this means that you can withdraw the hand in upper direction, and it won't make the robot follow it. It ignores the hand once it's at a safe height.

The first task, however, is slightly more complex: you need to build an array with all the positions of the pointer stored. Declare it somewhere at the beginning of the program, outside of the `setup()` and `draw()` functions.

```
PVector [] path = new PVector[100]; //you're declaring array of 100 points
```

Then allocate memory for it at the end of `setup()` function.

```
for(int i = 0; i < 100; i++){
   path[i] = new PVector();
}
```

Once this is done, you need to store pointer X and Y values in them. This is done by adding the following lines after the main scanning loop, but before you draw the robot:

```
path[frameCount % 100].x = pointer.x;
path[frameCount % 100].y = pointer.y;
```

`frameCount` is the number of frames rendered from the beginning. The modulo operator (%) makes sure this never goes out of bounds, as the array is only 100 records long. (The last position in the array is number 99, actually, as you're counting from 0.)

Next, inside the robot drawing routine (inside of the `pushMatrix()` block, and after `translate()` command), add the following code:

```
for (int i = 0; i < 100; i++){
  ellipse(path[i].x, path[i].y, 5, 5);
}
```

This displays your path in the form of small dots, following the pointer's trajectory. You can test the code now to see how the trajectory shows. Make sure to wave your hand above the table; you can also use a stick or a paper tube.

Finally, you need to average some values. Do this by taking the last 20 positions (you can use more or less, depending on robustness of the installation). So, instead of these two lines

```
    float penX = pointer.x; //you're capturing coordinates of your pen tip (mouse)
    float penY = pointer.y;
```

write these two lines

```
    float penX = 0; //you're resetting pen coordinates
    float penY = 0;
//and adding fractions of 20 last positions to them
    for (int i = 0; i < 20; i++) {
      penX += path[(frameCount + 100 - i) % 100].x/20.0;
      penY += path[(frameCount + 100 - i) % 100].y/20.0;
    }
```

Test the changes and observe how smooth the movement is now (Figure 8-28). This definitely makes it less prone to damage.

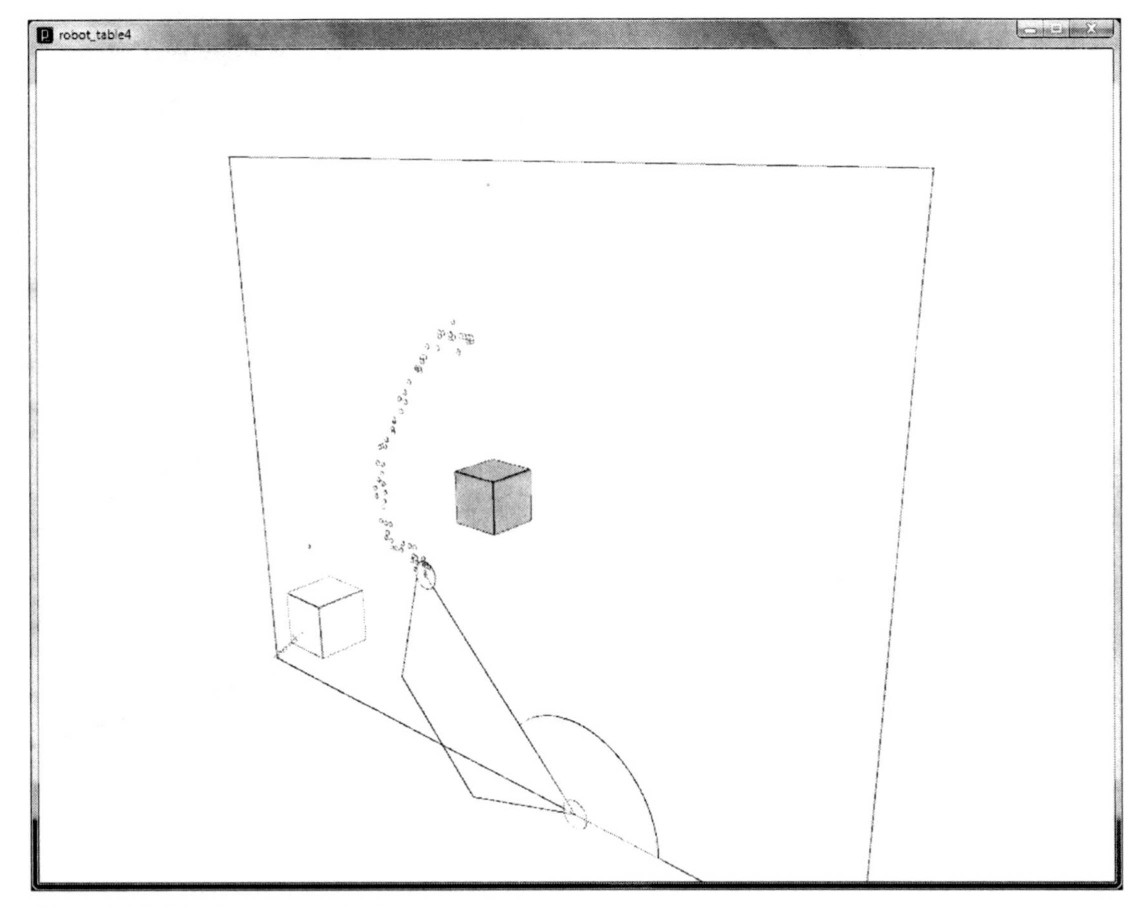

Figure 8-28. *User input smoothed*

The Drawing Robot in Action

The last step is to add a few extra instructions to send the robot angles to the Arduino and thus drive the robot. First, declare the libraries and the Arduino instance and then initialize it. Next, declare the variables defining the servo pins. The following block goes at the very beginning of the program:

```
import processing.serial.*;
import cc.arduino.*;
Arduino arduino;

int servo1pin = 11;
int servo2pin = 10;
int servo3pin = 9;
```

Within setup(), initialize the Arduino object and set the pin modes, after the size() command.

```
arduino = new Arduino(this, Arduino.list()[1]);
arduino.pinMode(servo1pin, Arduino.OUTPUT);
arduino.pinMode(servo2pin, Arduino.OUTPUT);
arduino.pinMode(servo3pin, Arduino.OUTPUT);
```

At the very end of the draw() function, add three lines of code that send the angles of the robot to the Arduino board using the Arduino library. Remember the reasoning behind the angle transformations from the "Driving the Physical Robot" section.

```
arduino.analogWrite(servo1pin, 90-round(degrees(angle1*1.0))); // move servo 1
arduino.analogWrite(servo2pin, 90-round(degrees(angle2*1.0))); // move servo 2
arduino.analogWrite(servo3pin, round(degrees(angle3))); // move servo 3
```

And voila! After some initial play (you might still recalibrate it a bit), you can add a pen (by attaching it with a paper clip) and start sketching! Your small gestural control of servo-powered mechanism is finally working!

Figures 8-29 and 8-30 show the robot at work. You can see a video of the piece at www.vimeo.com/34672180.

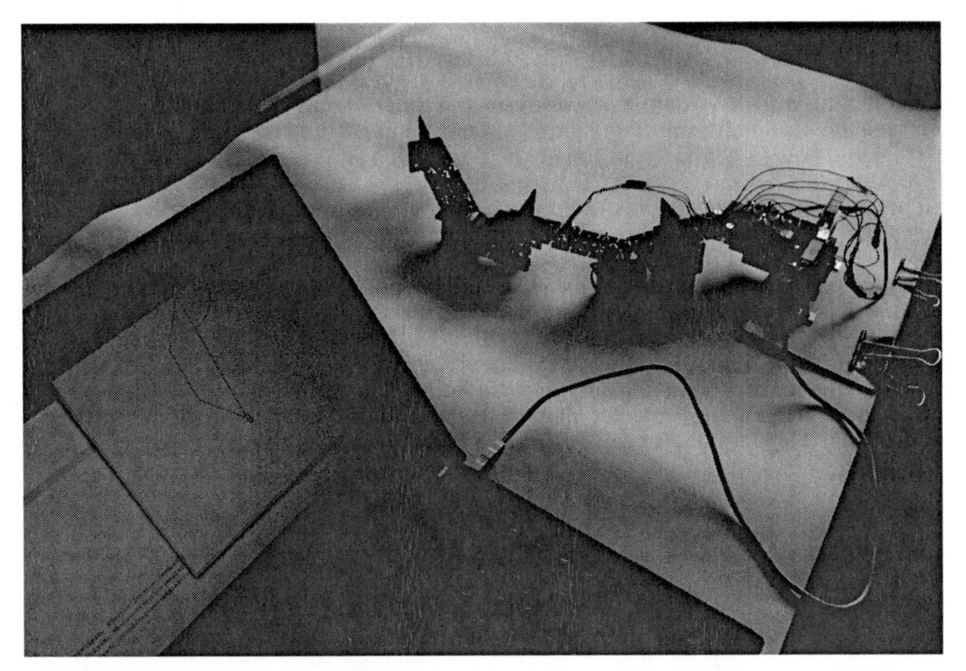

Figure 8-29. *The drawing robot at work*

Figure 8-30. *The robot drawing patterns on a sheet of paper*

Summary

This project had you building your own robotic drawing arm and programming it to follow the movements of your hand on a tangible table. You figured out how to parse the Kinect point cloud data using geometrical rules and how to acquire your finger position, which you used as the guide for the movements of the robotic arm.

Unlike the rest of the projects in this book, this project didn't require you to write any Arduino code at all. Instead, you learned how to use the Arduino Processing library and the Firmata Arduino firmware to control your motors directly from Processing.

Further alterations and improvements on this project could include adding an external power supply for better responsiveness of the servos or adding a microservo to lift and lower the pen to control the drawing flow.

I recommend spending some time calibrating this robot properly for a good match between the coordinate readouts from the Kinect's scanning and the positioning of the arm. For advanced experimentators, you can try to hang the projector above the table and project an image of the virtual robot onto the physical one. It leads to proper calibration and adds a new layer of augmented information.

CHAPTER 9

Kinect Remote-Controlled Vehicles

by Ciriaco Castro

In Chapter 5, you learned how to hack a remote control in order to control different devices via your body movements. In this chapter, you are going a step further in the control of devices by hacking an RC vehicle's electronics. You're going to use a RC car, but the same logic can be applied to any RC vehicle (helicopter, boat, etc.).

You are going to learn how to use an H-Bridge to control motors. Using the Kinect, you are going to produce an easy interface that connects body movements with the vehicle's movements (Figure 9-1) and you'll apply what you learned from the XBee technology to do it wirelessly. Finally, you are going to create an automated routine that will avoid crashes by using a proximity sensor and certain programmed behaviors. Figure 9-2 shows the parts you'll need for this chapter and Table 9-1 lists the specifics.

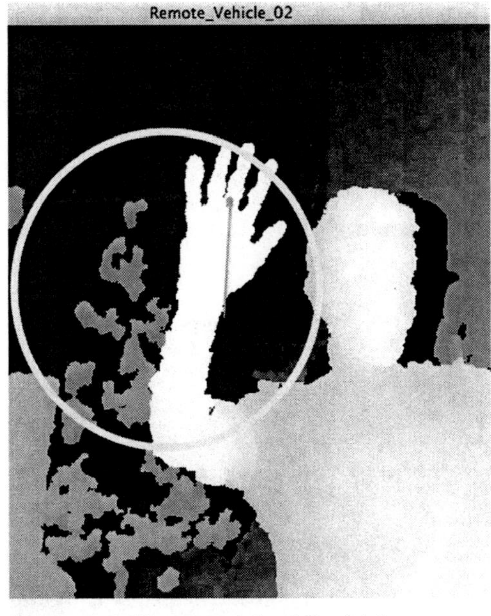

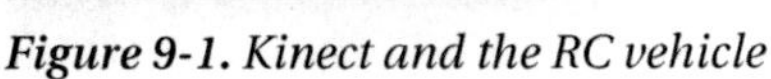

Figure 9-1. Kinect and the RC vehicle

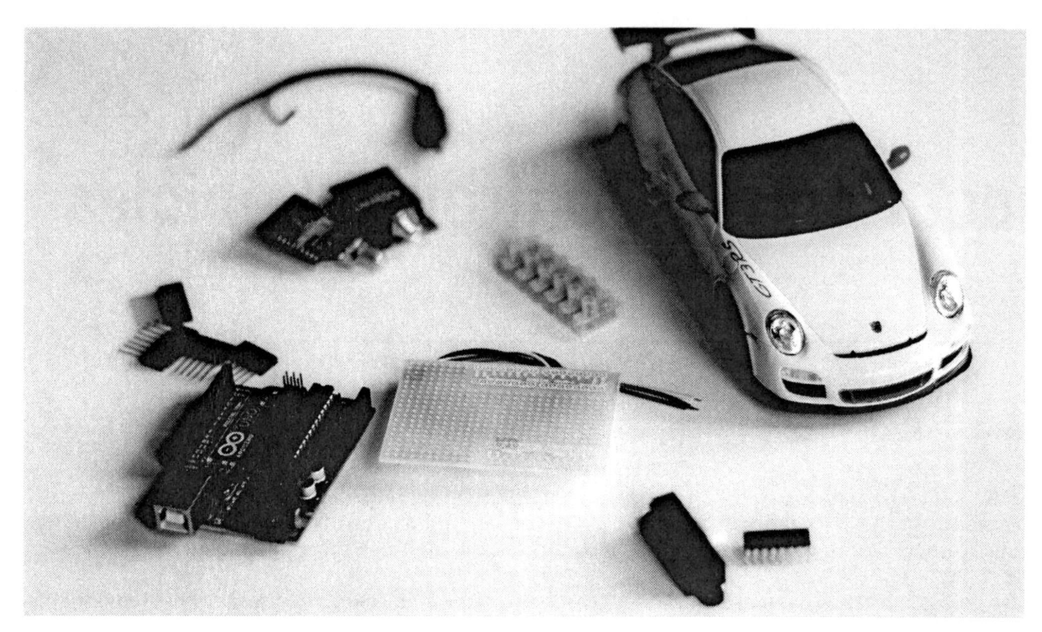

Figure 9-2. RC vehicle components

Table 9-1. The Parts List

Part	Description	Price
1 RC vehicle	Pick a RC car big enough to carry an Arduino board (approximately 20 x 10 cm). Make sure it has two motors.	Our cost $23.99
2 XBee 1mW Chip Antenna - Series 1	XBee 1mW Chip Antenna - Series 1	$22.95 each
XBee Explorer USB	You will need a mini USB cable to connect it.	$24.95
XBee Explorer Regulated	To interface the XBee with the Arduino	$9.95
1 prototype shield	SparkFun DEV-07914 or similar	$16.95
23 Breakaway headers-straight	You should have a bunch of these already from previous chapters.	$1.50
1 infrared distance sensor	From SparkFun or similar	$8.75
1 motor driver	Dual H-Bridge from SparkFun or similar	$5.89

Electrical Motors and the H-Bridge

Before hacking and programming the car, let's talk about electrical motors and how you are going to drive them using an H–Bridge.

An electric motor converts electrical energy into mechanical energy. Most electric motors operate through the interaction of magnetic fields and current conductors to generate the force. The kind of motor in most toys is a DC motor (Figure 9-3). A DC motor is an electric motor that runs on direct current (DC) electricity, usually from batteries.

Figure 9-3. *RC motor and H-Bridge*

To control a motor, you can turn it on and off using only one switch (or transistor). But controlling the direction is a bit more difficult. It requires a minimum of four switches (or transistors) arranged in a clever way. This can be simplified using an H-Bridge.

H-Bridges are four switches (or transistors) arranged in a shape that resembles an H. Each H-Bridge can control two motors. Each H-Bridge has 16 pins (eight per side), as shown in Figure 9-4. Figure 9-5 shows the schematic of an H-Bridge.

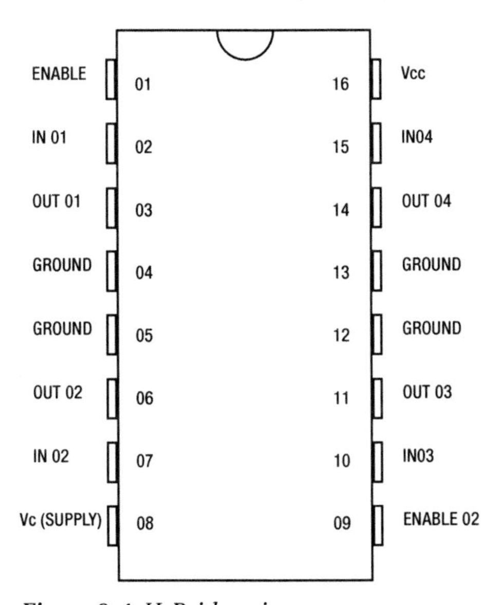

Figure 9-4. *H-Bridge pins*

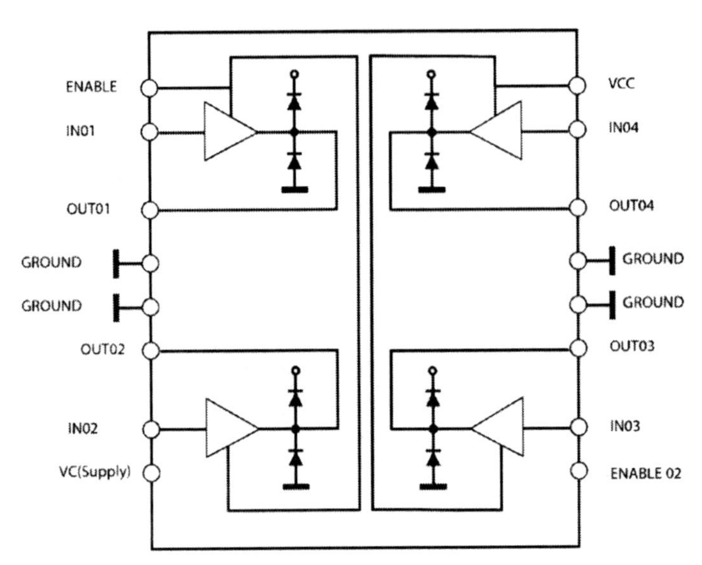

Figure 9-5. *H-Bridge schematic*

You are going to use the H-Bridge to drive two motors. Each pair of channels is equipped with an ENABLE input (pins 1 and 9). A separate supply input is provided in pins 8 and 16, allowing operation at a lower voltage. Pins 4, 5, 12, and 13 are connected to ground. Each side of the H-Bridge has two transistors (pins 2 and 7, and pins 10 and 15). One is responsible for pushing this side to HIGH and the other for pulling this side to LOW. When one side is pulled HIGH and the other LOW, the motor spins in one direction. If you want the motor to spin in the opposite way, you just have to reverse the transistors (the first side LOW and the latter HIGH).

Pins 3 and 6 are connected to the poles of one motor, and pins 11 and 14 are connected to the poles of a second motor. According to the direction of the current, the motor will spin in one direction or the opposite one.

Hacking a Car

Now that you understand these basic concepts of the motor, you can start to build your RC vehicle. The first step is to open up and observe what's inside (Figure 9-6). You should see a circuit and a series of cables. Looking at the circuit, you can see a RC receiver that controls the behavior of the motors. If you follow the cables, there are two that connect the circuit to the batteries. Another two connect the circuit to the front motor, and another two connect the circuit to the back motor. RC cars normally have two motors. The front one turns the vehicle left and right. The back one provides traction, which is the moving of the vehicle frontwards and backwards.

Figure 9-6. RC car opened

You're not going to use the existing circuit because you're going to build your own. So cut the cables that connect the actual circuit to the motors and the cables that connect to the batteries. Unscrew the board and leave the car ready for your new circuit (Figure 9-7).

Figure 9-7. RC car, just motors and electrical cables

You're going to drive the car using your Arduino. As a first step, you will be using serial data coming from a simple Processing sketch. Later, once everything is up and running, you will use the Kinect and a XBee module to drive the car with your body movements.

Start by joining the power and ground cables coming from the battery pack to a power jack so you can plug your Arduino using just the external batteries (Figure 9-8).

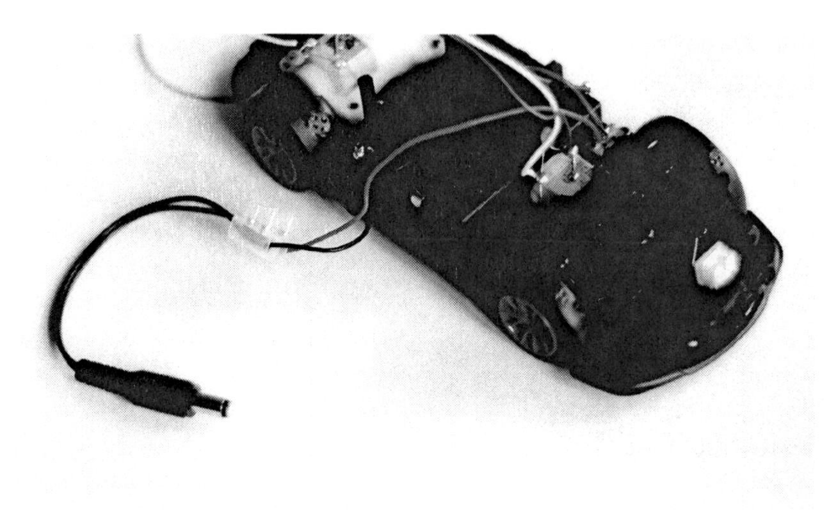

***Figure 9-8.** RC car with the electrical connection to battery pack*

So far, you've had an overview of electrical motors and the H-Bridge. You have opened an RC vehicle, looked inside, and extracted the circuit in order to start to building a new one.

Building the Circuit

The next step is to prepare your Arduino board and connect it to an H-Bridge in order to control the two motors (the back one that provides the traction and the front one that provides the direction). In this project, you're going to use a prototype shield instead of fusing the strip panel. A prototype shield is a platform for building a new project; it makes it easier to connect an Arduino to all the external bits and pieces. Shields are normally designed to sit on top of an Arduino, facilitating the connections. There are some pins that connect to the Arduino pins and there's a prototyping area where you solder your external components.

Start by soldering your breakaway headers in order to plug them into your Arduino board. Then solder two cables to 5V Arduino and ground and plug your H Bridge (Figure 9-9).

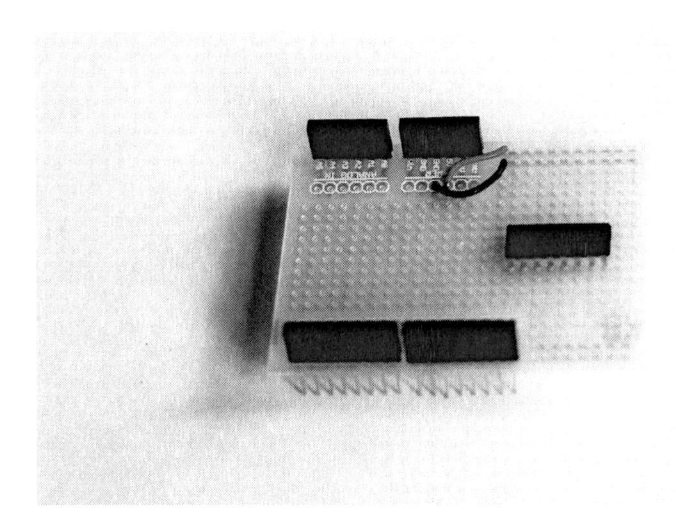

Figure 9-9. *Prototype board and the H-Bridge*

Next, solder the H-Bridge and start making all the connections. The H-Bridge chip is going to be used as two bridges controlling two separate motors. Connect the H-Bridge's pins 8 and 16 to 5V voltage. You can power the motors and the H Bridge using any external power device (connecting it to these pins), but remember to share the ground with the Arduino board. In this case, everything is powered through the Arduino (Figure 9-10).

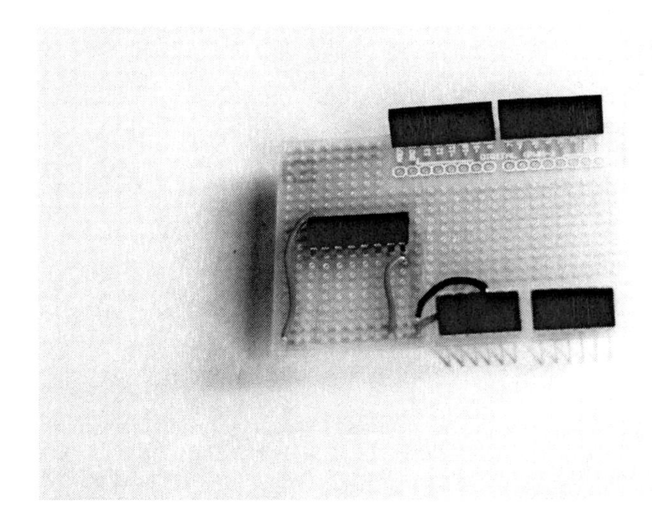

Figure 9-10. *H-Bridge connected with power*

Then connect the H-Bridge pins 4, 5, 12, and 13 to ground. The next set of cables that you are going to solder are the transistors to Arduino pins. Link Arduino pins 11 and 5 with H-Bridge pins 10 and 15, and Arduino pins 6 and 10 with H-Bridge pins 2 and 7. The circuit will look like the one in Figures 9-11 and 9-12.

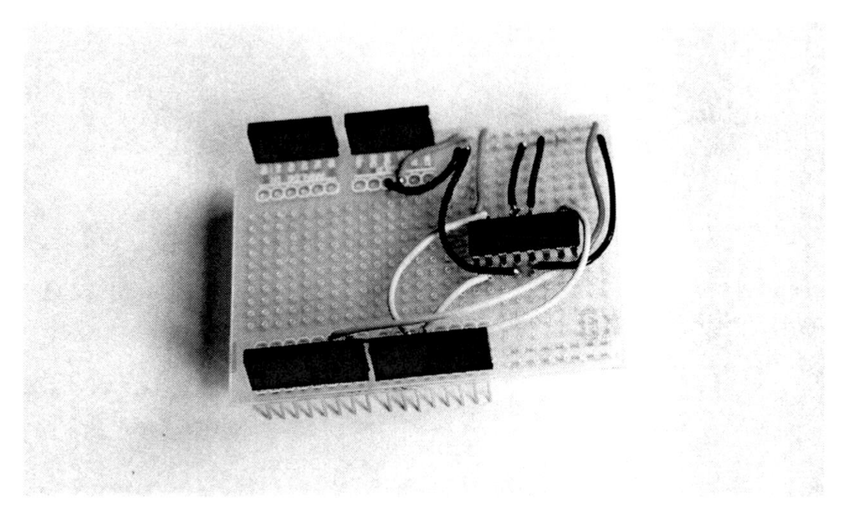

Figure 9-11. *H-Bridge with inputs for power and ground*

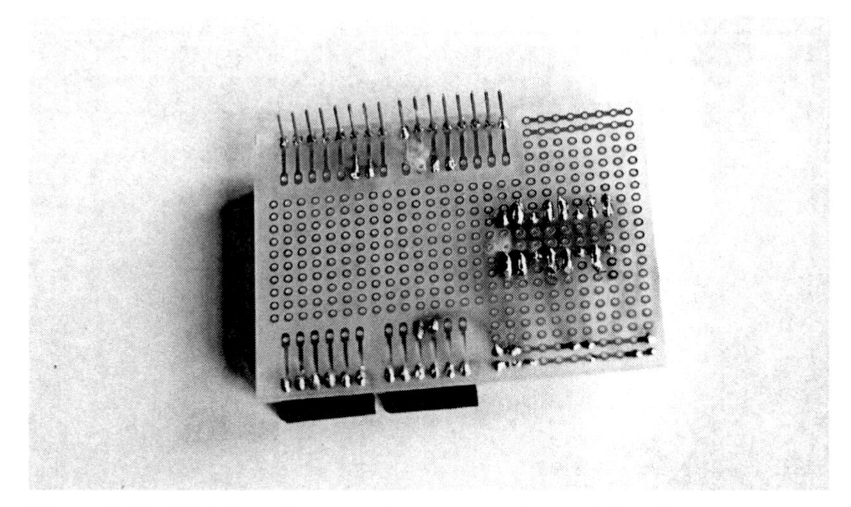

Figure 9-12. *H Bridge with inputs for power and ground (back)*

The next step is to connect the ENABLE inputs. Use a cable that connects H-Bridge pins 1 and 9 with Arduino pins 9 and 8. Remember to have a line in your code that allows them to be HIGH in order to activate the H-Bridge. The last step is to solder the H-Bridge OUT pins (pins 3, 6, 11, and 14) to the motor poles. Figures 9-13 and 9-14 show the circuit from the front and the back.

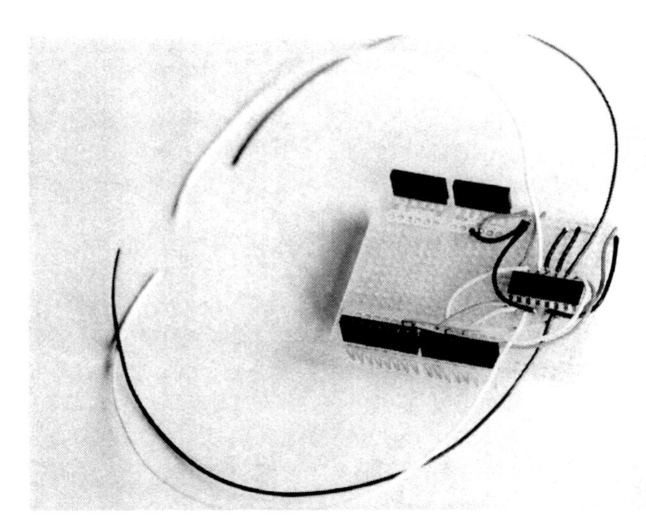

Figure 9-13. *H-Bridge ready for connecting motors*

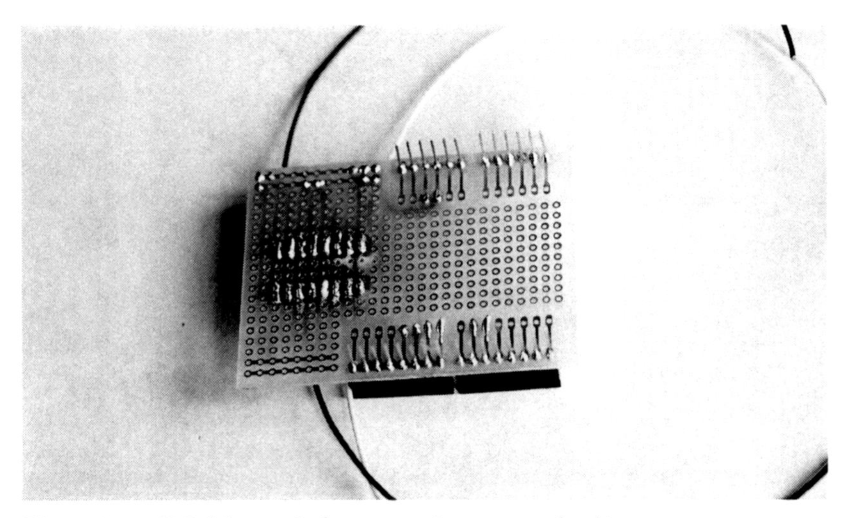

Figure 9-14. *H-Bridge ready for connecting motors (back)*

The diagram in Figure 9-15 shows the whole circuit.

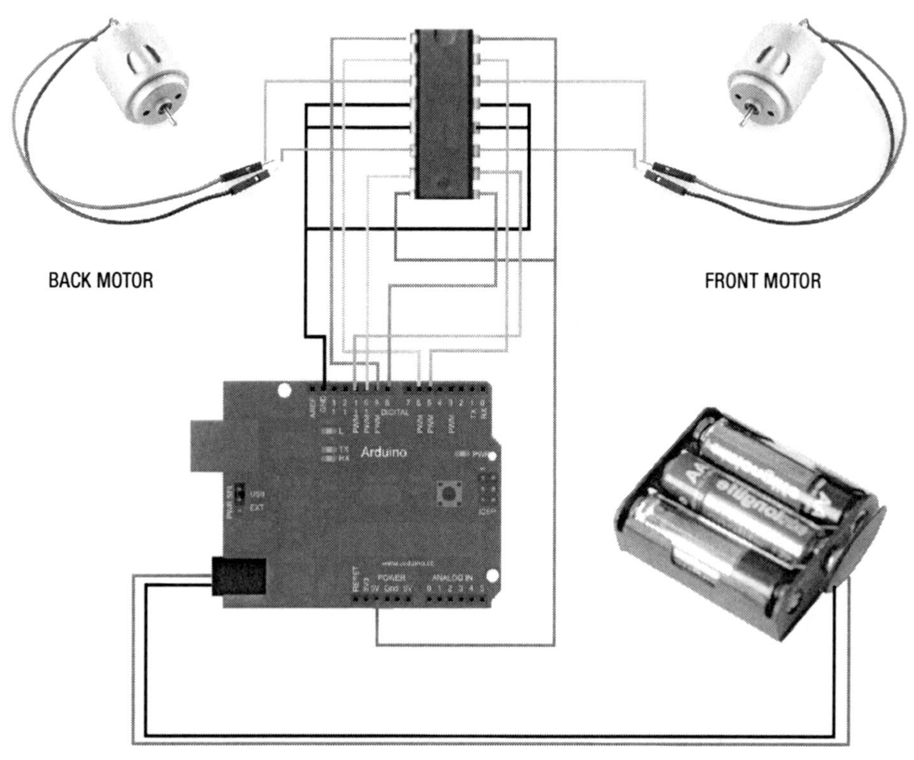

Figure 9-15. Circuit diagram

Testing the Circuit

It's time to test the circuit. You are going to use a simple Processing sketch that will activate each motor according to the position of the mouse, using serial communication with Arduino (Figure 9-16).

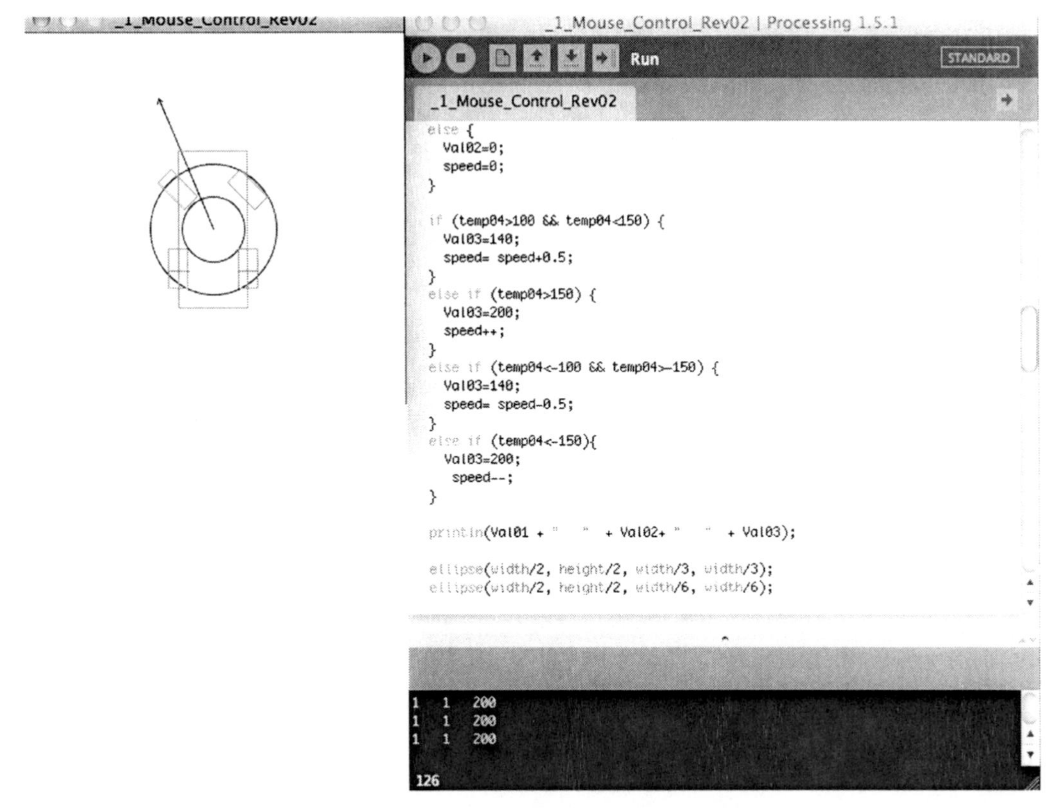

Figure 9-16. *Processing sketch for testing*

The code imports the Serial library in order to create serial communication with Arduino. After that, a series of variables are defined in order to store the values that are going to be transmitted. Finally, other variables are used for building a little car simulation that shows a visual description of what are you transmitting. These variables are used inside a `car()` function.

```
import processing.serial.*;
Serial myPort;
boolean serial = true;

int Val01, Val02, Val03;
float temp01, temp02, temp03, temp04;

//car variables
int carwidth= 70;
int carheight =120;
int wheels=30;
float angle;
float speed, wLine, wLinePos;
```

Setup Function

The setup() function just declares the size of your sketch and defines the port used for transmitting the data.

```
void setup() {
  size(300, 300);
  smooth();

  if (serial) {
    String portName = Serial.list()[0];
    myPort = new Serial(this, portName, 9600);
  }
}
```

Draw Function

The draw() function defines the arrow from a series of PVectors. One is defined from the mouse position (end of the arrow) and the other from the center of the canvas. Once you know these positions, you create another PVector that is the subtraction of the first two. After that, you call the function drawVector() that is defined by two PVectors (and will be explained later).

```
void draw() {
  background(255);
  PVector mouse = new PVector(mouseX, mouseY);
  PVector center = new PVector(width/2, height/2);
  PVector v = PVector.sub(mouse, center);

  //Draw  vector between 2 points
  drawVector(v, center);
  noFill();
```

The next part of the code defines the values sent via serial communication. You map the X coordinates in a range of values from 0 (left) to 255 (right). In order to do this, you define a temporary value as the difference between the mouse's X position and the canvas' center. This value is remapped in the range in a temporary float. Then you introduce a series of if conditionals so the final value that you to transmit is 0 (no movement), 1 (turn left), and 2 (turn right). These values activate an action in your virtual car simulation.

For the Y coordinate you also use a temporary value that refers to the mouse's Y coordinate and the center of the canvas. You remap this value between -255 (bottom area), 0 (center), and 255 (top area). This is the value that you pass.

The back motor is going to have two values: one that marks the direction (Val02) and other that marks the speed (Val03). The Val02 variable respond to 0 (don't activate the motor), 1 (activate it forward), and 2 (activate it backward). So according to the position of the mouse, it will change; if it's closer to the center, the motor won't be activated; if it's far, the car will go forward or backwards.

Another condition that you want to introduce is speed; this value depends on the distance of the mouse to the center. At the moment, just two speeds are defined: one of 140 and another of 200.

```
  temp01 = mouseX-width/2;
  temp02 = height/2- mouseY;
  temp03= int(map(temp01, -width/2, width/2, 0, 255));

//turn left
if (temp03<100) {
```

```
    Val01=1;
    angle=-PI/4;
  }

//turn right
  else if (temp03>150) {
    Val01=2;
    angle=PI/4;
  }

//no turn
  else {
    Val01=0;
    angle=0;
  }

// decide where to move
  temp04= int(map(temp02, -height/2, height/2, -255, 250));
//move front
  if (temp04>0 && temp04>50) {
    Val02=1;
  }

//move back
  else if (temp04<0 && temp04<-50) {
    Val02=2;
  }

//don't move
  else {
    Val02=0;
    speed=0;
  }

//decide speed
  if (temp04>100 && temp04<150) {
    Val03=140;
    speed= speed+0.5;
  }
  else if (temp04>150) {
    Val03=200;
    speed++;
  }
  else if (temp04<-100 && temp04>-150) {
    Val03=140;
    speed= speed-0.5;
  }
  else if (temp04<-150){
    Val03=200;
     speed--;
  }
```

The next lines of code just draw a couple of circles to define the motor area where the motor won't work. This area is introduced in case you want to stop the car and not keep it constantly in movement.

Then you call the car() function that represents a virtual car on the screen. The last line of the draw() function just sends serial data.

```
  ellipse(width/2, height/2, width/3, width/3);
  ellipse(width/2, height/2, width/6, width/6);

car();
  if (serial){
    sendSerialData();
  }
}
```

DrawVector Function

The drawVector() function represents an arrow starting from the center of your screen towards your mouse position (an arrow between two points). You mark the points using PVectors, but you also want to represent an arrowhead that points toward the direction that the car will move.

The arrowhead is represented by two lines. But you have to do a bit of math in order to have the rotation. You translate locally the coordinate system using pushmatrix and translate. These functions locate a new local origin to be the reference for the rotation of the arrowhead.

The angle that you rotate your coordinate system is defined by the PVector function heading2d. This function returns a float that defines the angle of rotation for the vector. So once your coordinate system has been rotated, you define the lines that form your arrow: a vertical line (remember that you have just rotate your UCS) that goes from 0,0 to the length (or magnitude) of your PVector, and for the head, you just define two lines from the extreme to both sides.

The last step is to use popmatrix to close the translations and rotations.

```
void drawVector(PVector v, PVector loc) {
  pushMatrix();
  float arrowsize = 4;
  translate(loc.x, loc.y);
  stroke(0);
  rotate(v.heading2D());
  float len = v.mag();
  line(0, 0, len, 0);
  line(len, 0, len-arrowsize, +arrowsize/2);
  line(len, 0, len-arrowsize, -arrowsize/2);
  popMatrix();
}
```

Car Function

The car() function represents a diagramatic verison of your car. The central rectangle defines the body and the four small ones represent the wheels. Select a blue color and use the functions pushMatrix and popMatrix to represent each rectangle on the screen, allowing rotation for the front wheels. So, according to the values received (or the mouse position), the wheels turn in one direction or the other. Finally, to represent the speed, you define a red line that crosses the back wheels, simulating the rotation of the wheels.

```
void car() {
rectMode(CENTER);
  //body
```

```
  noFill();
  stroke(30, 144, 255);
  rect(width/2, height/2, carwidth-wheels/2, carheight);

  //front wheels
  pushMatrix();
  translate(width/2-carwidth/2+wheels/4, height/2-carheight/2+wheels);
  rotate(angle);
  rect(0, 0, wheels/2, wheels);
  popMatrix();
  pushMatrix();
  translate(width/2+carwidth/2-wheels/4, height/2-carheight/2+wheels);
  rotate(angle);
  rect(0, 0, wheels/2, wheels);
  popMatrix();

  //back wheels
  pushMatrix();
  translate(width/2, height/2);
  rect(-carwidth/2+wheels/4, carheight/4, wheels/2, wheels);
  rect(carwidth/2-wheels/4, carheight/4, wheels/2, wheels);
//line simulating speed
  stroke(255, 0, 0);
  line(-carwidth/2, wLine, -carwidth/2+wheels/2, wLine);
  line(carwidth/2, wLine, carwidth/2-wheels/2, wLine);
  wLine=(carheight/4+wLinePos+speed);

  if (wLine<carheight/4-wheels/2) {
    wLine=carheight/4+wheels/2;
    speed=0;
    wLinePos=wheels/2;
  }
  else if (wLine>carheight/4+wheels/2) {
    wLine=carheight/4-wheels/2;
    speed=0;
    wLinePos=-wheels/2;
  }
  popMatrix();
}
```

SendSerial Function

The sendSerial() function is the same as in previous chapters; it just sends data with a letter before it in order to recognize and avoid any loss of data.

```
void sendSerialData() {
  myPort.write('S');
  myPort.write(Val01);
  myPort.write(Val02);
  myPort.write(Val03);
}
```

With this code, you send four values of serial data to drive your motors. The first value is just to identify the car, the second value is a range for moving the front motor, the third value defines a direction, and fourth value defines a speed. Now you have a sketch that sends serial data to your Arduino according to your mouse position. You are really close to your first test but first you need to program your Arduino.

Arduino Testing Sketch

The code starts by declaring floats and the Arduino pins that are connected to your H-Bridge. Note that for the back motor you use analog pins.

```
float Val01, Val02, Val03;
int motorFrontLeft = 11;   // H-bridge leg 10
int motorFrontRight = 5;  // H-bridge leg 15
int motorBackUp = 6;     // H-bridge leg 2 (PWM)
int motorBackDown = 10;  // H-bridge leg 7 (PWM)
int enablePinFront = 8; // H-Bridge leg 9
int enablePinBack = 9; // H-Bridge leg 1
```

Setup Function

The setup() function just declares the start of serial communication and the Arduino pin mode; in this sketch all pins are outputs.

```
void setup() {
  // initialize the serial communication:
  Serial.begin(9600);
  // Set the pin modes
  pinMode(motorFrontLeft, OUTPUT);
  pinMode(motorFrontRight, OUTPUT);
  pinMode(motorBackUp, OUTPUT);
  pinMode(motorBackDown, OUTPUT);
  pinMode(enablePinFront, OUTPUT);
  pinMode(enablePinBack, OUTPUT);
}
```

Loop Function

Start by reading the serial data. If any data is available, the first piece of information should match the letter S. If this is the case, continue reading and store the values as floats.

After reading the values, you introduce a series of if conditionals that will drive the motors according to the values obtained. For the front motor drive, you get values 1 (turn left), 2 (turn right), and 0 (no action in case you want to drive in a straight line). If the values are 1 or 2, the H-Bridge's ENABLE pin is active, meaning that the motor starts to work and the direction is defined by the functions TurnLeft() and TurnRight().

For the back motor, you receive two values (Val02 and Val03); one defines the direction and the other defines the speed.

Variable Val02 defines the direction. If its value is 1, it puts the H-Bridge ENABLE pin as HIGH and calls the MoveUp() function that is defined by a value (speed). If variable Val02 is 2, the ENABLE pin is

HIGH and calls the function `MoveDown()` runs, moving the car in the opposite direction. Finally, if variable `Val02` is 0, the ENABLE pin is LOW, meaning that this motor is disconnected and the car won't move.

Variable `Val03` defines the speed of the motor. Potentially, you could have values from 0 to 255, but according to the Processing sketch you just have two speeds, normal and fast, with values of 140 and 200.

```
void loop() {
  // check if data has been sent from the computer:
  if (Serial.available()>4) { // If data is available to read,
    char val = Serial.read();
    if(val == 'S') {
      Val01 = Serial.read();
      Val02 = Serial.read();
      Val03 = Serial.read();
    }
  }

  // turning left and right
  if (Val01==1) {
    enable(enablePinFront);
    TurnRight();
  }
  else if (Val01==2) {
    enable(enablePinFront);
    TurnLeft();
  }
  else if(Val01==0)disable(enablePinFront);
   if (Val02==1) {
    enable(enablePinBack);
    MoveUp(Val03);
  }
  else if (Val02==2) {
    enable(enablePinBack);
    MoveDown(Val03);
  }
  else if(Val02==0) {
   disable(enablePinBack);
   }

}
```

Turning Functions

The functions for turning right and turning left are similar. Each function sets one pin to HIGH and one to LOW. The direction of the current flow, and therefore the direction of the motor, is determined by which pin is in which state. The function is straightforward: a `digitalWrite` is applied to the Arduino pins.

```
//function to turn Right
void TurnRight(){
  digitalWrite(motorFrontRight,HIGH);
  digitalWrite(motorFrontLeft,LOW);
```

```
}

//function to turn Left
void TurnLeft(){
  digitalWrite(motorFrontLeft,HIGH);
  digitalWrite(motorFrontRight,LOW);
}
```

Move Functions

The move functions are similar to the turning functions; the main difference is that you use `analogWrite` with an external value defined in the void. This value marks the speed of rotation of the motor. So instead of using HIGH and LOW, you use speed: 255 as a maximum value and 0 to indicate the car has stopped.

```
//function to move
void MoveUp(int speedD){
  analogWrite(motorBackUp,speedD);
  analogWrite(motorBackDown,0);
}
//function to move reverse
void MoveDown(int speedD){
  analogWrite(motorBackDown,speedD);
  analogWrite(motorBackUp,0);
}
```

Enable/Disable Functions

The enable and disable functions just write HIGH or LOW to the pin marked when they are called. You use them for the H-Bridge ENABLE pins, activating or disconnecting the motors.

```
void enable(int pin){
  digitalWrite(pin, HIGH);
}
void disable(int pin){
  digitalWrite(pin, LOW);
}
```

Now you're ready to test your car for the first time! Upload your Arduino code (leaving the USB cable plugged into the car; you need it for the serial communication). Run the Processing sketch and move your mouse. The car should move and turn according to your mouse position!

Proximity Sensors

Now let's introduce a bit more complexity. You are going to use a proximity sensor in the front part of the car to check the distance from another object. Using the values it returns, you are going to introduce a routine of "stopping the car" inside your Arduino code. The idea is to avoid collisions. The code will let you control the car but if it is going to crash with an object, it will override your movements and automatically stop (and actually move backwards a little).

The routine will check the value coming from the proximity sensor and it will compare it with a threshold (a value that we have defined previously in the code).

If this value is higher than the threshold it will mean that an object is really close. If that is the case, the back motor will start to turn backwards for a short period of time and after that it will check again the value coming from the sensor, repeating this routine if the value is still bigger than the threshold.

If the value from the sensor is smaller than the threshold it means that there is not any object close enough, and there is no risk of crashing, so the car will move normally.

An analog infrared (IR) distance sensor is a really simple component (Figure 9-17). It shines a beam of IR light from an LED and measures the intensity of light that is bounced back using a phototransistor. If you film this sensor with a cellphone or a camera and you look at the sensor through the screen, you will see the IR LEDs glowing.

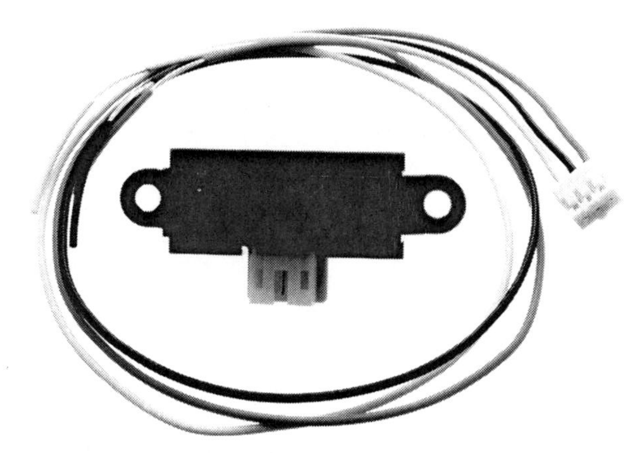

Figure 9-17. Infrared proximity sensor

To make it work, just connect +5v and ground, and an analog signal will be returned (Figure 9-18). This signal is in a voltage proportional to the distance between the sensor and an object in front of it. The sensor you are using has a range of 80 cm (0.4 volts) and 10 cm (3 volts).

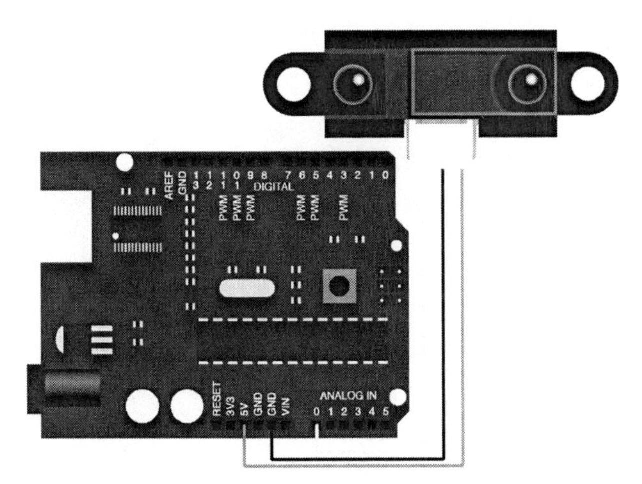

Figure 9-18. Proximity sensor connected to Arduino

You can work with the values that come from the sensor if you choose an appropriate maximum value to act as threshold in order to activate your stop car function.

The IR proximity sensor has been used in many interactive applications. Even if you are just using the outcome voltage coming from the sensor, it's worthwhile to know how the sensor works and how you can measure real distances with it. (You can skip this explanation if you want.)

The voltage returned is not linear, as you can see in Figure 9-19.

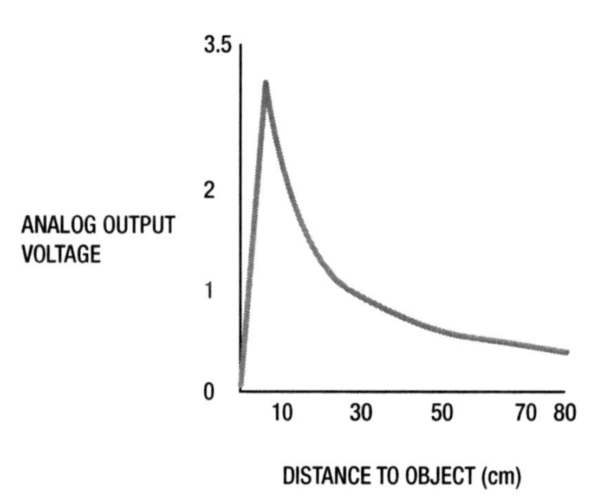

Figure 9-19. Proximity sensor's voltage/distance graph

However, you can convert the voltage returned to distance using an equation. Depending of the model of sensor, this equation will vary. For the Arduino code, the equation for transforming inputs to distance is

```
float distance = 12343.85 * pow(analogRead(sensorPin),-1.15);
```

Before adding it to your circuit, do a quick test by connecting a sensor to your Arduino board using a pigtail or an infrared sensor jumper. Use this code for the test:

```
int sensorPin = 0;
void setup(){
  Serial.begin(9600);
}

void loop(){
  int val = analogRead(sensorPin);
  //float val = 12343.85 * pow(analogRead(sensorPin),-1.15);
  Serial.println(val);
  delay(100);
}
```

Use Arduino analog Pin 0 for receiving the values, and start the serial communication. Next, choose which values you want out of the Arduino: the voltage (`int val`) or the distance (`float val`). Finally, just add a delay to slow down the output.

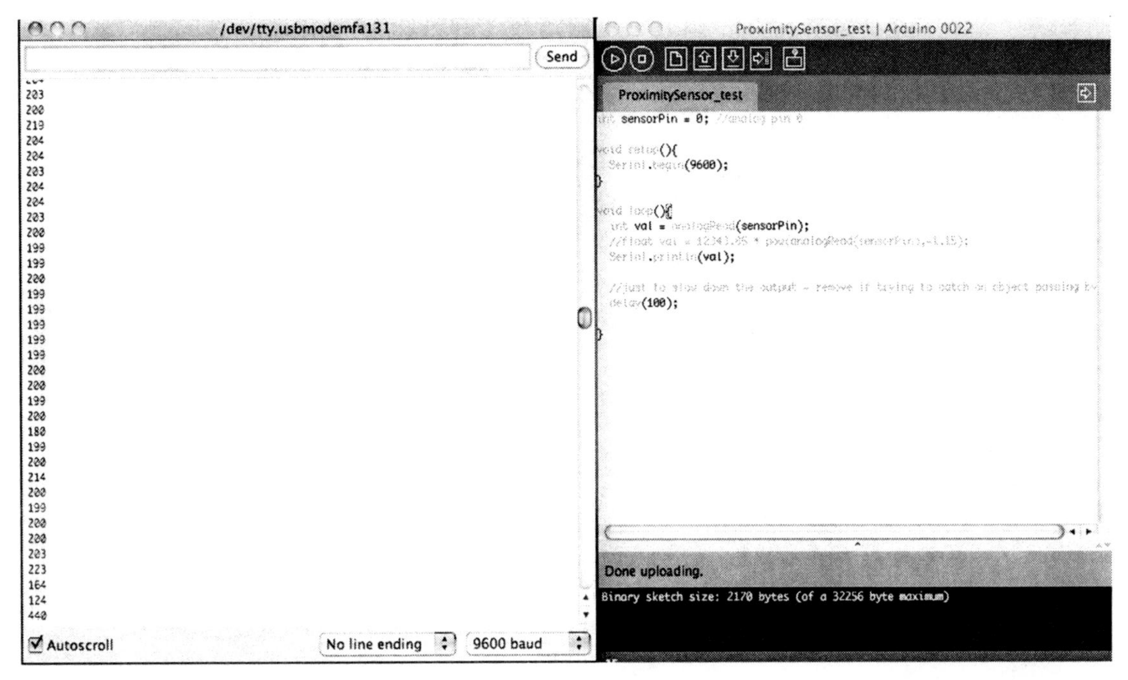

Figure 9-20. Proximity sensor's output values

The values that you obtain in the Serial Monitor, depending on the code you use, are between 650 (really close) and 20 (object far away), as shown in Figure 9-20. If we comment the *int val* and uncomment the line *float val*, we will use the equation previously described in the code. If we upload this "modified" code to Arduino we will obtain values between 70 and 7, meaning real distance in centimeters. Using this equation is great because we have real values, but it also gives us a big error when the object is really far away, due to the math that you applied.)

Enough theory! Let's get back to the RC vehicle. To keep things simple, use the voltage values. In your case, just knowing the threshold value that indicates that your car is close to an object is enough to activate your stop car routine.

After testing, add the sensor to your car circuit. Just solder the cables to power, ground, and analog pin 0 in your prototype board. The circuit looks like the one in Figure 9-21.

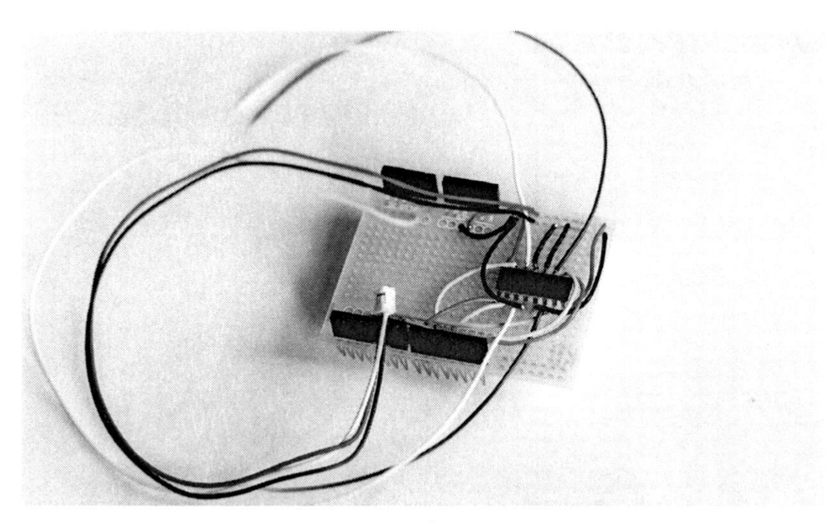

Figure 9-21. Proximity sensor connected

Fix the sensor on the front area of the car using some removable adhesive (in case you want to use the sensor for another application), as shown in Figure 9-22.

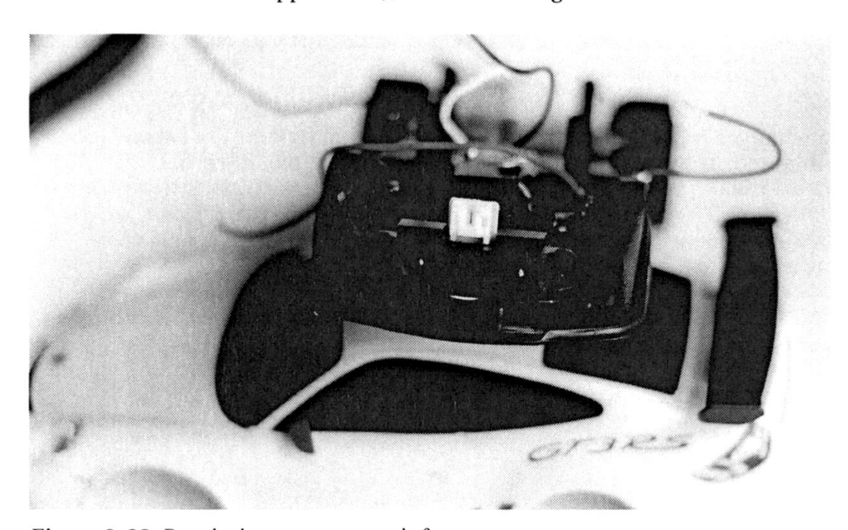

Figure 9-22. Proximity sensor on car's front

Now change the Arduino code in order to get an automated response according to the data coming from the sensor. As mentioned, the idea is to stop the car and reverse it for long enough to avoid a collision. If the data coming from the sensor is high (meaning an object is really close), the back motor will automatically reverse.

Add a few variables to the beginning of the code. A new variable (`Val04`) activates or deactivates the automatic response. The sensor is incorporated with an analog IN pin (pin0), an initial value, and a threshold value to mark the distance from which the automated response will be active.

```
float Val01, Val02, Val03, Val04;
(....)
int sensorPin =  0;
int sensorValue=0;
int Value =500; //threshold for detecting object
```

Setup Function

The setup() function remains the same.

Loop Function

The loop() function incorporates this new value from the serial data. According to the value of Val04, Arduino reads the data coming from the sensor (automatic response on) or just marks this value as 0 (automatic response off).

Then you add a general if before all the functions, marking the value coming from the sensor and comparing it to the threshold. In other words, if no object is close, you drive the vehicle with your Processing sketch, but if an object is nearby, the code jumps directly to the new function called stopCar().

```
void loop() {
if (Serial.available()>5) {
    char val = Serial.read();
    if(val == 'S'){
      Val01 = Serial.read();
      Val02 = Serial.read();
      Val03 = Serial.read();
      Val04 = Serial.read();
    }
  }

if (Val04= 1)sensorValue = analogRead(sensorPin);
else sensorValue=0;

  // turning left and right
  if (sensorValue<Value) {
    if (Val01==1){
      enable(enablePinFront);
      TurnRight();
    }
    else if (Val01==2) {
      enable(enablePinFront);
      TurnLeft();
    }
    else if(Val01==0)disable(enablePinFront);
    if (Val02==1) {
      enable(enablePinBack);
      MoveUp(Val03);
    }
```

```
      else if (Val02==2) {
        enable(enablePinBack);
        MoveDown(Val03);
      }
      else if(Val02==0) {
         disable(enablePinBack);
         }
    }

    else {
      enable(enablePinBack);
      stopCar();
    }
}
```

Move Functions

All move functions remain the same. You just add a new function `stopCar()` that exists only to call the previous function `MoveDown()` and disable the back motor once it has moved backwards. This function acts as a brake, avoiding collisions.

```
void stopCar(){
for(int i=0; i<2000; i++){
    MoveDown(200);
}
    disable(enablePinBack);
}
```

Once you have connected the proximity sensor, you can test your car. Upload this new Arduino code (you are still not wireless, so you need to leave the USB cable plugged to your Arduino board). Run the previous Processing sketch again and move your mouse. The car should move as before, but if you put an object in front of it, it should stop and then move backwards.

XBee and Wireless

The last step is to add a wireless module. You learned how to use XBee modules for wireless communication in pervious chapters. (Have a look at those chapters for a refresher on the concepts and how to install drivers in case you haven't done it before.)

As you have seen, you can use an XBee shield or just connect the cables that you need. In this case, you want to everything as compact as possible, so solder the cables coming from the XBee Explorer as described in Table 9-2 and shown in Figure 9-23.

***Table 9-2.** XBee Explorer Regulated to Arduino Connections*

XBee Explorer Regulated Pins	Arduino Pins
XBee 3.3V	Arduino 3.3V
XBee GND	Arduino GND

XBee Explorer Regulated Pins	Arduino Pins
XBee DIN	Arduino TX (Digital Pin 1)
XBee DOUT	Arduino RX (Digital Pin 0)
XBee DIO3	Arduino RESET

Figure 9-23. XBee Explorer

If you solder the cables on your prototype shield, the circuit will look like that in Figure 9-24.

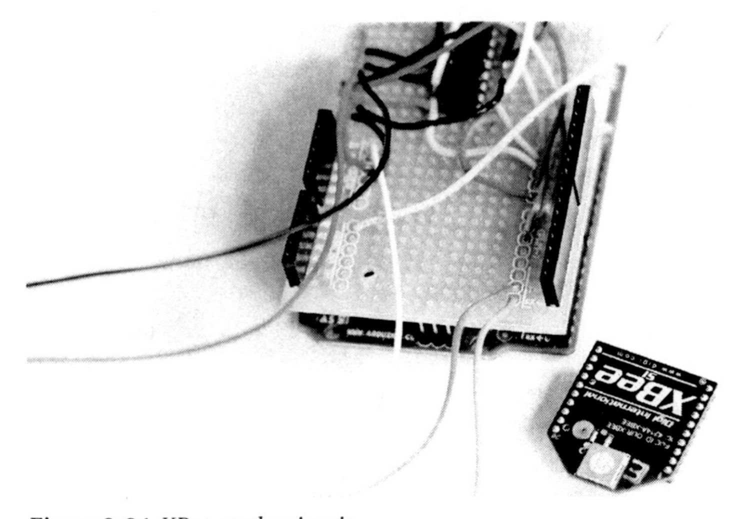

Figure 9-24. XBee on the circuit

Now just tidy up the cables and fix them to the vehicle to avoid disconnections when the vehicle is in movement. The final vehicle will look like Figure 9-25.

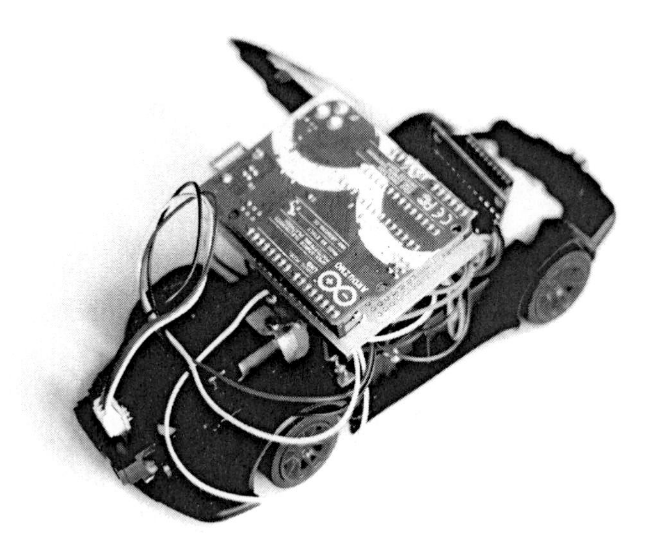

***Figure 9-25.** The complete hacked RC car*

The *very* last step is to prepare a Kinect sketch and start driving your car. But first, test the car. Plug your XBee Explorer USB to your computer, run the Processing sketch, and make sure that data is getting to your Arduino. You should see a flashing Arduino LED RX. If that's happening, you should be able to drive your car just using your mouse.

Kinect RC Interface

Now you're going to use the movement of your hand to drive the vehicle.

- According to its relative position to the center of the screen, you can drive the car, moving it towards this direction.
- If your hand is in the central position, the vehicle won't move.
- If you draw a circle with your hand, you activate a virtual switch, turning on and off the automatic response, as shown in Figure 9-26.

  The automatic response marks if Arduino is going to read the data coming form the sensor, and therefore it will react automatically when an object is really close using the function *stopcar()*.

  In the case this automatic response is off, the data coming from the sensor will be overwrite to 0, meaning that the car will be "blind" respect the objects that are in front and won't have any automatic reaction.

  Note that this circle should be drawn close to the center of the screen where the car is not in movement so the gesture of doing a circle won't drive the car over the place.

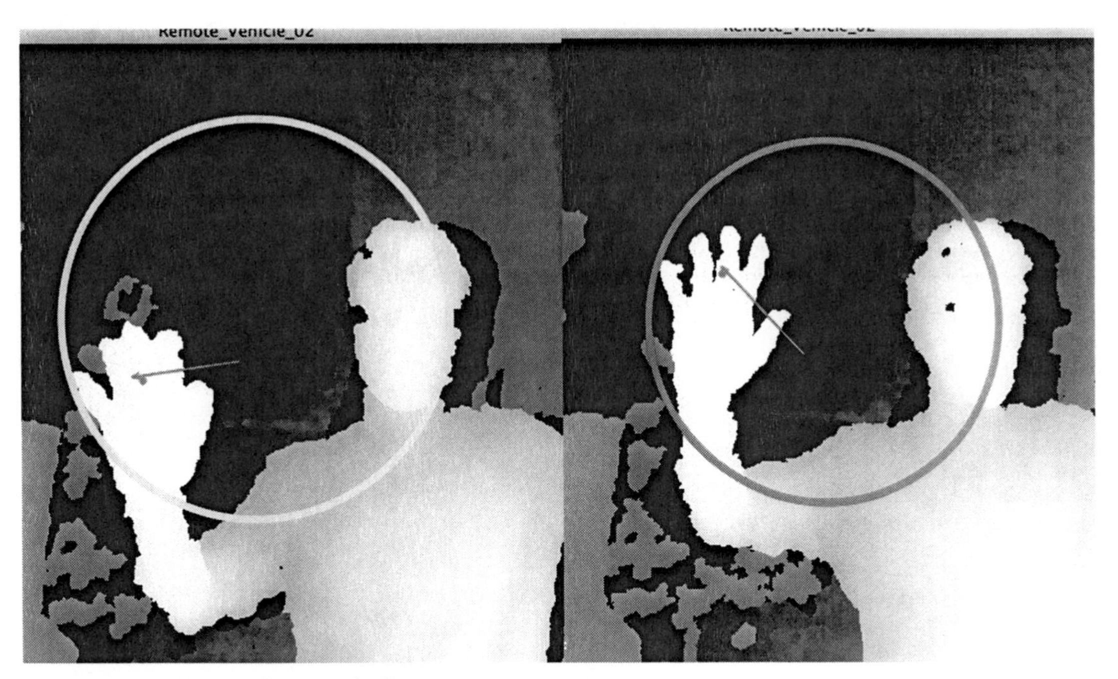

Figure 9-26. Safety mode on and off

Similar code has been used in the previous chapters. You're just creating a few new functions and putting some text on the screen in order to display live data. The code is based on hand and circle recognition. Start by importing the libraries.

```
import SimpleOpenNI.*;
import processing.opengl.*;
import processing.serial.*;

SimpleOpenNI kinect;
Serial myPort;
```

Next, declare the NITE objects explicitly, using the NITE functions for hand tracking, gesture recognition, and circle detection. You declare the session manager (that deal with NITE gesture recognition), control points (that deal with hand points), and circle detector.

```
// NITE
XnVSessionManager sessionManager;
XnVPointControl pointControl;
XnVCircleDetector circleDetector;
```

You're going to work with text on the screen so you declare a font and some strings. (You declare them here in order to avoid null values once you start your Processing sketch. They update their value once the hand has been detected.)

```
// Font for text on screen
PFont font;
String Ctext;
String dir;
```

The other variable that you declare is handTrackFlag, which refers to a hand being detected. The same principle is applied to the boolean circleTrackFlag. Other variables that you declare are PVectors that represent the positions of the hand on the screen, and the position of the hand in the real world. Then other PVectors will represent the center of a circle when we do the gesture and again we use two different variables: one representing the center on the screen and another representing the center in the real world. Other variables are radius and a counter t, that will serve for changing the state of a Boolean activating or deactivating the automated response. PVector v is the difference between the screen center and your hand position but in 2D. Boolean automated sends data about using the proximity sensor or not, and Boolean serial allows you to send serial data. The last set of values is a set of temporary ones that are remapped before being sent via serial communication to Arduino.

```
// Variables for Hand Detection
boolean handsTrackFlag = false;
boolean circleTrackFlag = false;
PVector screenHandVec = new PVector();
PVector handVec = new PVector();
int t = 0;
float rad;
PVector centerVec = new PVector();
PVector screenCenterVec = new PVector();

PVector v = new PVector();
boolean automated=true;

boolean serial = true;
int Val01, Val02, Val03, Val04;
float temp01, temp02, temp03, temp04;
```

Setup Function

In the setup() function, you initialize all the objects that you will be using later on, starting from the Kinect ones and then calling the NITE session manager where listeners will be added. As explained in previous chapters, you need to call methods create, destroy and update, so the methods XnVControl() and XnVDetector() invoke your callback functions in order to create, destroy and update one object (a circle or a hand detection).

```
void setup() {
// Simple-openni object
kinect = new SimpleOpenNI(this);
kinect.setMirror(true);
// enable depthMap generation, hands + gestures
kinect.enableDepth();
kinect.enableGesture();
kinect.enableHands();

// setup NITE
sessionManager = kinect.createSessionManager("Wave", "RaiseHand");
// Setup NITE.s Hand Point Control
pointControl = new XnVPointControl();
pointControl.RegisterPointCreate(this);
pointControl.RegisterPointDestroy(this);
pointControl.RegisterPointUpdate(this);
```

```
  // Setup NITE's Circle Detector
  circleDetector = new XnVCircleDetector();
  circleDetector.RegisterCircle(this);
  circleDetector.RegisterNoCircle(this);

// Add it to the session
sessionManager.AddListener(pointControl);
sessionManager.AddListener(circleDetector);
```

After calling the Nite methods, add the size of your canvas and initialize the font (which should be added to the data folder). You also initialize the Strings that you defined before and finally start your serial communication (depending on a Boolean, so it's easy to connect and disconnect).

```
// Set the sketch size to match the depth map
size(kinect.depthWidth(), kinect.depthHeight());
smooth();

// Initialize Font
font = loadFont("SansSerif-12.vlw");
Ctext="Automated mode ON";
dir = "-";

//Initialize Serial Communication
  if (serial) {
    String portName = Serial.list()[0]; // This gets the first port
    myPort = new Serial(this, portName, 9600);
  }
}
```

Draw Function

The purpose of this sketch is to send data to Arduino, according to the hand position and the gesture that you perform in front of the Kinect camera. You start by calling a black background and a PVector to mark the center of the screen, as shown in Figure 9-27.

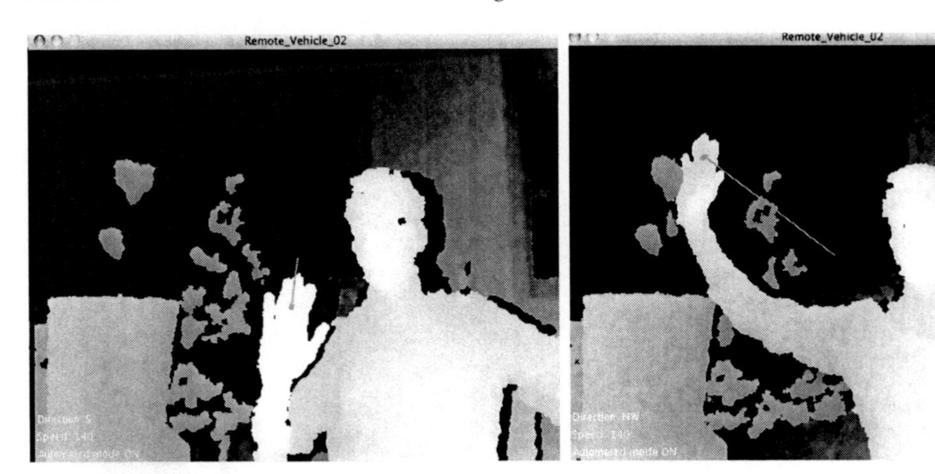

Figure 9-27. *Driving the car*

After that, you update the Kinect object so you can see what your Kinect is seeing.

```
void draw()
{
  background(0);
  PVector centerL = new PVector(width/2, height/2);
  // Update Kinect data
  kinect.update();
  // update NITE
  kinect.update(sessionManager);

  // draw depthImageMap
  image(kinect.depthImage(), 0, 0);
```

Next, you calculate the difference in 2D (just XY coordinates) between the projection of the hand and the center of the canvas. This operation should be familiar; remember your Processing test sketch? You are doing the same thing but using the data coming from the Kinect.

```
    //displacement between the centre and the hand in 2D
  v.x = screenHandVec.x-centerL.x;
  v.y = screenHandVec.y-centerL.y;
```

Next, call a series of functions depending on whether the Kinect is tracking a hand. If so, it draws a hand and an arrow and then drives the car. If the Kinect detects a circle, it draws on the screen. A series of functions is running and updated continuously: some text on the canvas and the data that you are sending via serial.

```
  if (handsTrackFlag){
    drawHand();
    drawArrow(v, centerL);
    controlCar();
  }
  if (circleTrackFlag){
    drawCircle();
  }
  textDisplay();
  if (serial){
  sendSerialData();
  }
}
```

Other Functions

The function `controlCar()` maps all the data in just a few values in order to transmit them to the Arduino. This function is similar to the one used previously in the Processing sketch. It just remaps the arrow vector according to the coordinates X and Y and then stores these values. It uses temporary values to remap them with a series of `if` conditions in order to have a clean output.

```
void controlCar() {
  temp01 = screenHandVec.x-width/2;
  temp02 = height/2- screenHandVec.y;
  temp03 = int(map(temp01, -width/2, width/2, 0, 255));

  if (temp03 < 75) {
    Val01 = 1;
  }
```

```
  if (temp03 > 175) {
    Val01 = 2;
  }
  if ((temp03 > 75) && (temp03 < 175)) {
    Val01 = 0;
  }
  temp04= int(map(temp02, -height/2, height/2, -255, 250));
  if ((temp04 > 0) && (temp04 > 50)) {
    Val02 = 1;
  }
  else if ((temp04 < 0) && (temp04 < -50)) {
    Val02 = 2;
  }
  else {
    Val02 = 0;
  }
  if ((temp04 > 100) && (temp04 < 150)) {
    Val03 = 140;
  }
  if (temp04 > 150) {
    Val03 = 200;
  }

  if ((temp04 < -100) && (temp04 > -150)) {
    Val03 = 140;
  }
  if (temp04 < -150) {
    Val03 = 200;
  }
  //println(Val01 + "    " + Val02+ "    " + Val03+ "    " +Val04 );
}
```

The Other function represents the hand on the screen. Once the hand is detected, it appears as a red dot that makes the same movements as your hand. There's also a function for representing a circle in case the gesture is detected. With the counter `t`, you can change the Boolean `automated` every time the new circle is produced. So the automated behavior is activated and deactivated with the same function. If you draw a circle, the automated behavior is turned off, and if you repeat the gesture, it's on again. These gestures also change the color of the circle and the text referring to the automated mode as well as `Val04`, which is sent via serial.

```
// Draw the hand on screen
void drawHand() {

  stroke(255, 0, 0);
  pushStyle();
  strokeWeight(6);
  kinect.convertRealWorldToProjective(handVec, screenHandVec);
  point(screenHandVec.x, screenHandVec.y);
  popStyle();

}
void drawCircle() {
if(t == 1)automated=!automated;
    if (automated == true){
```

```
        Val04 = 0;
        noFill();
        strokeWeight(6);
        stroke(0, 255, 0);
        ellipse(screenCenterVec.x, screenCenterVec.y, 2*rad, 2*rad);
        textAlign(LEFT);
        Ctext = "Automated mode ON";
      }
      if (automated==false){
        Val04 = 1;
        noFill();
        strokeWeight(6);
        stroke(255, 0, 0);
        ellipse(screenCenterVec.x, screenCenterVec.y, 2*rad, 2*rad);
        textAlign(LEFT);
        Ctext = "Automated mode OFF";
      }

    // println(automated);

}
```

The function `drawArrow` represents an arrow between two PVectors. It was used at the beginning of the chapter and it is pretty straightforward.

```
void drawArrow(PVector v, PVector loc){
  pushMatrix();
  float arrowsize = 4;
  translate(loc.x, loc.y);
  stroke(255, 0, 0);
  strokeWeight(2);
  rotate(v.heading2D());
  float len = v.mag();
  line(0, 0, len, 0);
  line(len, 0, len-arrowsize, +arrowsize/2);
  line(len, 0, len-arrowsize, -arrowsize/2);
  popMatrix();
}
```

`textDisplay` represents live text (meaning that the values update automatically) on the screen. The function `text` asks for a value or string plus X and Y positions. In this sketch, you are showing the car's direction, speed of the back motor, and whether the automated mode is activated or not. According to the values you are sending to Arduino, you make the rules for representing these values using a series of `if` conditionals. The last function is `sendSerialData`, which as usual uses a letter to identify each message.

```
void textDisplay(){
   text(Ctext, 10, kinect.depthHeight()-10);
   int value;
   if(Val02 == 0) {
     value=0;
     }
   else {
     value = Val03;
    }
   text("Speed: "+value, 10, kinect.depthHeight()-30);
```

```
    text("Direction: "+dir, 10, kinect.depthHeight()-50);

  if ((Val02 == 1) && (Val01 == 0)) {
    dir ="N";
  }
  if ((Val02 == 2) && (Val01 == 0)) {
    dir="S";
  }
  if ((Val02 == 0) && (Val01 == 1)) {
    dir="W";
  }
  if ((Val02 == 0) && (Val01 == 2)) {
    dir="E";
  }
  if ((Val02 == 1) && (Val01 == 2)) {
    dir="NE";
  }
  if ((Val02 == 2) && (Val01 == 2)) {
    dir="SE";
  }
  if ((Val02 == 2) && (Val01 == 1)) {
    dir="SW";
  }
  if ((Val02 == 1) && (Val01 == 1)) {
    dir="NW";
  }
}

void sendSerialData() {
  // Serial Communcation
  myPort.write('S');
  myPort.write(Val01);
  myPort.write(Val02);
  myPort.write(Val03);
  myPort.write(Val04);
}
```

The last part is the NITE callbacks. As shown in previous chapters, they refer to the events of creating an object (hand or circle) once the gesture is detected, destroying it when the hand or the circle has gone off the screen, and updating them. Hand events and circle events have different callbacks. The only lines that are new are the ones that refer to the counter `t`. Each time a new circle is created, the counter starts to add, meaning that just once it will have the value of 1 (which is necessary for changing a Boolean). Each time that circle is destroyed, this counter comes back to 0.

```
// XnVPointControl callbacks
void onPointCreate(XnVHandPointContext pContext) {
  println("onPointCreate:");
  handsTrackFlag = true;
  handVec.set(pContext.getPtPosition().getX(),
              pContext.getPtPosition().getY(),
              pContext.getPtPosition().getZ());

}
```

```
void onPointDestroy(int nID) {
  println("PointDestroy: " + nID);
  handsTrackFlag = false;
}

void onPointUpdate(XnVHandPointContext pContext) {
  handVec.set(pContext.getPtPosition().getX(),
              pContext.getPtPosition().getY(),
              pContext.getPtPosition().getZ());
}

// XnVCircleDetector callbacks
void onCircle(float fTimes, boolean bConfident, XnVCircle circle) {
  println("onCircle: " + fTimes + " , bConfident=" + bConfident);
  circleTrackFlag = true;
  t++;
  centerVec.set(circle.getPtCenter().getX(),
                circle.getPtCenter().getY(), handVec.z);
  kinect.convertRealWorldToProjective(centerVec, screenCenterVec);
  rad = circle.getFRadius();
}

void onNoCircle(float fTimes, int reason) {
  println("onNoCircle: " + fTimes + " , reason= " + reason);
  circleTrackFlag = false;
  t = 0;
}
```

Now, test this code with your vehicle. The Arduino code hasn't changed, so just plug the Kinect and the XBee USB Explorer. Run the new Processing sketch and start to drive, as shown in Figure 9-28!

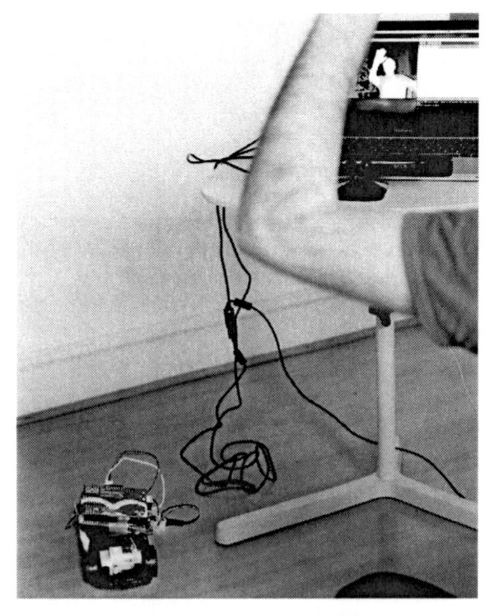

Figure 9-28. *Driving your RC vehicle*

Summary

If you use this code (make sure that the Boolean `serial` is set to `true`), you will be sending serial data to drive a hacked RC car. Using the position of your hand, you will be able to drive it; using a gesture, you can activate a simple anti-collision system.

You have learned several new things in this chapter. Regarding hardware, you learned about RC motors and how to drive them using a simple H-Bridge. You opened a remote control car and, using the components, reprogrammed its behavior using Arduino. You also learned how proximity sensors work and the kind of data you can obtain from them.

This application is only the principle; the same logic can be applied to any RC vehicle. Next time, hack a helicopter or a boat! Just open it up, investigate what's inside, and start to have fun!

CHAPTER 10

Biometric Station

by Enrique Ramos

In 1487, Leonardo da Vinci created a drawing that he entitled "The Vitruvian Man," depicting what he believed to be the ideal human proportions. But these proportions are not only a matter of beauty. Height, weight, and other body measures can actually be an indicator of appropriate development and general health in both children and adults. Kinect has a device that can precisely measure all your body proportions in real time. If you combine this ability with other sensors, you can create a general-purpose biometric station for your personal use and delight.

Throughout this chapter, you will learn how to hack a bathroom scale and connect it wirelessly to your computer, how to combine the incoming data with the Kinect point cloud and skeleton data, and how to compare this information to external databases for different purposes. Table 10-1 shows the parts required for this project. You will be able to identify people, automatically get body mass indexes, and check whether they fall within desirable ranges, as shown in Figure 10-1. You will store that information so you can create graphs of the evolution.

This is an advanced chapter in terms of programming techniques, and it will introduce some Java file parsing techniques and the use of Java AWT components. Welcome to Kinect biometrics!

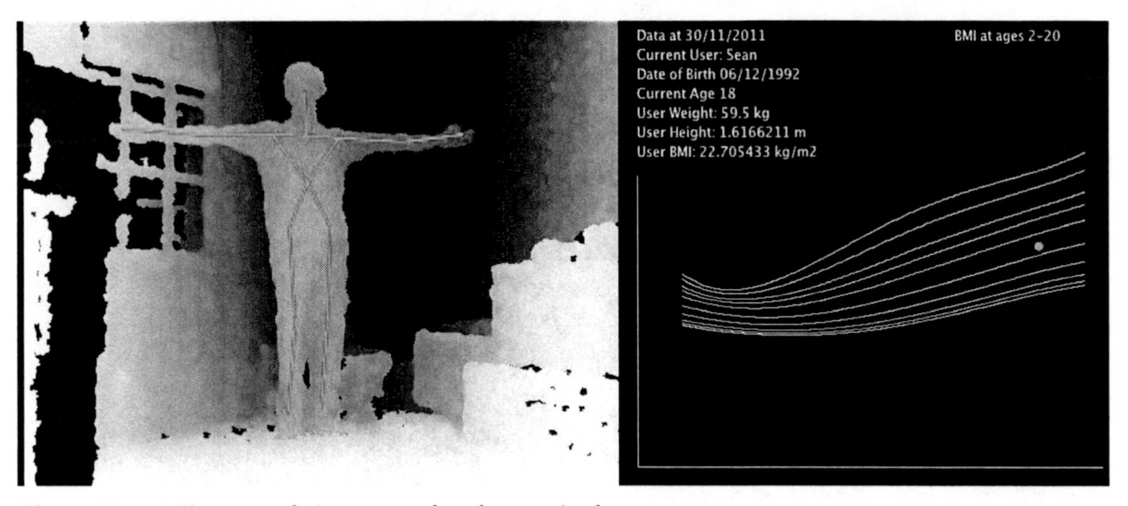

Figure 10-1. A Kinect user being scanned and recognized

Table 10-1. Parts list

Part	Description	Price
Electronic bathroom scale	With LCD display!	$20-40
XBee Shield	To interface the XBee with your Arduino	$24.95
XBee Explorer USB	You will need a mini USB cable to connect it	$24.95
2 XBee modules	XBee 1mW Chip Antenna - Series 1	$22.95 each
Stackable headers	6-pin and 4-pin. At least two of each	$0.50 each
1 Prototype Shield	SparkFun DEV-07914 or similar.	$16.95

Hacking a Bathroom Scale

In this project you will combine weight information with volume information from the Kinect sensor. You learned how to acquire Kinect data in a previous chapter, so now you need to figure out how to get the weight information into the computer. Well, one option is to type the information in, but that would be too easy, wouldn't it? You're going to go a little further and hack a standard bathroom scale to communicate wirelessly with your computer, so you can get the volume and weight information at the same time (Figure 10-2).

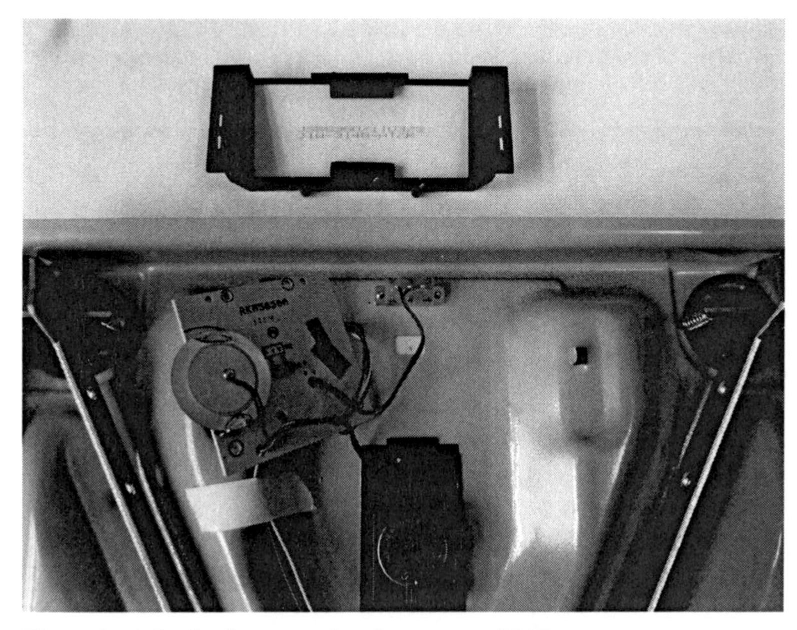

Figure 10-2. Scale after removing the cover and LCD

Hacking a bathroom scale would seem to be a simple task. If you have previously worked with pressure sensors, you might think that it's just a matter of finding the analog output, reading it from your Arduino, and mapping the values to the real weight. Well, it happens to be slightly more complicated than that. Most modern scales use the combined output of one to four strain gauges. The main PCB processes this information and displays it on the LCD.

If, after a close inspection of your PCB, you don't find any analog output that makes sense regarding weight data, look at the back of the PCB. There's often a three-pin breakaway header (Figure 10-3) that provides an analog signal that changes depending on the state of the scale (off, weighing, weight found). You will see how to use this information later.

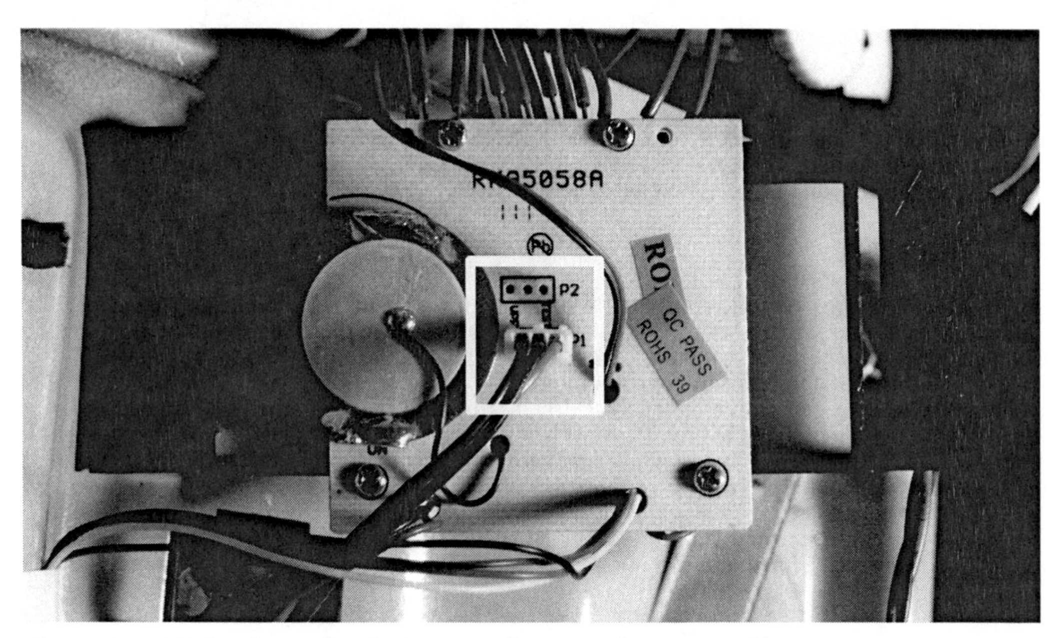

Figure 10-3. *PCB breakaway headers connected to an analog sensor cable*

Trying to get information from the strain gauges directly seems downright impossible, so why not let the PCB do its job and just sniff the weight information as it's passed to the LCD? This lets the PCB do all computation and means that you get the same numbers as are shown on the LCD.

LCD hacking is a very handy technique indeed; the same approach can be applied to any other appliance or device with a built-in LCD (which is pretty much every single device you can buy today). You might as well be getting the minutes remaining before you can eat your microwaved meal or the temperature in your home boiler. But for the moment, let's stick to the weight on your bathroom scale.

Seven-Segment LCD

Most bathroom scales, and actually most built-in LCDs, are seven-segment displays (Figure 10-4). This includes calculators and all kind of home appliances. They are widely used because this is a very optimized way of displaying decimals compared to dot-matrix displays, as every number can be encoded in one byte of information; seven bits defining the number, and one extra for the decimal point.

Figure 10-4. Individual segments (from Wikipedia), and the LCD on your scale.

The information is usually encoded in the order of *gfedcba* or *abcdefg*, but you can find other encodings, as you will see in this chapter. You have to check the signals from the PCB to the LCD to figure out the encoding of your scale, so first let's hack the scale.

Hacking the LCD

Once you have removed the cover and exposed the guts of your scale, you're probably looking at an LCD display attached to a printed circuit board (PCB). Unscrew the LCD from the PCB. There will be some form of connection between the LCD and the 16 pins on the PCB (Figure 10-5). In our case, it was connective rubber that we kept for future repositioning.

The ultimate goal here is to acquire and decode the information sent from the PCB to the LCD, so you solder one wire to each of the LCD pins. (Later you will try to figure out which ones define the eight bits of information relevant to you.) Use flexible wrap wire or a 16-wire ribbon cable, preferably; hook-up wire is too sturdy and could strip the PCB connectors off.

If there are connections on the PCB that you can use without impeding the repositioning of the LCD, use those so you can still read the numbers on your scale as depicted in Figure 10-6. This will be extremely helpful when analyzing the information from the pins.

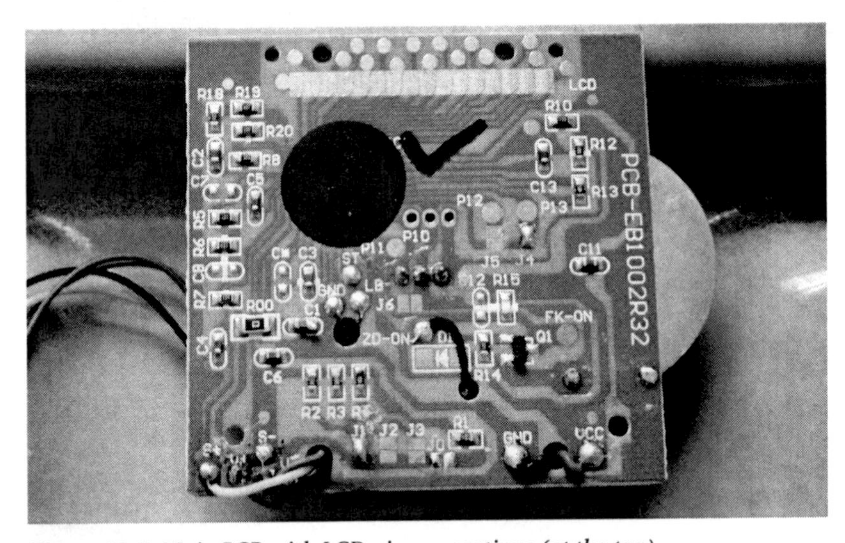

Figure 10-5. Main PCB with LCD pin connections (at the top)

Test that all pins are properly connected and there is no electrical flow between them. Then replace the LCD and the scale cover, leaving all the LCD wires and the analog cable you previously connected outside. You will probably need to carve an indent on the plastic cover with a scalpel so you can pass the wires through. Connect all the cables to an Arduino prototyping shield so they are ready to be easily plugged into your board (Figures 10-6 through 10-8).

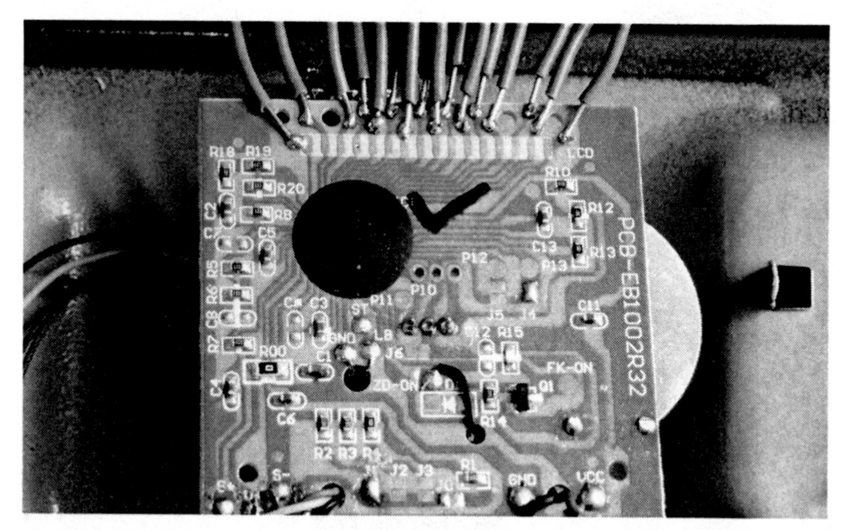

Figure 10-6. *Wires soldered to LCD pins. The rubber connection can still be plugged in.*

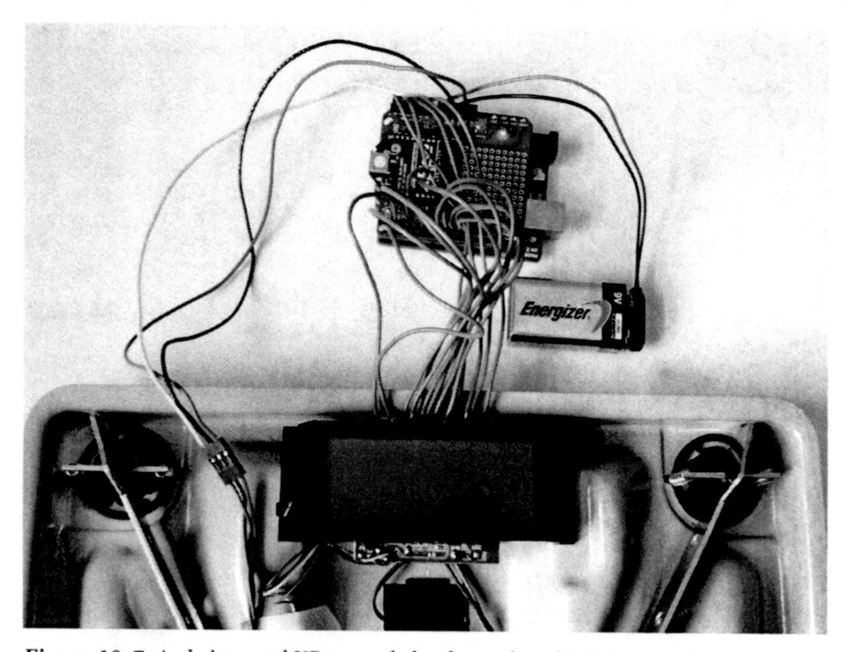

Figure 10-7. *Arduino and XBee module plugged and LCD repositioned*

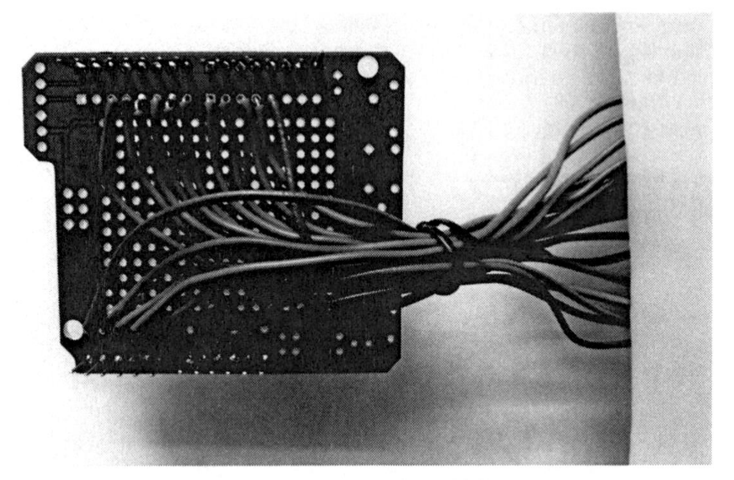

Figure 10-8. Arduino Prototype Board connected to all the LCD pins and the PCB

Add an XBee module to the board, as explained in Chapter 7, and put everything into a transparent plastic box along with an external power supply (Figure 10-9). This way, you won't need to depend on the USB cable's length to use your scale.

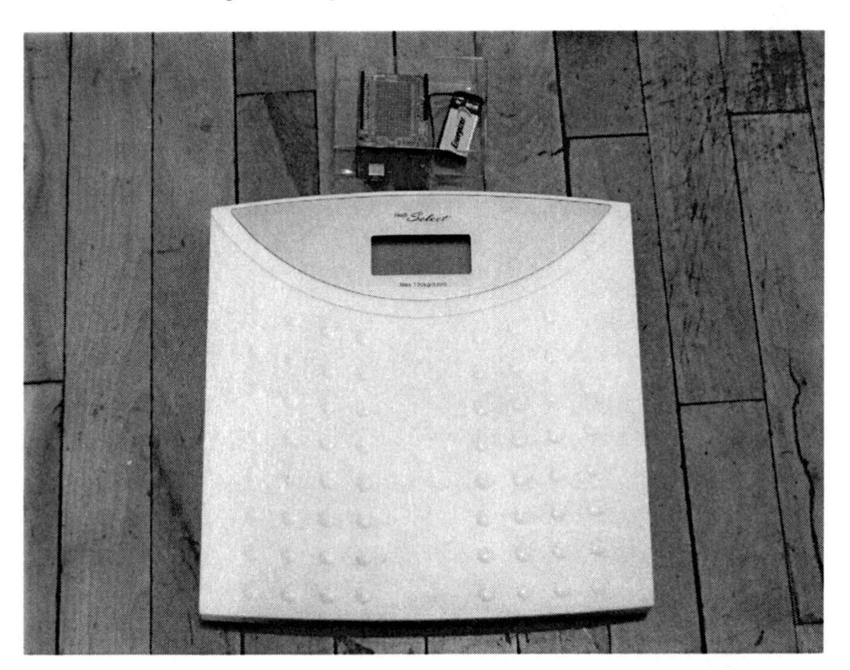

Figure 10-9. Scale finished and plugged to the Arduino and XBee module

Acquiring the LCD Signal

You have a pretty exciting cryptographic exercise ahead. When you connect your shield to the Arduino and read the values coming from the LCD pins, you will be overwhelmed by a constant flow of ones and zeros passing through your Serial Monitor. Don't despair; you are going to make sense of all this in the next few pages!

Connect your Arduino to your computer and start a new Arduino sketch. If you found a breakout header like the one shown previously, you might be able to power your scale from the Arduino; connect the pins to ground and 3.3V so you can get by without using a battery.

By now you should be acquainted with Arduino code, so you should have no problem reading through the following code.

```
#define dataSize 16
int data [dataSize];

void setup() {
  Serial.begin(9600);
  pinMode(2,INPUT);
  pinMode(3,INPUT);
  pinMode(4,INPUT);
  pinMode(5,INPUT);
  pinMode(6,INPUT);
  pinMode(7,INPUT);
  pinMode(8,INPUT);
  pinMode(9,INPUT);
  pinMode(10,INPUT);
  pinMode(11,INPUT);
  pinMode(12,INPUT);
}

void loop() {
  for(int i = 2; i<13; i++){
    data[i-2] = digitalRead(i);
  }
  data[11] = digitalRead(A5);
  data[12] = digitalRead(A4);
  data[13] = digitalRead(A3);
  data[14] = digitalRead(A2);
  data[15] = digitalRead(A1);

  for(int i = 0; i<dataSize; i++){
    Serial.print(data[i]);
  }
  Serial.println();
}
```

You are defining your pins as inputs (only the digital ones) and then storing their current digital readings in the array of integers you have called `data`. Note that the order is defined by the connections you used. In your Arduino, you connected the LCD pins left to right from Arduino pin `2` to `12` and then from analog `5` to `1`, hence the order of the readings.

You then print the resulting array to your serial port. If you open the Serial Monitor, you will be confronted with something similar to the following lines:

```
0000000011100000
1000000000000000
0000000001010000
0010000000000000
0000000011110000
0000000011100000
0100000000000000
```

This won't really make any sense to you yet, but you can start to identify a pattern. There are a series of sparse ones written on the left side of the string, and then some more tightly grouped at the center-right. Not much, but it's a start.

Reading the LCD Signal

Every line of code seems to be different, but an LCD should keep the pins high for a while, turn them off, and them on again for a certain time. Arduino should be faster than that, so you should get at least a couple of consecutive lines that look the same every time. You might have guessed the answer already: Arduino can indeed loop faster than the LCD cycle, but the serial printing slows it down considerably, so you don't get the same reading twice. Let's test this with some more sophisticated code.

You defining a bi-dimensional array named `serial` that will store fifty lines of LCD signals. Whenever it acquires the fifty lines, it prints them to the Serial Monitor. This way, you're not slowing down your Arduino every cycle, so you should be able to gather the information faster. When you print the information out, you will, of course, spend some time and certainly skip some cycles, but you will still have a consistent stream of data in your serial array.

```
#define dataSize 16
int data [dataSize];
#define lines 50
int serial [lines][dataSize];
int index;

void setup() {
  Serial.begin(9600);
  pinMode(2,INPUT);
  pinMode(3,INPUT);
  pinMode(4,INPUT);
  pinMode(5,INPUT);
  pinMode(6,INPUT);
  pinMode(7,INPUT);
  pinMode(8,INPUT);
  pinMode(9,INPUT);
  pinMode(10,INPUT);
  pinMode(11,INPUT);
  pinMode(12,INPUT);
}

void loop() {
  for(int i = 2; i<13; i++){
    data[i-2] = digitalRead(i);
```

```
  }
  data[11] = digitalRead(A5);
  data[12] = digitalRead(A4);
  data[13] = digitalRead(A3);
  data[14] = digitalRead(A2);
  data[15] = digitalRead(A1);

  for(int i = 0; i<dataSize; i++){
    serial[index][i] = data[i];
  }
  index++;
  if(index==lines){
    index=0;
    for(int i = 0; i<lines; i++){
      for(int j = 0; j<dataSize; j++){
        Serial.print(serial[i][j]);
      }
      Serial.println();
    }
    Serial.println("_________________");
  }
}
```

If you upload this sketch and open the Serial Monitor, you will get a surprisingly different panorama:

```
0000000001010000
0000000001010000
0000000001010000
0000000011110000
0000000011110000
________________
1000000000000000
1000000000000000
0000000000000000
0010000000000000
0010000000000000
```

You can see that the LCD pins are sending a pattern for a time and then changing to another pattern. You can't synchronize your Arduino to the LCD, but you can write code that reads the numbers only if they have changed and if they are maintained for at least two cycles (to avoid misreading at the transition point).

```
#define dataSize 16
int data [dataSize];
int previousData [dataSize];
#define lines 50
int serial [lines][dataSize];
int index;

// Booleans defining the change
boolean changed;
```

```
boolean scanning;

void setup() {
  Serial.begin(9600);
  pinMode(2,INPUT);
  pinMode(3,INPUT);
  pinMode(4,INPUT);
  pinMode(5,INPUT);
  pinMode(6,INPUT);
  pinMode(7,INPUT);
  pinMode(8,INPUT);
  pinMode(9,INPUT);
  pinMode(10,INPUT);
  pinMode(11,INPUT);
  pinMode(12,INPUT);
}

void loop() {
  for(int i = 2; i<13; i++){
    data[i-2] = digitalRead(i);
  }
  data[11] = digitalRead(A5);
  data[12] = digitalRead(A4);
  data[13] = digitalRead(A3);
  data[14] = digitalRead(A2);
  data[15] = digitalRead(A1);
```

Check if the numbers have changed from the previous reading

```
  int same = 0;
  for(int i = 0; i<dataSize; i++){
    if(data[i]==previousData[i])same++;
    previousData[i] = data[i];
  }
  if(same<dataSize){
    changed=true;
    scanning=true;
  }
  else{
    changed=false;
  }
```

If the numbers have NOT changed and if they have occurred at least twice (so scanning is `true`), store the the data into the `serial[]` array.

```
  if(changed == false && scanning == true){
   for(int i = 0; i<dataSize; i++){
      serial[index][i] = data[i];
    }
    index++;
    scanning=false;
  }
```

Finally, if you have reached the buffer limit, reset the index to 0 and run a `for` loop to print each line in your `serial[]` array to `serial`.

```
  if(index==lines){
    index=0;
    for(int i = 0; i<lines; i++){
      for(int j = 0; j<dataSize; j++){
        Serial.print(serial[i][j]);
      }
      Serial.println();
    }
    Serial.println("_________________");
  }
}
```

This will actually give you the key to reading the scale's LCD signals. If you open your Serial Monitor, you now see a clear pattern in the flow of data streaming from the scale.

```
1000000000000000
0010000000000000
0001000000000000
0000000011100000
0000000011110010
0000000001010000
0000000011110000
0100000000000000
```

This is your data stream when the scale is showing a 0.0 kg signal. If you settle on another weight, the signal will look differently. Let's try 67.4kg.

```
1000000000000000
0010000000000000
0001000000000000
0000001001000000
0000001101010010
0000001001110000
0000001111100000
0100000000000000
```

Well, that's good; you can now see that there is a pattern that repeats itself at the far-left of your strings, and then some numbers that change at the center. The ones appearing on the right side are actually defining if the kg or lb symbol is turned on.

After quite a lot of testing, we were able to identify the pins that carried the information on the digits and the ones that dealt with other elements. You should run a similar trial and error process to find out the order in your particular scale. This was the one in ours.

Pin	0	1	2	3	4	5	6	7	8	9	10	11	12	13	14	15
	1	0	0	0	0	0	0	0	0	0	0	0	0	0	0	0
	0	0	**1**	0	0	0	0	0	0	0	0	0	0	0	0	0
	0	0	0	**1**	0	0	0	0	0	0	0	0	0	0	0	0

The first three rows don't change with the weight, so they can be used to identify the start of the message when pin 3 is turned on.

```
0  0  0  0  |  0  0  |  1  0  |  0  1  |  0  0  |  0  0  0  0
0  0  0  0  |  0  0  |  1  1  |  0  1  |  0  1  |  0  0  1  0
0  0  0  0  |  0  0  |  1  0  |  0  1  |  1  1  |  0  0  0  0
0  0  0  0  |  0  0  |  1  1  |  1  1  |  1  0  |  0  0  0  0
```

After the three first rows have passed, the fifth to eighth rows carry the important information. Columns 4-5 each consist of 8 bits, defining the hundreds. Columns 6-7 define the tens, 8-9 the units, and 10-11 the decimals. Column 14 provided information on the unit used (kg or lbs).

```
0  1  0  0  |  0  0  |  0  0  |  0  0  |  0  0  |  0  0  0  0
```

The last row is always constant and can be used to identify the end of the message.

Sending the Signal to Processing

You have acquired the information sent to the LCD display. You now want to decode it so you can work with the weight. You are going to send the raw information to Processing and write a decoder to transform this information into numbers.

Stop scanning columns 12, 13, and 15 because they don't add any useful information. Write a routine that checks the start of the message (one on column 3) and the end of the message (one on column 2) and sends to serial all the rows in between. Also send the information coming from analog pin 0 connected to the PCB's analog output that changes with the state of the weighing process.

```
#define dataSize 13
int data [dataSize];
int previousData [dataSize];
#define lines 4
int serial [lines][dataSize];
int index;
// Booleans defining the change
boolean changed;
boolean scanning;
boolean start;
int state;
int previousState;

void setup() {
  Serial.begin(9600);
  pinMode(2,INPUT);
  pinMode(3,INPUT);
  pinMode(4,INPUT);
  pinMode(5,INPUT);
  pinMode(6,INPUT);
  pinMode(7,INPUT);
  pinMode(8,INPUT);
  pinMode(9,INPUT);
  pinMode(10,INPUT);
  pinMode(11,INPUT);
  pinMode(12,INPUT);
}

void loop() {
```

```
  for(int i = 2; i<13; i++){
    data[i-2] = digitalRead(i);
  }
  data[11] = digitalRead(A5);
  data[12] = digitalRead(A2);
  state = analogRead(A0);

  // Check if the numbers have changed from the previous reading
  int same = 0;
  for(int i = 0; i<dataSize; i++){
    if(data[i]==previousData[i])same++;
    previousData[i] = data[i];
  }
  if(same<dataSize){
    changed=true;
    scanning=true;
  }
  else{
    changed=false;
  }

  if(changed == false && scanning == true){
    if(data[1] == 1){
      start=false;
      Serial.println('S');
      for(int i = 0; i<4; i++){
        for(int j = 0; j<dataSize; j++){
          Serial.print(serial[i][j]);
        }
        Serial.println();
      }
      Serial.print('X');
      Serial.println(state);
    }
    if(start){  // If you have started the message
      for(int i = 0; i<dataSize; i++){
        char number[1];
        // Serial.print(itoa(data[i], number, 10));
        //        Serial.print(" ");
        serial[index][i]=data[i];
      }
      index++;
    }
    scanning=false;
    if(data[3] == 1){
      start=true;
      index = 0;
    }
  }
}
```

This code will output the following data stream to the serial buffer:

```
S
0000000001100
0000000001111
0000000011110
0000000010110
X532
S
0000000001100
0000000001111
0000000011110
0000000010110
X532
```

You have enough data to identify the start and end of the information package, and all the necessary data to recompose the weight information at the other end of the line.

Decoding the LCD Signal

The decoding process is far simpler than the acquisition of the information. At this point, you can easily write a Processing program that receives the information from Arduino and decodes it into numbers you can understand. You are going to pack this process into a class so you can reuse the implementation in different applications.

This class is called LCD. You input a two-dimensional array defining the individual number on the seven-segment display and you get a float defining the weight.

```
public class LCD {
  private float data;
  int lcdArray[][] = new int[4][13];
  private String units = "kg";
  private int state;
  boolean foundWeight;
  boolean scanning;

  int[] zero =          { 1, 1, 1, 0, 1, 1, 1 };
  int[] one =           { 0, 0, 1, 0, 1, 0, 0 };
  int[] two =           { 1, 1, 0, 1, 1, 0, 1 };
  int[] three =         { 1, 0, 1, 1, 1, 0, 1 };
  int[] four =          { 0, 0, 1, 1, 1, 1, 0 };
  int[] five =          { 1, 0, 1, 1, 0, 1, 1 };
  int[] six =           { 1, 1, 1, 1, 0, 1, 1 };
  int[] seven =         { 0, 0, 1, 0, 1, 1, 1 };
  int[] eight =         { 1, 1, 1, 1, 1, 1, 1 };
  int[] nine =          { 1, 0, 1, 1, 1, 1, 1 };
```

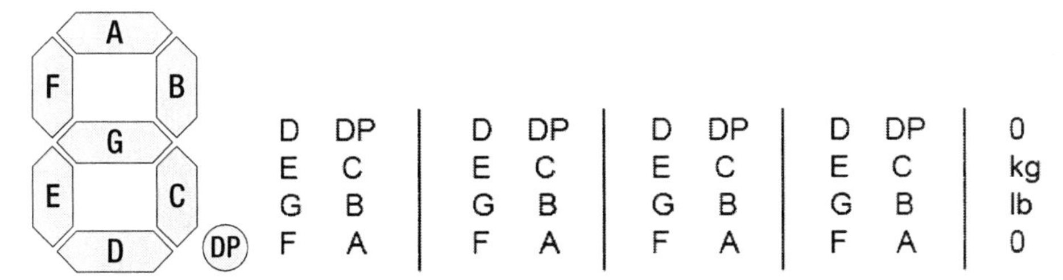

***Figure 10-10.** Seven-segment display code*

The arrays you just defined are a blueprint to check your data against. They are based on the distribution of the seven-segment bits coming from Arduino. After, again some intensive testing sessions, we found the pattern depicted in Figure 10-10.

The method getReading() returns the real value of the weight.

```
  public float getReading() {
    return data;
  }
```

Use the method setLcdArray() to input the new values from Arduino into your LCD class to keep the numbers updated. You convert every character to an integer, store them in the lcdArray, and run the update() method to update your real value.

```
  public void setLcdArray(char[][] stringData) {
    if (stringData[1][12] == '1') {
      units = "kg";
    }
    else if (stringData[2][12] == '1') {
      units = "lb";
    }
  for (int i = 0; i < stringData.length; i++) {
        for (int j = 0; j < stringData[0].length; j++) {
          lcdArray[i][j] = Character.digit(stringData[i][j], 10);
        }
      }
    this.update();
  }
```

Within the update function, you reorganize your main two-dimensional array into a three-dimensional array that stores the four two-dimensional arrays defining each of the digits.

```
public void update() {
    int[] digits = new int[4];
    int[][][] segments = new int[4][4][2];

    for (int i = 0; i < lcdArray.length; i++) {
      for (int j = 4; j < 6; j++) {
        segments[0][i][j - 4] = lcdArray[i][j];
      }
      for (int j = 6; j < 8; j++) {
        segments[1][i][j - 6] = lcdArray[i][j];
      }
      for (int j = 8; j < 10; j++) {
        segments[2][i][j - 8] = lcdArray[i][j];
```

```
      }
      for (int j = 10; j < 12; j++) {
        segments[3][i][j - 10] = lcdArray[i][j];
      }
    }
```

Then you use the function getNumber(), passing each of the bi-dimensional arrays and thus getting an array of the individual numbers displayed on the LCD.

```
    for (int i = 0; i < digits.length; i++) {
      digits[i] = getNumber(segments[i]);
    }
```

Lastly, you recompose the individual digits into the real value of the weight by multiplying the hundreds by 100, the tens by 10, the units by 1, and the decimals by 0.1. This real value is stored in the float variable data.

```
    data = digits[0] * 100 + digits[1] * 10 + digits[2] + digits[3] * 0.1;
  }
```

You use the function getNumber() to get the decimal number represented by each of your seven-segment bi-dimensional arrays. This process starts by extracting the seven relevant bits of the incoming array and storing them in a linear array. You compare this array to each one of the number blueprints that you defined previously. If it matches one of them, the corresponding number will be returned.

```
  public int getNumber(int segments[][]) {
    int flatSegment[] = new int[7];
    for (int i = 0; i < segments.length; i++) {
      for (int j = 0; j < segments[0].length; j++) {
        if (i + j == 0) {
          flatSegment[i * 2 + j] = segments[i][j];
        }
        else if (!(i == 0 && j == 1)) {
          flatSegment[i * 2 + j - 1] = segments[i][j];
        }
      }
    }
    if (Arrays.equals(flatSegment, zero)) { return 0; }
    else if (Arrays.equals(flatSegment, one))   { return 1; }
    else if (Arrays.equals(flatSegment, two))   { return 2; }
    else if (Arrays.equals(flatSegment, three)) { return 3; }
    else if (Arrays.equals(flatSegment, four))  { return 4; }
    else if (Arrays.equals(flatSegment, five))  { return 5; }
    else if (Arrays.equals(flatSegment, six))   { return 6; }
    else if (Arrays.equals(flatSegment, seven)) { return 7; }
    else if (Arrays.equals(flatSegment, eight)) { return 8; }
    else if (Arrays.equals(flatSegment, nine))  { return 9; }
    else {
      return 0;
    }
  }
```

The setState() function is used to set the state of the scale from the main sketch. You perform a check of the state of the scale and set the scanning and foundWeight Boolean values accordingly.

```
  public void setState(int state) {
    this.state = state;
```

```
    if (state < 518 && state > 513) {
      scanning = true;
    }
    if (state < 531 && state > 526 && scanning) {
      foundWeight = true;
    }
  }
```

The following function displays the raw data in array form on screen.

```
  public void displayArray(int x, int y) {
    pushMatrix();
    translate(x, y);
    for (int i = 0; i < lcdArray[0].length; i++) {
      for (int j = 0; j < lcdArray.length; j++) {
        text(lcdArray[j][i], 20 * i, j * 20);
      }
    }
    popMatrix();
  }
}
```

Note that the last curly brace the closing curly brace for the whole class.

Using the Weight Data

You are going to write a first simple sketch to test the decoding process. Then you will use it with your Kinect data, so let's check that it works properly before starting to make things more complex.

Start a new sketch and create a separate tab in which you paste the LCD class. Then go back to the main tab and add all the necessary code to acquire serial communication. Then declare and initialize the LCD object, and declare a character array to store the incoming data.

```
import processing.serial.*;
Serial myPort;
PFont font;
LCD lcd;
char[][] stringData = new char[4][13];
int currentLine;

public void setup() {
  size(400, 300);
  font = loadFont("SansSerif-14.vlw");
  textFont(font);
  String portName = Serial.list()[0]; // This gets the first port on your computer.
  myPort = new Serial(this, portName, 9600);
  myPort.bufferUntil('\n');
  lcd = new LCD();
}
```

The main draw() function updates the LCD with the latest data from Arduino, displays the array data, and draws the result weight on the screen, as shown in Figure 10-11.

```
public void draw() {
  background(0);
  lcd.setLcdArray(stringData);
```

```
  lcd.displayArray(50, 50);
  text(lcd.data + " " + lcd.units, 50, 200);
}
```

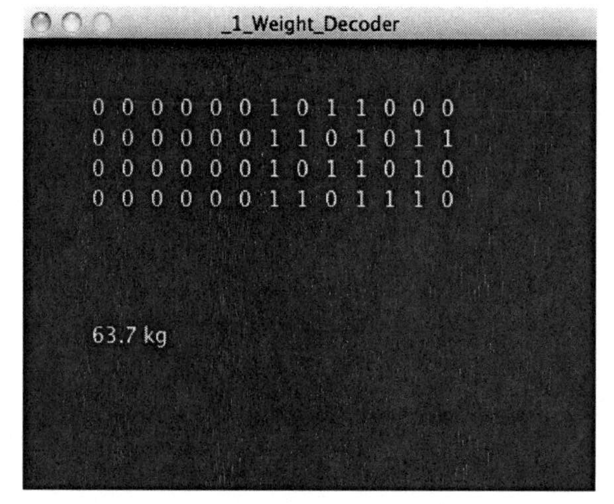

Figure 10-11. *Testing the LED data decoder*

You have worked with the serialEvent() function previously in this book, so it shouldn't be completely unknown to you. Trim the newline character from your incoming string, and then check whether you have received a start or end of communication character.

```
public void serialEvent(Serial myPort) {
String inString = myPort.readString();
inString = inString.substring(0, inString.indexOf('\n') - 1);
```

If you receive an S, set the current line to 0, so all new data is stored on the first line of your data array.

```
  if (inString != null) {
    if (inString.equals("S")) {
      currentLine = 0;
    }
```

If the first character of your incoming string is an X, this means you are done with this reading. The numbers following the X are the reading of your LCD state. Pass that value to the LCD object.

```
    else if (inString.charAt(0) == 'X') {
      String newState = inString.substring(1);
      lcd.setState(Integer.valueOf(newState));
    }
```

Otherwise, check if the length of the array is correct, that you are not beyond the fourth line, and store the values of the incoming characters into the stringData array.

```
    else {
      if (inString.length() == 13 && currentLine<4) {
        for (int i = 0; i < stringData[0].length; i++) {
          stringData[currentLine][i] = inString.charAt(i);
```

```
        }
        currentLine++;
      }
    }
  }
  else {
    // increment the horizontal position:
    println("No data to display");
  }
}
```

And that's all! When you run this you should see the array of ones and zeros on the screen defining the LCD pins information and the real value of the weight displayed on the scale. You're done with the weight data acquisition; let's have fun with it and your Kinect data!

Implementing the Biometric Recognition

You have probably had enough of playing with your scale by now, so you're going to plunge straight into the final piece of code that will use skeleton tracking to recognize users and the weight input to generate body mass index (BMI) values. The basic functioning of the program is the following; whenever a user enters the field of vision of Kinect, you start tracking him. Find the highest and lowest points defining your user and extract the height of the person from those points. At the same time, you skeletonize the user and extract the proportions of their limbs to use as a footprint to match the user with previously saved information. If the footprint is similar to an existing user, you recognize him and retrieve the data from previous readings. You also have an input interface that allows you to create new users, stating their names and dates of birth. Once you have all of the skeleton and height data, the user steps on the bathroom scale and you read his weight. You then save all the current information in an external CSV file and calculate the person's BMI and display it on screen.

■ **Note** A CSV (comma-separated values) file is a file format that stores tabular data separated by commas. It is used extensively in the transfer of data between applications.

Controlling the height and weight of children from early life up to adulthood can be a good way to keep track of adequate development, so for the sake of this exercise, compare your data to the male BMI-for-age chart for children less than 20 years old. You can find this chart at `http://www.cdc.gov/growthcharts/clinical_charts.htm`. This chart applies to US citizens; you can probably find equivalent charts from your government's health program.

Imports and Variable Declaration

The first thing you need to do in your Processing sketch is to import the external libraries that you need for this process. You are building a basic user interface, so import Java AWT and `AWT.event`, which are part of the Java Foundation Classes and allow you to interact with your sketch in runtime by introducing new users.

```
import processing.serial.*;
import SimpleOpenNI.*;
import java.awt.*;
```

```
import java.awt.event.*;

SimpleOpenNI kinect;
Serial myPort;
PFont font;
PFont largeFont;
LCD lcd;
```

You are going to be juggling quite a few external files, so next you define an array called `files` that stores all the previous user data files that exist in the data folder. You also declare a current user file, two AWT `TextFields`, and one button for the user interface.

```
// User Input and Files
File[] files; // Existing user files
File userFile; // Current user
TextField name;
TextField birthdate;
Button b = new Button("New User");
```

The serial variables are similar to previous examples. The `stringData` array stores the incoming information from Arduino and one integer, `currentLine`, which you use in the process of parsing the incoming data. The Boolean `scan` is set to `true`, and then it is changed to `false` once you have acquired the final weight information.

```
// Serial Variables
boolean serial = true;
char[][] stringData = new char[4][13];
int currentLine;
boolean scan = true;
```

The numeric user data is stored in the array of floats `userData`. You also store some user information in `String` format, such as the name and date of birth.

```
// User Data
float[] userData = new float[11];
int user; // Current user ID
String userName = "new User";
String userBirthDate = "00/00/0000";
int userAge;
float userBMI;
float userHeightTemp;
```

You declare a `double` array to keep the data in the chart that you want to display, and a `File` object defining the chart CSV file. The String `today` contains the current date.

```
// Charts
float[][] chart;
File chartFile;
String today;
```

Setup Function

The `setup()` function is slightly longer than usual due to the user interface elements and the objects managing the external files. The first half of the function deals with the Simple-OpenNI object and the Processing fonts.

Along with the depthMap and the user capabilities, you enable the scene from NITE. The scene allows you to separate the background from the foreground user. This means you can select the 3D points that make up the user's volume and extract dimensions from it, namely the user height (Figure 10-12).

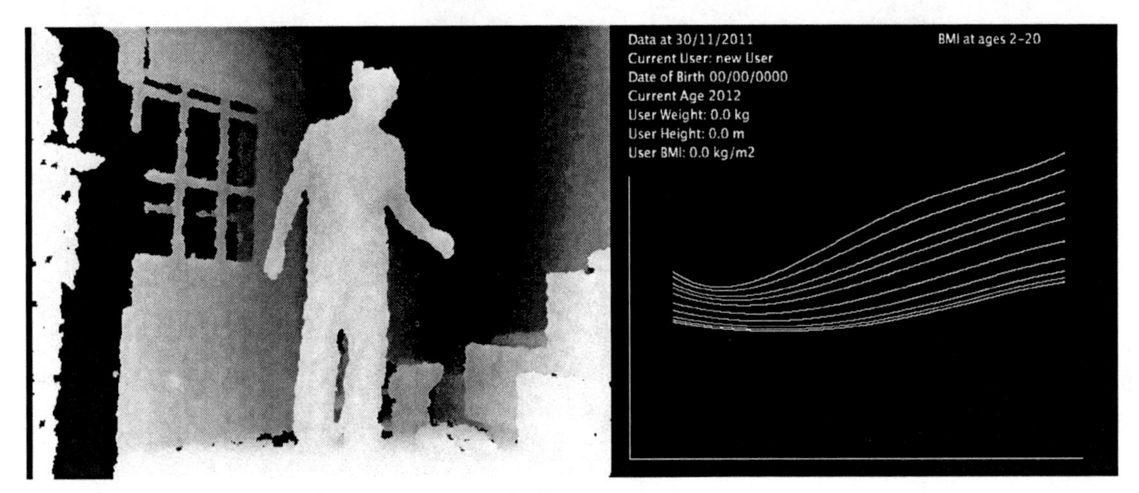

Figure 10-12. *Kinect scene: the user is automatically recognized.*

```
public void setup() {
  size(1200, 480);
  smooth();

  kinect = new SimpleOpenNI(this);
  kinect.setMirror(true);
  kinect.enableDepth();
  kinect.enableScene();
  kinect.enableUser(SimpleOpenNI.SKEL_PROFILE_ALL);

  font = loadFont("SansSerif-12.vlw");
  largeFont = loadFont("SansSerif-14.vlw");
  textFont(largeFont);
  lcd = new LCD();

  if (serial) {
    String portName = Serial.list()[0]; // This gets the first port
    myPort = new Serial(this, portName, 9600);
    myPort.bufferUntil('\n');   // don't generate a serialEvent() unless you get a newline
  }
```

User Interface

Now you initialize the AWT objects that constitute your user interface (Figure 10-13). The input is based on two Java AWT TextFields and one button. You actually position these three elements out of the Processing frame, right under it. When one user is detected, you resize the height of the Processing frame, but not the applet, so you then expose the part of the frame containing the interface. The two text fields contain the name and date of birth of the new user.

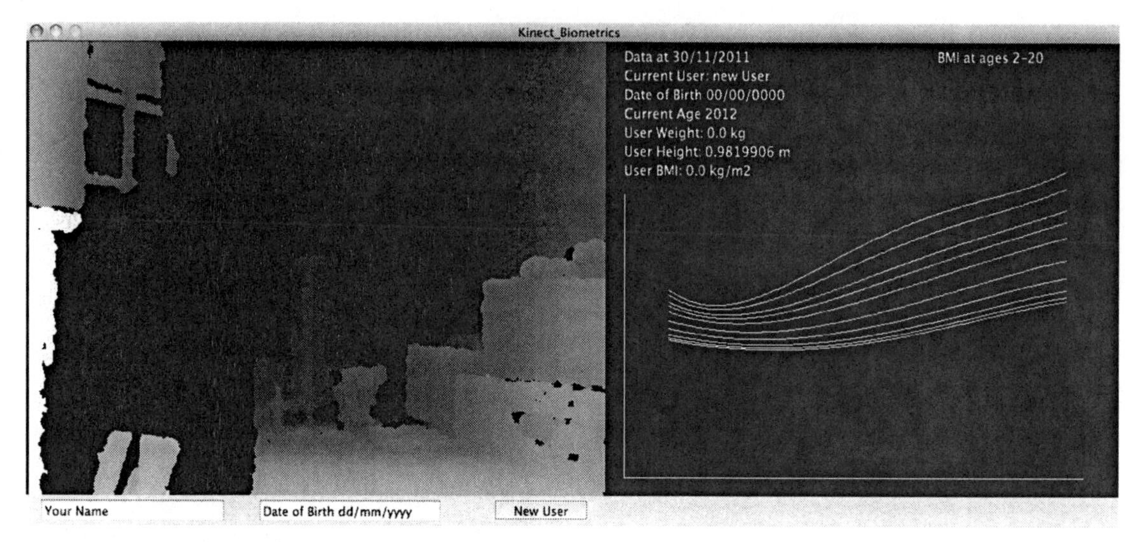

Figure 10-13. User interface

```
// Setting Up the User Input Elements
name = new TextField(40);
birthdate = new TextField(40);
frame.add(name);
frame.add(birthdate);
frame.add(b);
name.setBounds(20, height + 27, 200, 20);
name.setText("Your Name");
birthdate.setBounds(260, height + 27, 200, 20);
birthdate.setText("Date of Birth dd/mm/yyyy");
b.setBounds(520, height + 27, 100, 20);
```

You need to add an action listener to your button so an action is triggered when you press it. In your case, you store the user name and date of birth and you call the function `newUser()`, passing these strings as parameters.

```
b.addActionListener(new ActionListener() {
  public void actionPerformed(ActionEvent e) {
    userName = name.getText();
    userBirthDate = birthdate.getText();
    newUser(userName, userBirthDate);
  }
}
);
```

You want to be able to recognize users created in previous sessions. For this, you need to scan the data folder in search of CSV files starting with the characters "user_". Use Java's `FilenameFilter` for this purpose. The following lines first get a list of all files in the data folder and then apply the filter and store the resulting files into the File array `files`. You then print the file names to the console for debugging.

```
// Acquire the user files already on the folder
File directory = new File(dataPath(""));
FilenameFilter filter = new FilenameFilter() {
```

```
    public boolean accept(File dir, String name) {
      return name.startsWith("user_");
    }
  };
  files = directory.listFiles(filter);
  println("List of existing user files");
  for (File f : files) {
    println(f);
  }
```

To finish the setup() function, you choose the chart you are going to be comparing your results to and trigger the function readChart(), which retrieves all the data from the file. You also get the current date with the function getDate().

```
  // Acquire the chart file
  chartFile = new File(dataPath("charts/Male_2-20_BMI.CSV"));
  readChart(chartFile);
  getDate();
}
```

Draw Function

The draw() loop, in contrast to the setup() function, is quite simple because you have implemented the main routines into separate functions that are detailed next. The first thing you do in each loop is to update the data from the Simple-OpenNI and LCD objects.

```
public void draw() {
  kinect.update();
  background(0);
  lcd.setLcdArray(stringData);
  image(kinect.sceneImage(), 0, 0);
```

Then, if you have detected a user, you update the height data, as you are taking it from the user point cloud.

```
  if (kinect.getNumberOfUsers() != 0) {
    updateUserHeight();
  }
```

If you are tracking a skeleton as well, you draw it on screen and update the user data from his limbs proportions.

```
if (kinect.isTrackingSkeleton(user)) {
    drawSkeleton(user);
    updateUserData(user);
  }
```

Then, if you are still scanning the weight (scan is true), and the LCD object's variable foundWeight is true (meaning you got a steady weight), you try to recognize the user from previous users, and then you store the first piece of data (the weight from the scale) in the first position of your userData array and the height in its second position. You then update the BMI value and save the user for future retrieval. You set the scan Boolean to false so you stop updating the values.

```
   if (lcd.foundWeight && scan) {
    userRecognition();
    userData[0] = lcd.data;
```

```
    userData[1] = userHeightTemp;
    userBMI = userData[0]/pow(userData[1] /1000, 2);
    saveUser();
    scan = false;
  }
```

And finally, you take care of the visualization by printing the chart and the user data on the right side of your Processing screen.

```
  printUserData();
  drawChart();
}
```

Additional Functions

You probably guessed it: a short draw loop means a quite extensive list of additional functions! Some of these functions are very similar to code used in previous examples, so we won't go into detail on those. Others are more complex and will be described exhaustively, explaining what they do and when they are called.

Before you get into the user functions, there is a helper function called getDate() that uses the Java Calendar class to fetch the current date and store it in the variable today. It also calculates the age of the current user by subtracting the user's date of birth from the current date.

```
public void getDate() {
  Calendar currentDate = Calendar.getInstance();
  SimpleDateFormat formatter = new SimpleDateFormat("dd/MM/yyyy");
  today = formatter.format(currentDate.getTime());
  Date todayDate = currentDate.getTime();
  Date userDate = null;
  try {
    userDate = formatter.parse(userBirthDate);
  }
  catch (ParseException e) {
    e.printStackTrace();
  }
  long diff = todayDate.getTime() - userDate.getTime();
  userAge = (int) (diff/31556926000l);
}
```

New User

The newUser() function is called whenever you press the New User button in your user interface. The user name and date of birth are passed as arguments.

You first need to check if you have pressed the button without entering a name or a date, in which case you prompt the user to insert the data. Then, you use the Java function DateFormat() to check whether the date that has been inserted corresponds to the format you are expecting. If it doesn't, a parse exception is thrown and caught by your try-catch statement, and you prompt the user to insert a properly formatted date of birth (which doesn't mean it is the real one anyway...).

If the data is correct, you proceed to check whether there is already a user with that name in the system, in which case you simply use that file for the rest of the time. If it doesn't exist, you create it and add the user name and date of birth to the first line in the file. You finally flush and close the PrintWriter.

```
public void newUser(String name, String birth) {
  // Start the user file
  if (name.equals("Your Name") || birth.equals("Date of Birth dd/mm/yyyy")) {
    println("Please insert your name and date of Birth");
  }
  else {
    DateFormat format = new SimpleDateFormat("dd/MM/yyyy");
    try {
      format.parse(birthdate.getText());
      PrintWriter output;
      userFile = new File(dataPath("user_" + userName + ".csv"));
      if (!userFile.exists()) {
        output = createWriter(userFile);
        output.println(name + "," + birth);
        output.flush(); // Write the remaining data
        output.close(); // Finish the file
        getDate();
        println("New User " + name + " created successfully");
      }
      else {
        setPreviousUser(userFile);
      }
    }
    catch (ParseException e) {
      System.out
        .println("Date " + birthdate.getText()
        + " is not valid according to "
        + ((SimpleDateFormat) format).toPattern()
        + " pattern.");
    }
  }
}
```

Updating User Data

You call updateUserData() from the draw() loop in every cycle if you are tracking a skeleton. This function first stores all the real world coordinates of the user's joints; then it calculates the dimensions of arms, forearms, thighs, shins, and shoulder breadth; and then it stores all of this data in the userData array, starting from the third position (the first two are occupied by the weight and height).

This array serves later as an identifier for the user. The body proportions stored in the userData array can be matched and compared to other users' identifiers and thus the person tracked can be recognized. The first step is to initialize all the PVectors that store the location of the user's joints.

```
void updateUserData(int id) {
  // Left Arm Vectors
  PVector lHand = new PVector();
  PVector lElbow = new PVector();
  PVector lShoulder = new PVector();

  // Left Leg Vectors
  PVector lFoot = new PVector();
  PVector lKnee = new PVector();
  PVector lHip = new PVector();
```

```
// Right Arm Vectors
PVector rHand = new PVector();
PVector rElbow = new PVector();
PVector rShoulder = new PVector();

// Right Leg Vectors
PVector rFoot = new PVector();
PVector rKnee = new PVector();
PVector rHip = new PVector();
```

Then you use the getJointPositionSkeleton() function to get the real world coordinates into the vectors.

```
// Left Arm
kinect.getJointPositionSkeleton(id, SimpleOpenNI.SKEL_LEFT_HAND, lHand);
kinect.getJointPositionSkeleton(id, SimpleOpenNI.SKEL_LEFT_ELBOW, lElbow);
kinect.getJointPositionSkeleton(id, SimpleOpenNI.SKEL_LEFT_SHOULDER, lShoulder);
// Left Leg
kinect.getJointPositionSkeleton(id, SimpleOpenNI.SKEL_LEFT_FOOT, lFoot);
kinect.getJointPositionSkeleton(id, SimpleOpenNI.SKEL_LEFT_KNEE, lKnee);
kinect.getJointPositionSkeleton(id, SimpleOpenNI.SKEL_LEFT_HIP, lHip);
// Right Arm
kinect.getJointPositionSkeleton(id, SimpleOpenNI.SKEL_RIGHT_HAND, rHand);
kinect.getJointPositionSkeleton(id, SimpleOpenNI.SKEL_RIGHT_ELBOW, rElbow);
kinect.getJointPositionSkeleton(id, SimpleOpenNI.SKEL_RIGHT_SHOULDER, rShoulder);
// Right Leg
kinect.getJointPositionSkeleton(id, SimpleOpenNI.SKEL_RIGHT_FOOT, rFoot);
kinect.getJointPositionSkeleton(id, SimpleOpenNI.SKEL_RIGHT_KNEE, rKnee);
kinect.getJointPositionSkeleton(id, SimpleOpenNI.SKEL_RIGHT_HIP, rHip);
```

And finally, you store the distances between certain joints in the userData array.

```
  userData[2] = PVector.dist(rHand, rElbow);           // Right Arm Length
  userData[3] = PVector.dist(rElbow, rShoulder);               // Right Upper Arm Length
  userData[4] = PVector.dist(lHand, lElbow);           // Left Arm Length
  userData[5] = PVector.dist(lElbow, lShoulder);               // Left Upper Arm Length
  userData[6] = PVector.dist(lShoulder, rShoulder);    // Shoulder Breadth
  userData[7] = PVector.dist(rFoot, rKnee);            // Right Shin
  userData[8] = PVector.dist(rHip, rKnee);             // Right Thigh
  userData[9] = PVector.dist(lFoot, lKnee);            // Left Shin
  userData[10] = PVector.dist(lHip, lKnee);            // Left Thigh
}
```

Updating User Height

The detection of the user's height is not based on skeleton tracking but directly on the point cloud, so you implement a separate function for it. This function is called from draw in case you're tracking a user in the scene.

You have previously worked with functions that parsed and printed on screen the Kinect point cloud. This function is similar, but you won't print any of the points. What you want to do is to find out the higher and lower points from the user point cloud so you can subtract their Y coordinates and thus find the real height of the user in millimeters (Figure 10-14).

Figure 10-14. User height from highest and lowest points

You will only use one of every three points to speed up the process; if you want a more precise measurement, you can decrease the steps to 1. Use userMap from Simple-OpenNI to get a flat image of the points that define the user.

```
void updateUserHeight() {
  int steps = 3;
  int index;
  int[] userMap = kinect.getUsersPixels(SimpleOpenNI.USERS_ALL);
  PVector userCenter = new PVector();
  PVector[] realWorldPoint = new PVector[kinect.depthHeight() * kinect.depthWidth()];
```

Then you initialize the userTop and userBottom PVectors to the center of the mass of the user, so you can start looking for a top and bottom pixel from a neutral ground.

```
  kinect.getCoM(1, userCenter); // Get the Center of Mass
  PVector userTop = userCenter.get();
  PVector userBottom = userCenter.get();
```

And you run a loop checking every point in the depth map looking for user points. You determine that a point belongs to a user if the equivalent pixel in the userMap is not black. If you find a user point, you check if it's higher than the higher pixel found or lower than the lowest pixel found; in positive case, you set it as the highest or lowest point.

```
  for (int y = 0; y < kinect.depthHeight(); y += steps) {
    for (int x = 0; x < kinect.depthWidth(); x += steps) {
      index = x + y * kinect.depthWidth();
      realWorldPoint[index] = kinect.depthMapRealWorld()[index].get();
      if (userMap[index] != 0) {
        if (realWorldPoint[index].y > userTop.y)
          userTop = realWorldPoint[index].get();
        if (realWorldPoint[index].y < userBottom.y)
          userBottom = realWorldPoint[index].get();
      }
    }
  }
```

Once you have found the highest and lowest points, you store the value of the difference of their Y coordinates into `userHeightTemp`. But you won't do it straight away. You want to smooth the noisy data to avoid continuous oscillations of the value, so you use Processing's `lerp()` function to progressively make the transition from one value to the next one.

```
  userHeightTemp = lerp(userHeightTemp, userTop.y - userBottom.y, 0.2f);
}
```

Saving the User

If you have successfully created a user and tracked his weight, height, and proportions, now you want to store this information into the previously created CSV file (Figure 10-15).

For this, you create a `File` object with the name of the current user, and if you find that it doesn't exist yet, you prompt the user to insert his name and birth date. If the file does exist, use a Java `BufferedWriter` to append information to the file without wiping it out.

Enrique	12/11/1982										
29/11/2011	63.0	1686.2014	268.18143	227.60263	263.1212	236.44228	279.46335	305.47165	396.49283	270.1577	393.8149

Figure 10-15. *Structure of the user information CSV file*

```
private void saveUser() {
  userFile = new File(dataPath("user_" + userName + ".csv"));
  if (!userFile.exists()) {
    println("Please Enter your Name and Date of Birth");
  }  else {
    try {
      FileWriter writer = new FileWriter(userFile, true);
      BufferedWriter out = new BufferedWriter(writer);
      out.write(today);
      for (int i = 0; i < userData.length; i++) {
        out.write("," + userData[i]);
      }
      out.write("\n");
      out.flush(); // Write the remaining data
      out.close(); // Finish the file
    }
    catch (IOException e) {
      e.printStackTrace();
    }
    println("user " + userName + " saved");
  }
}
```

User Recognition

Now you have all the data you need from your current user, and you want to know if the same user was previously registered in the system. This means that there is a CSV file containing very similar limb lengths to the ones of the current user.

In order to find this out, you make use of the generalization of the concept of vector subtraction to a higher-dimensional space. What? Well, the next few lines are a technical explanation of how to measure the similitude between two users based on the skeleton data. If you are not that interested in multidimensional vector operations, simply skip the math and use the code provided next!

Vector Subtraction in Higher Dimensions

You are using the Cartesian coordinate system with basis vectors.

$$\vec{e_1} = (1,0,0) \quad \vec{e_2} = (0,1,0) \quad \vec{e_3} = (0,0,1)$$

If the subtraction of two two-dimensional vectors is

$$\vec{a} - \vec{b} = (a_1 - b_1)\vec{e_1} + (a_2 - b_2)\vec{e_2}$$

and the subtraction of two three-dimensional vectors is

$$\vec{a} - \vec{b} = (a_1 - b_1)\vec{e_1} + (a_2 - b_2)\vec{e_2} + (a_3 - b_3)\vec{e_3}$$

then the subtraction of n-dimensional vectors will be

$$\vec{a} - \vec{b} = (a_1 - b_1)\vec{e_1} + (a_2 - b_2)\vec{e_2} + \ldots\ldots + (a_n - b_n)\vec{e_n}$$

This means that you can treat your user data array as a multidimensional vector and calculate the distance or difference between two users as the magnitude of the vector resulting from subtracting the two.

$$\vec{a} - \vec{b} = \vec{c} \qquad \left\|\vec{c}\right\| = \sqrt{c_1^2 + c_2^2 + c_3^2 + \ldots + c_n^2}$$

This is performed by the function getUserDistance(), which takes an array as parameter and returns the magnitude of the vector resulting from the subtraction of the vector "pattern" and the one formed by the user data, excluding the two first parameters, so the recognition doesn't get affected by changes in weight or height.

```
float getUserDistance(float[] pattern) {
  float result = 0;
  for (int i = 0; i < pattern.length; i++) {
    result += pow(pattern[i] - userData[i + 2], 2);
  }
  return sqrt(result);
}
```

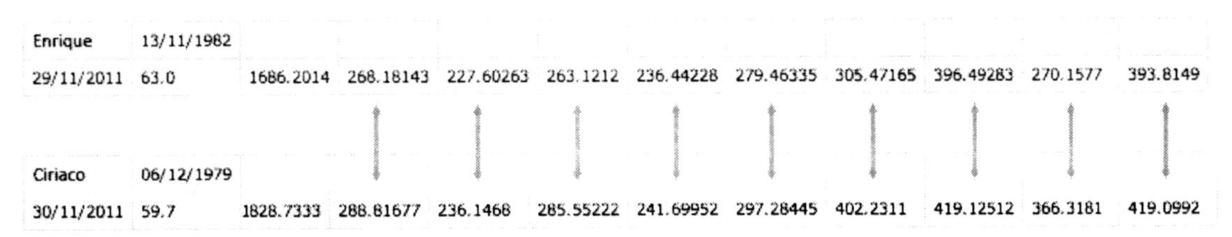

Enrique	13/11/1982										
29/11/2011	63.0	1686.2014	268.18143	227.60263	263.1212	236.44228	279.46335	305.47165	396.49283	270.1577	393.8149
Ciriaco	06/12/1979										
30/11/2011	59.7	1828.7333	288.81677	236.1468	285.55222	241.69952	297.28445	402.2311	419.12512	366.3181	419.0992

***Figure 10-16.** Users being compared to previously stored information*

The function userRecognition() is the bit of code that deals with comparing the current user to the information previously stored in CSV files (Figure 10-16). First, you create an array of floats that stores the distances of the current user to the various previously stored users, hence the size of it being files.length.

Then, for each user file, you read its content; if the line contains more than two chunks of data (so it's not the header), you store all the relevant floats in the array pattern, and you get its distance to your current user data using the previously defined function getUserDistance().

```
private void userRecognition() {
  float[] distances = new float[files.length];
  for (int i = 0; i < files.length; i++) {
    String line;
    try {
      BufferedReader reader = createReader(files[i]);
      while ( (line = reader.readLine ()) != null) {
        String[] myData = line.split(",");
        if (myData.length > 2) {
          float[] pattern = new float[9];
          for (int j = 3; j < myData.length; j++) {
            pattern[j - 3] = Float.valueOf(myData[j]);
          }
          distances[i] = getUserDistance(pattern);
        }
      }
    }
    catch (IOException e) {
      e.printStackTrace();
      println("error");
    }
  }
```

Once you have all the distances, or differences, you only need to find the user that is closest to your current user; if this distance is small enough (after some testing, we set the minimum to 100), you set the user in the file as the current user (Figure 10-17).

```
  int index = 0;
  float minDist = 1000000000;
  for (int j = 0; j < distances.length; j++) {
    if (distances[j] < minDist) {
      index = j;
      minDist = distances[j];
    }
  }
  if (minDist < 100)
    setPreviousUser(files[index]);
  println("Recognising User");
}
```

The function setPreviousUser() simply retrieves the user data from the selected file and updates the user name and birth date.

```
private void setPreviousUser(File file) {
  String line;
  try {
    BufferedReader reader = createReader(file);
```

```
    line = reader.readLine();
    String[] myData = line.split(",");
    userName = myData[0];
    userBirthDate = myData[1];
    getDate();
  }
  catch (IOException e) {
    e.printStackTrace();
    println("error");
  }
  println("User set to " + "userName, born on" + userBirthDate);
}
```

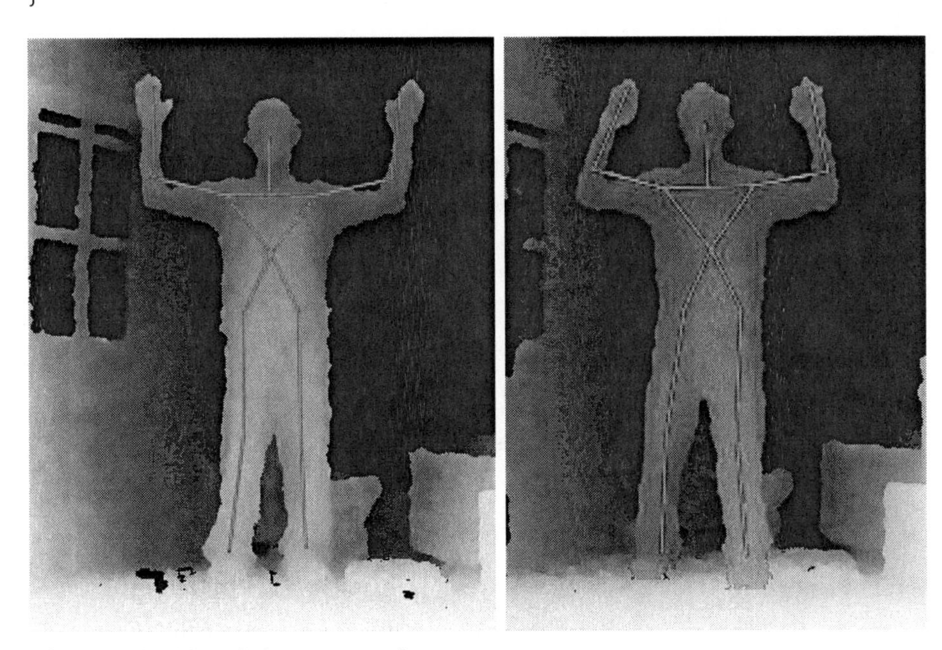

Figure 10-17. Users being compared

Reading and Drawing the Chart

The following functions follow the same principles and methods you used to read the user data from external files, but this time you read the chart files and transform the numeric data into graphs representing the percentiles of BMI.

```
private void readChart(File file) {
  String line;
  int lineNumber = 0;

  try {
    int lines = count(file);
    chart = new float[lines][11];
    BufferedReader reader = createReader(file);
```

```
    while ( (line = reader.readLine ()) != null) {
      String[] myData = line.split(",");
      if (lineNumber == 0) {
      }
      else {
        for (int i = 0; i < myData.length; i++) {
          chart[lineNumber - 1][i] = Float.valueOf(myData[i]);
        }
      }
      lineNumber++;
    }
  }
  catch (IOException e) {
    e.printStackTrace();
    println("error");
  }
}
```

The function count() is a helper function used to find out how many lines you will be reading from a CSV file.

```
public int count(File file) {
  BufferedReader reader = createReader(file);
  String line;
  int count = 0;
  try {
    while ( (line = reader.readLine ()) != null) {
      count++;
    }
  }
  catch (IOException e) {
    e.printStackTrace();
    println("error");
  }
  return count;
}
```

The function drawChart() simply runs through the chart array and draws the lines defining the percentiles. You can apply this method to every chart you have included in the data folder of the sketch.

```
private void drawChart() {
  float scaleX = 2;
  float scaleY = 10;

  pushStyle();
  stroke(255);
  noFill();
  pushMatrix();
  translate(0, height);
  scale(1, -1);
  translate(660, 20);
  line(0, 0, 500, 0);
  line(0, 0, 0, 300);
  for (int i = 1; i < chart[0].length; i++) {
    beginShape();
```

```
    for (int j = 0; j < chart.length - 1; j++) {
      vertex(chart[j][0] * scaleX, chart[j][i] * scaleY);
    }
    endShape();
  }
  strokeWeight(10);
  stroke(255, 0, 0);
  point(userAge*12*scaleX, userBMI*scaleY);
  popMatrix();
  popStyle();
}
```

Graphic Output

The following functions handle the on-screen representation of the user data in text form and the drawing of the user's skeleton.

```
private void printUserData() {
  fill(255);

  text("Data at " + today, 660, 20);
  text("Current User: " + userName, 660, 40);
  text("Date of Birth " + userBirthDate, 660, 60);
  text("Current Age " + userAge, 660, 80);

  text("User Weight: " + userData[0] + " " + lcd.units, 660, 100);
  text("User Height: " + userHeightTemp/1000 + " m", 660, 120);
  text("User BMI: " + userBMI + " kg/m2", 660, 140);
  fill(255, 255, 150);
  text("BMI at ages 2-20", 1000, 20);
}

void drawSkeleton(int userId) {
  pushStyle();
  stroke(255, 0, 0);
  strokeWeight(3);
  kinect.drawLimb(userId, SimpleOpenNI.SKEL_HEAD, SimpleOpenNI.SKEL_NECK);

  kinect.drawLimb(userId, SimpleOpenNI.SKEL_NECK, SimpleOpenNI.SKEL_LEFT_SHOULDER);
  kinect.drawLimb(userId, SimpleOpenNI.SKEL_LEFT_SHOULDER, SimpleOpenNI.SKEL_LEFT_ELBOW);
  kinect.drawLimb(userId, SimpleOpenNI.SKEL_LEFT_ELBOW, SimpleOpenNI.SKEL_LEFT_HAND);

  kinect.drawLimb(userId, SimpleOpenNI.SKEL_NECK, SimpleOpenNI.SKEL_RIGHT_SHOULDER);
  kinect.drawLimb(userId, SimpleOpenNI.SKEL_RIGHT_SHOULDER, SimpleOpenNI.SKEL_RIGHT_ELBOW);
  kinect.drawLimb(userId, SimpleOpenNI.SKEL_RIGHT_ELBOW, SimpleOpenNI.SKEL_RIGHT_HAND);

  kinect.drawLimb(userId, SimpleOpenNI.SKEL_LEFT_SHOULDER, SimpleOpenNI.SKEL_TORSO);
  kinect.drawLimb(userId, SimpleOpenNI.SKEL_RIGHT_SHOULDER, SimpleOpenNI.SKEL_TORSO);
  kinect.drawLimb(userId, SimpleOpenNI.SKEL_TORSO, SimpleOpenNI.SKEL_LEFT_HIP);
  kinect.drawLimb(userId, SimpleOpenNI.SKEL_LEFT_HIP, SimpleOpenNI.SKEL_LEFT_KNEE);
  kinect.drawLimb(userId, SimpleOpenNI.SKEL_LEFT_KNEE, SimpleOpenNI.SKEL_LEFT_FOOT);

  kinect.drawLimb(userId, SimpleOpenNI.SKEL_TORSO, SimpleOpenNI.SKEL_RIGHT_HIP);
```

```
    kinect.drawLimb(userId, SimpleOpenNI.SKEL_RIGHT_HIP, SimpleOpenNI.SKEL_RIGHT_KNEE);
    kinect.drawLimb(userId, SimpleOpenNI.SKEL_RIGHT_KNEE, SimpleOpenNI.SKEL_RIGHT_FOOT);
    popStyle();
  }
```

Serial Event

The serialEvent() function is entirely derived from the previous example implemented in the "Using the Weight Data" section. This function acquires the data from Arduino and sets the LCD state to the incoming data.

```
public void serialEvent(Serial myPort) {
  // get the ASCII string:
  String inString = myPort.readString();
  inString = inString.substring(0, inString.indexOf('\n') - 1);
  if (inString != null) {
    if (inString.equals("S")) {
      currentLine = 0;
    }
    else if (inString.charAt(0) == 'X') {
      String newState = inString.substring(1);
      lcd.setState(Integer.valueOf(newState));
    }
    else {
      if (inString.length() == 13 && currentLine < 4) {
        for (int i = 0; i < stringData[0].length; i++) {
          stringData[currentLine][i] = inString.charAt(i);
        }
        currentLine++;
      }
    }
  }
  else {
    println("No data to display");
  }
}
```

Simple-OpenNI User Events

These are the same good old Simple-OpenNI callbacks as usual. The only hint of novelty is the last line in the onNewUser() function, which increases your Processing frame height by 54 pixels so the user interface is displayed.

```
public void onNewUser(int userId) {
  println("onNewUser - userId: " + userId);
  println("  start pose detection");
  kinect.startPoseDetection("Psi", userId);
  frame.setSize(width, height + 54);
}

public void onLostUser(int userId) {
  println("onLostUser - userId: " + userId);
```

```
}

public void onStartCalibration(int userId) {
  println("onStartCalibration - userId: " + userId);
}

public void onEndCalibration(int userId, boolean successfull) {
  println("onEndCalibration - userId: " + userId + ", successfull: "
    + successfull);

  if (successfull) {
    println("  User calibrated !!!");
    kinect.startTrackingSkeleton(userId);
    user = userId;
  }
  else {
    println("  Failed to calibrate user !!!");
    println("  Start pose detection");
    kinect.startPoseDetection("Psi", userId);
  }
}

public void onStartPose(String pose, int userId) {
  println("onStartdPose - userId: " + userId + ", pose: " + pose);
  println(" stop pose detection");  kinect.stopPoseDetection(userId);
  kinect.requestCalibrationSkeleton(userId, true);
}

public void onEndPose(String pose, int userId) {
  println("onEndPose - userId: " + userId + ", pose: " + pose);
}
```

Summary

This has been quite a heavy project on the programming side. You had to acquire rather noisy data from a bathroom scale's LCD, and you made sense of it by reverse-engineering its logic until you had the numeric data you wanted. You then built a whole framework for human biometric scanning and you implemented a user interface plus a data storage and retrieval system. You even managed to perform user recognition using NITE's skeleton tracking capabilities, and you output all the gathered data on screen for visualization.

You used some Kinect biometrics in this project; obviously, there are innumerable spin-off projects to be generated from this. You could start by building a system to show all the available growth and weight charts on screen, or you could extract only the user-recognition implementation and build your own home security system. You could also implement a fitness evaluator from the user shape, and try to classify people according to body shapes and dimensions. If you have some commercial vision, you should definitely give a go to an online fitted clothes try-on service based on Kinect's three-dimensional scanning!

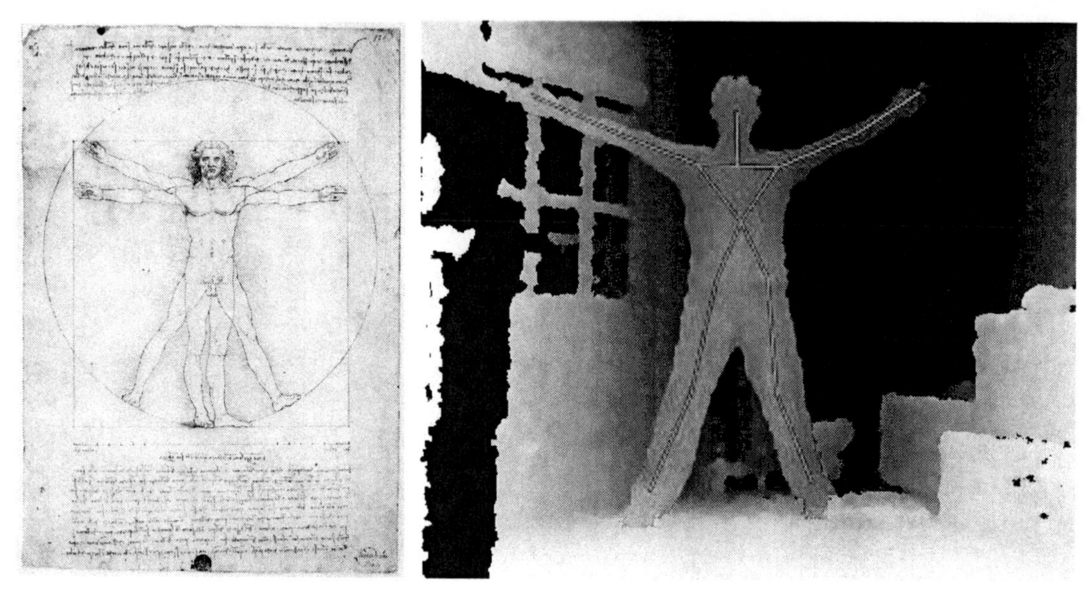

Figure 10-18. *Leonardo's Vitruvian Man, and a not-so-Vitruvian Kinect user*

CHAPTER 11

3D Modeling Interface

by Enrique Ramos

So far you've used Kinect to accurately track your movements in space, but when it comes to "clicking," you can't seem to dispense with your mouse. In this project, you're going to build a wearable user interface that will complement Kinect's tracking capabilities with accurate multi-clicking and you will use it to build 3D geometries by drawing directly in space.

Your user interface will be mounted on a standard glove and will use an Arduino LilyPad as its computational core. You will learn how to use the bending of your fingers as an input device via flex sensors. You will then learn how to develop your own simple CAD program to translate the position of your hand and the bending of your fingers into 3D geometries (as shown in Figure 11-1) that you will then export as `.dxf` files readable by other CAD and 3D-modeling packages.

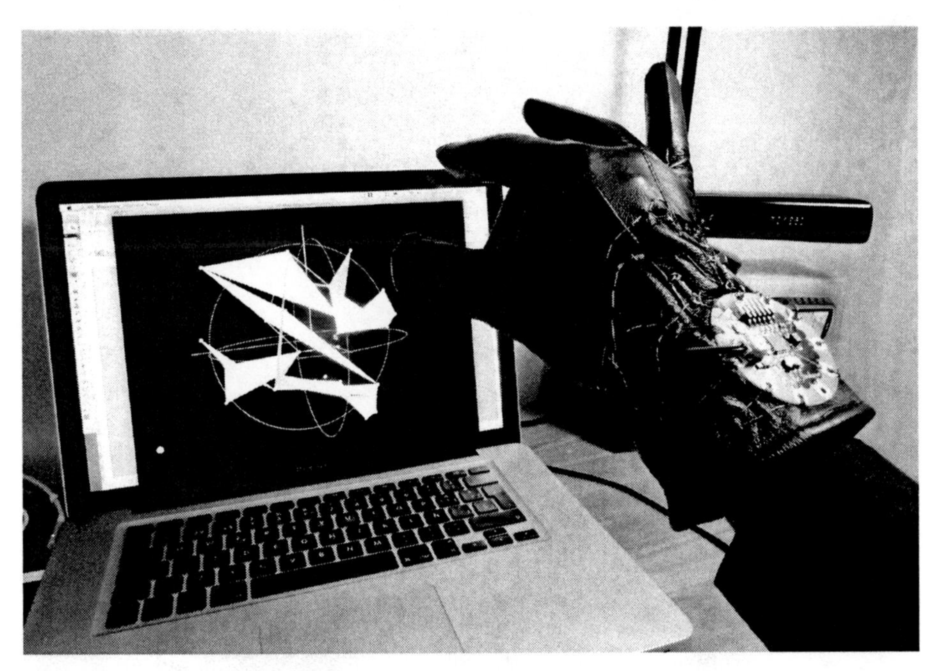

Figure 11-1. 3D modeling with the glove interface

The Interface

In previous chapters, you explored several ways of interacting with your computer using Kinect's 3D scanning plus NITE's hand and skeleton tracking capabilities. But the "clicking" precision using gesture recognition is rather inaccurate and can be discouraging for some applications.

In this chapter, you are going to bridge this gap by building a wearable wireless user interface that can be used as a 3D interface for several applications (Figure 11-2). You will take advantage of Kinect's excellent 3D hand-tracking capability, and you will complement it with a way to trigger different behaviors in your program without using conventional keyboard and mouse devices.

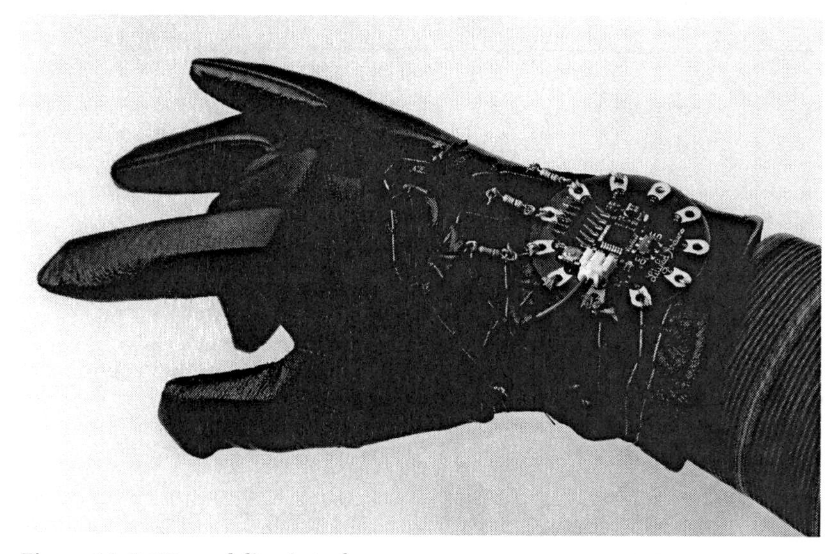

Figure 11-2. 3D modeling interface

Table 11-2 lists all the components necessary for this project and Figure 11-3 shows what they look like. If you need to buy all of the Arduino LilyPad stuff, we recommend the LilyPad Beginner's Kit from SparkFun, which includes most of the components for this project plus other parts you may find useful later.

***Table 11-1.** Parts List*

Part	Description	Price
One glove	(I guess you will have to buy two though...)	From $5
4 flex sensors	FLX-01-M from Images SI or similar	$10.00 each
4 10KΩ resistors	For the voltage dividers	$0.25 each
Arduino LilyPad Simple	You could also use the LilyPad main board	$19.95
LilyPad XBee	XBee breakout for LilyPad	$14.95

Part	Description	Price
XBee Explorer USB	You will need a mini USB cable to connect it.	$24.95
2 XBee modules	XBee 1mW Chip Antenna - Series 1	$22.95 each
FTDI Basic Breakout 5V	You will need this to program your LilyPad.	$14.95
USB A to MiniB Cable	To connect the FTDI to the computer	$4.95
Conductive thread	You only need a 60ft bobbin for the project.	$2.95
Needle	To sew your own circuit	$1.00
LiPo Battery - 110mAh	Polymer lithium ion battery	$6.95

■ **Note** You are using the Arduino LilyPad Simple in this project. As a result, you also need to use Arduino's SofwareSerial library to communicate with the XBee radio module, as you can't access pins 0 and 1 on the LilyPad Simple. This way of transmitting information is actually much slower than using pins 0 and 1, but it's good enough for this project. If you are planning on adding more demanding transmission of information to this project, you should consider using the standard Arduino LilyPad board. For more information, see the "Going Wireless" section.

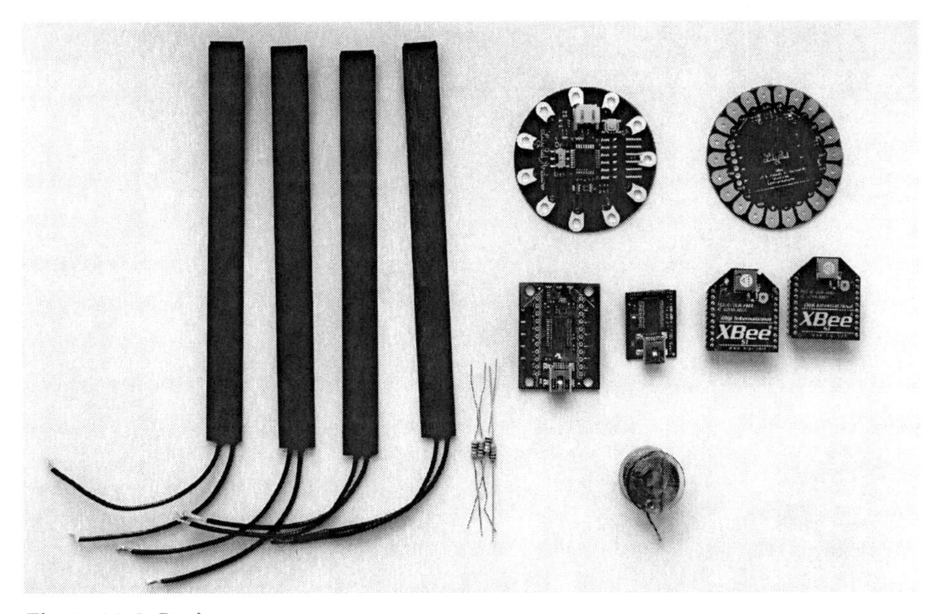

Figure 11-3. *Project parts*

Arduino LilyPad

When working on wearable technology with Arduino, it's difficult to find a better option than the Arduino LilyPad, an Arduino board developed cooperatively by Leah Buechley and SparkFun Electronics specifically to be sewn to fabric and clothing for wearable applications.

For this project, you are going to use the Arduino LilyPad Simple board. This board has fewer connections than the standard Arduino LilyPad, but it includes a built-in power socket for plugging in a LiPo battery plus an on/off switch. This will simplify the circuit by eliminating the need to sew an extra power supply to your glove.

Note that there's no USB connection on the LilyPad board (Figure 11-4). To connect the board to the computer, you will need to use an FTDI Basic Breakout that plugs into the 6-pin male header on the top of the LilyPad board. The FTDI Basic Breakout (Figure 11-5) can be plugged to the computer with a USB MiniB cable.

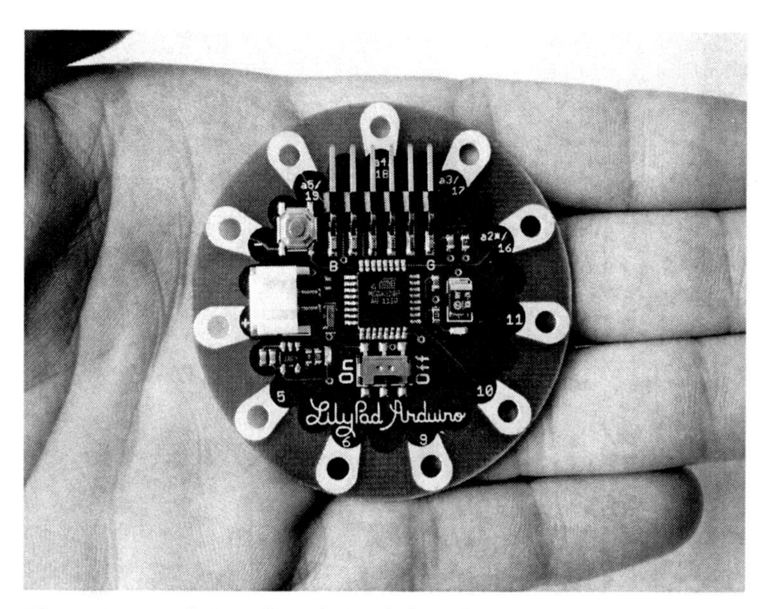

Figure 11-4. Arduino LilyPad Simple board

Figure 11-5. FTDI Basic Breakout

Flex Sensors

Flex sensors are electronic components that change resistance when bent. As with any other kind of variable resistors, they must be used with a voltage divider to get changing voltages into your analog input pins (Figure 11-6). This kind of sensors has been used extensively in the fabrication of virtual reality gloves.

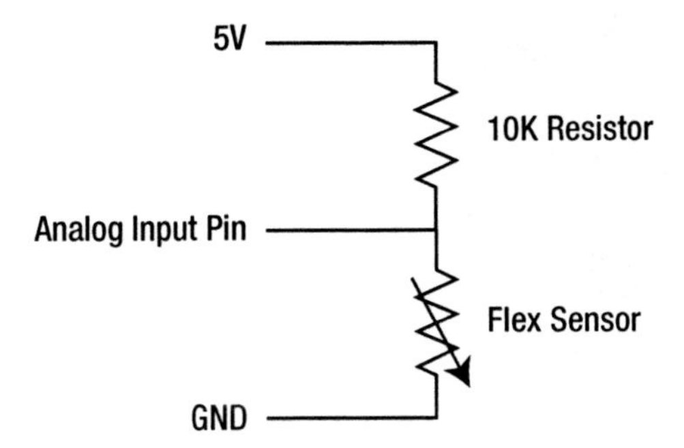

Figure 11-6. *Voltage divider for flex sensor*

For this project, you will use four two-directional Bi-Flex Sensors from Images Scientific Instruments. These sensors decrease their resistance as they are bent in either direction. (This feature is something you don't really need for the project, as your fingers only bend in one direction, so you can choose any other flex sensors as long as they fit nicely into the glove's fingers.)

There are three versions of the FLX-01 sensors, depending on the resistance ranges. This project calls for the FLX-01-M (medium resistance range of 20K-50K) and 10K resistors for each sensor, which provides a reasonable input range. The kind of circuit you are using for each sensor is called a *voltage divider* (see Figure 11-6). We discussed the calculation of output voltages and resistors for voltage dividers in the section "Feedback from Arduino" in Chapter 4. Go back to this section if you need a refresher.

The first thing you need to do is to install the flex sensors in the glove. There are many ways of doing this. Our gloves had a double skin, so we decided to insert the sensors between the fake-leather exterior skin and the fluffy interior one. After measuring the length of the sensor from the tip of each finger, cut some small incisions, just large enough for the sensors to slip into. Then insert one sensor in each finger, leaving the cables to protrude from the slits. Only use sensors for the thumb, index, middle, and ring fingers (Figure 11-7). Leave the little finger out because of the difficulty of controlling it independently (unless you are a piano player!) and because there are only four accessible analog pins on the LilyPad Simple.

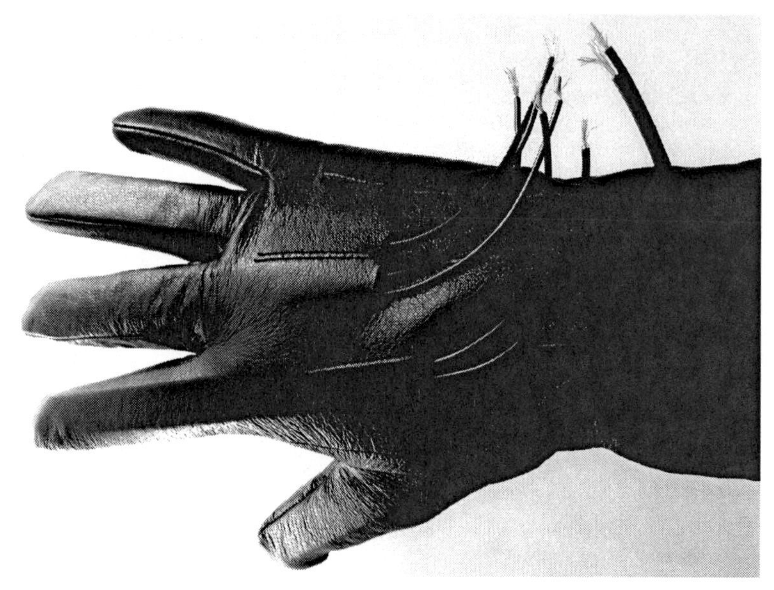

Figure 11-7. Flex sensors installed in the glove

Sewing the Circuit

Yes, that's the right word. This time you won't be "building" your circuit; you'll be "sewing" it (Figure 11-8). For wearable applications, you use conductive thread instead of normal wires to connect the components in the circuit; the result is a more flexible, stable, and lightweight design. A piece of advice: if you have never sewn a button in your life, you may want to get some help from more experienced relatives or friends before you put a needle in your hand!

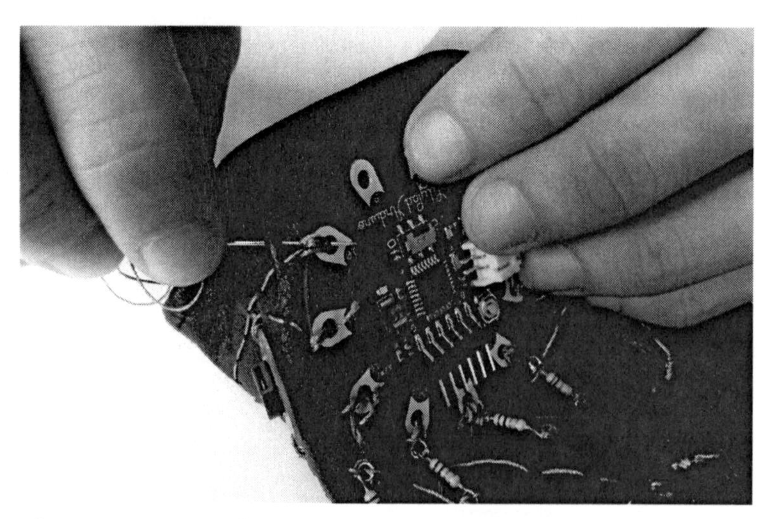

Figure 11-8. Sewing the circuit

The main idea is to link the LilyPad pins to the right components with the conductive thread. Note that you want to avoid the threads touching each other, just as you would do on a normal circuit. Sometimes you will need to cross threads but you don't want them to join electrically. The best way around this is to arrange for them to be on different sides of the skin at the crossing point.

Start by sewing a test circuit using your Arduino LilyPad and the four flex sensors. After proper testing, you can add other components.

First, sew the Arduino LilyPad to the glove, and then sew the connections to the flex sensors and the resistors (Figure 11-9). To connect the flex sensors, expose a section of the cables and loop around the exposed wires with your conductive thread. Then cut off the rest of the cable. If you look carefully at Figure 11-10, you can see that the ground thread connecting all the resistors together goes under the exterior skin most of the time so that it doesn't connect to the cables going from the flex sensors to the Arduino LilyPad pins.

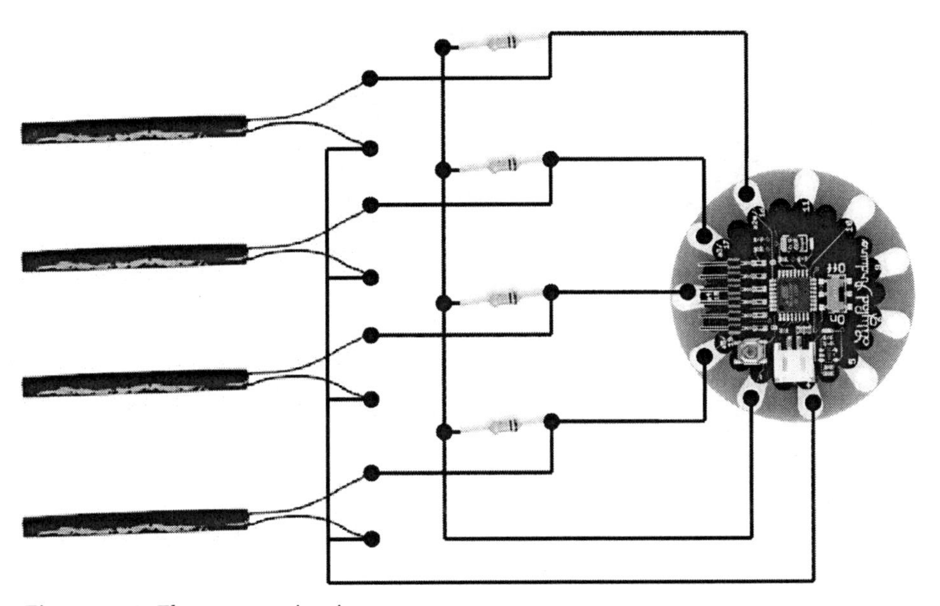

Figure 11-9. *Flex sensors circuit*

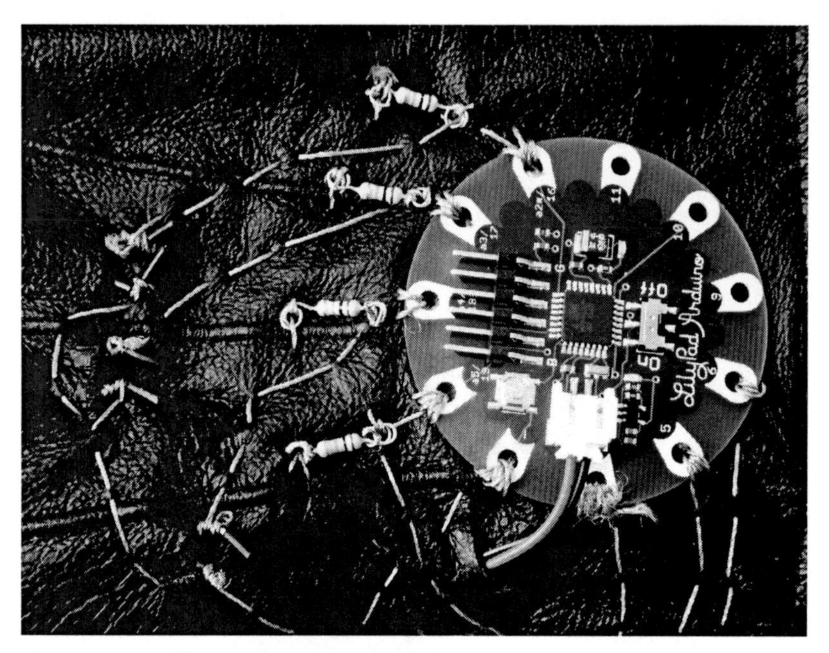

Figure 11-10. *The circuit sewn to the glove*

Test that the connections are correct using a multimeter and then connect the board to your computer using the FTDI Basic Breakout. Now you can load and test your first program (Figure 11-11).

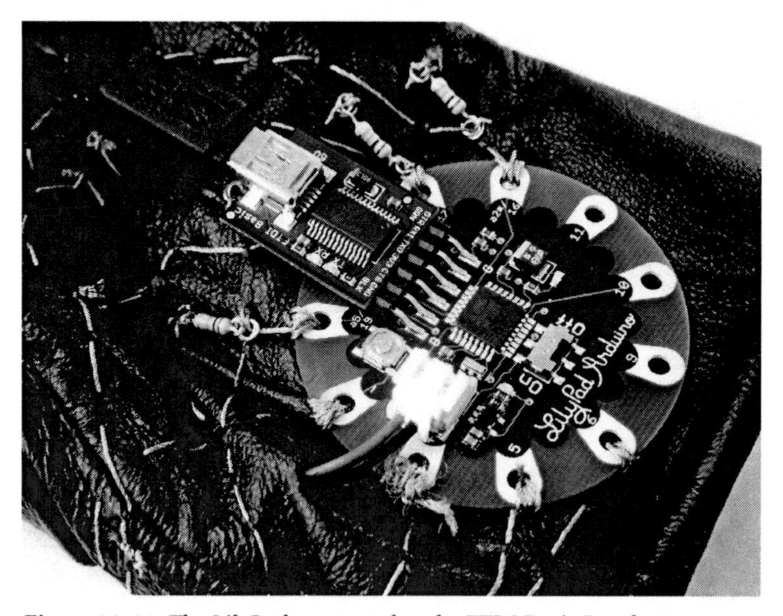

Figure 11-11. *The LilyPad connected to the FTDI Basic Breakout*

Testing the Circuit

Now that you have sewn the circuit, you need to check that everything works properly. To do so, you will upload an Arduino program to your board to read the input from the flex sensors and send it via serial so you can see the values on the screen.

Arduino Serial Sender

Open a new Arduino sketch and declare four integer variables, one for each finger. In the setup() function, start serial communication at 9600 baud.

```
int finger1, finger2, finger3, finger4;
void setup() {
  Serial.begin(9600);
}
```

In the loop() function, read the four sensors using analogRead() and send their values to the serial buffer, separating them by blank spaces and finishing with a println() function call to get a newline character at the end of each cycle.

```
void loop() {
  finger1 = analogRead(A5);
  finger2 = analogRead(A4);
  finger3 = analogRead(A3);
  finger4 = analogRead(A2);

  Serial.print(finger1);
  Serial.print(" ");
  Serial.print(finger2);
  Serial.print(" ");
  Serial.print(finger3);
  Serial.print(" ");
  Serial.println(finger4);
}
```

Upload this code and open the Serial Monitor. If you put the glove on and bend your fingers, you should see the numbers changing within a range of 100 to 700 or similar.

Processing Glove Data Receiver

It will be much easier to understand this data if you can get a visual representation of it, so you're going to build a simple Processing sketch that will receive the values and draw them on screen as variable-width rectangles. You will reuse some of these elements in the final project.

Open a new Processing sketch and import the Serial library. Then initialize the serial port and an array of four floats to contain the data from the flex sensor for each finger. The data from Arduino is actually integers, but you declare the fingerData array as a float data type because you are smoothing out the incoming data and the result may turn out to be floating point numbers. Then, within the setup() function, set the sketch size and initialize the serial communication.

```
import processing.serial.*;
Serial myPort;
float[] fingerData = new float[4];

void setup() {
  size(800, 300);
  // Serial Communication
  String portName = Serial.list()[0];
  myPort = new Serial(this, portName, 9600);
  myPort.bufferUntil('\n');
}
```

The draw() function begins by defining a black color for the background and then drawing a rectangle for each finger, the width based on the finger data from Arduino.

```
void draw() {
  background(0);
  for (int i = 0; i < fingerData.length; i++) {
    rect(0, 50+i*50, fingerData[i], 20);
  }
}
```

The serialEvent() function is triggered every time you get a newline character in the serial buffer. You have programmed the Arduino to send the four values from the flex sensors separated by blank spaces, so you need to get the string and split it at every blank space.

```
public void serialEvent(Serial myPort) {
String inString = myPort.readStringUntil('\n');
  if (inString != null) {
    String[] fingerStrings = inString.split(" ");
```

If you have the correct amount of String values (four), you convert every String to a float and then you add it to the fingerData[] array. You place the conversion inside a try{} … catch{} block to avoid the program stopping completely when it gets an anomalous character from Arduino, which happens occasionally. Use the Processing function lerp() to smooth out quick oscillations in the incoming data. Figure 11-12 shows what the data could look like.

```
    if (fingerStrings.length==4) {
      for (int i = 0; i < fingerData.length; i++) {
        try {
          float intVal = Float.valueOf(fingerStrings[i]);
          fingerData[i] = lerp(fingerData[i], intVal, 0.5);
        }
        catch (Exception e) {
        }
      }
    }
  }
}
```

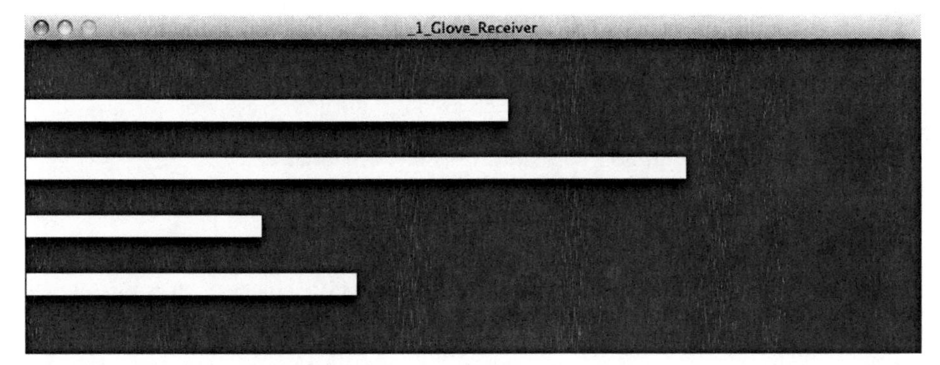

Figure 11-12. Finger-bending test

Going Wireless

Now that you have the basics of the glove interface working, you are going to improve your device by eliminating the cables. The ultimate goal of this project is an intuitive interface that works via your hand movements, so having a cable attached to it is not quite ideal.

To free yourself from the inconvenient cables, you need to add a power source to your glove design and an XBee radio module to transmit serial data wirelessly. You have used the XBee radio module in several other projects, so you should be acquainted with it already. If you skipped the previous chapters, read the introduction to XBee in Chapter 7.

LilyPad XBee

Happily, there's a LilyPad XBee breakout board that works on the same principles as the main LilyPad board. You are going to add this XBee breakout to your glove, and you will use an XBee Explorer USB connected to your computer to receive the information (Figure 11-13).

Figure 11-13. XBee Explorer USB

Connecting the XBee breakout to your LilyPad board isn't difficult. The way you normally do it is by attaching the + and - pins on the XBee breakout to 5V and ground on the LilyPad and the rx and tx to pins 1 and 0. But that's on a standard LilyPad board. You're using a LilyPad Simple, and after close inspection, you see that pins 0 and 1 don't exist. (Well, they do exist but the architecture of the LilyPad doesn't expose them.) So what should you do? Can you use the XBee module with the LilyPad Simple? The answer to this problem is one of the Arduino's core libraries: SoftwareSerial.

SoftwareSerial Library

Arduino's built-in support for serial communication in pins 0 and 1 occurs via a piece of hardware called UART (Universal Asynchronous Receiver/Transmitter), which is built into the ATmega microcontroller.

You can't access pins 0 and 1 on the LilyPad Simple, but you can replicate the serial reception and transmission functionality on other pins using the SoftwareSerial library, which is included with the Arduino IDE. This way of transmitting information is actually much slower than using pins 0 and 1, but it's still good enough to transmit the information from the glove in this project. If you are planning on adding some more demanding transmission of information to this project, you should definitely consider using the standard Arduino LilyPad board.

Wireless Arduino Code

Let's modify the previous Arduino code to work with the XBee through pins 5 and 6 using this library. First, you need to include the SoftwareSerial library and initialize a `SoftwareSerial` object, passing pin numbers 5 and 6 as parameters. Call the object `XBee`. This is the first time you're using an Arduino library in this book. You have previously included libraries in Processing by selecting Import Library in the Sketch menu or typing something like `import SimpleOpenNI.*;` directly in the text editor. In Arduino, the syntax is different because the language is based on C instead of Java. You can still go to Sketch > Import Library and choose SoftwareSerial; you'll see a new line of code in your text editor, `#include <SoftwareSerial.h>`, which is the Arduino equivalent of the Processing import statement.

So wherever you wrote `Serial` on your previous code, you now write `XBee.` The resulting code is the following:

```
#include <SoftwareSerial.h>
SoftwareSerial XBee(5,6);
int finger1, finger2, finger3, finger4;

void setup() {
  XBee.begin(9600);
}

void loop() {
  finger1 = analogRead(A5);
  finger2 = analogRead(A4);
  finger3 = analogRead(A3);
  finger4 = analogRead(A2);

  XBee.print(finger1);
  XBee.print(" ");
  XBee.print(finger2);
  XBee.print(" ");
  XBee.print(finger3);
```

```
  XBee.print(" ");
  XBee.println(finger4);
}
```

This is the Arduino code that you will be using for the rest of this chapter. Before trying this code with the previous Glove Receiver Processing sketch, you need to sew the XBee to the glove circuit. Figure 11-14 shows the final circuit for the project with the components arranged as in the physical device. It will take you some time to sew all the connections. Be sure to make the connections solid and reliable; otherwise, the movement of your hands will end up loosening them.

Apart from the XBee, you need to introduce the LiPo battery in order to make the glove completely autonomous. Just plug it in the LilyPad power supply socket and put the battery inside the glove (Figure 11-15). If your gloves have a double skin, this is the perfect place to hide it.

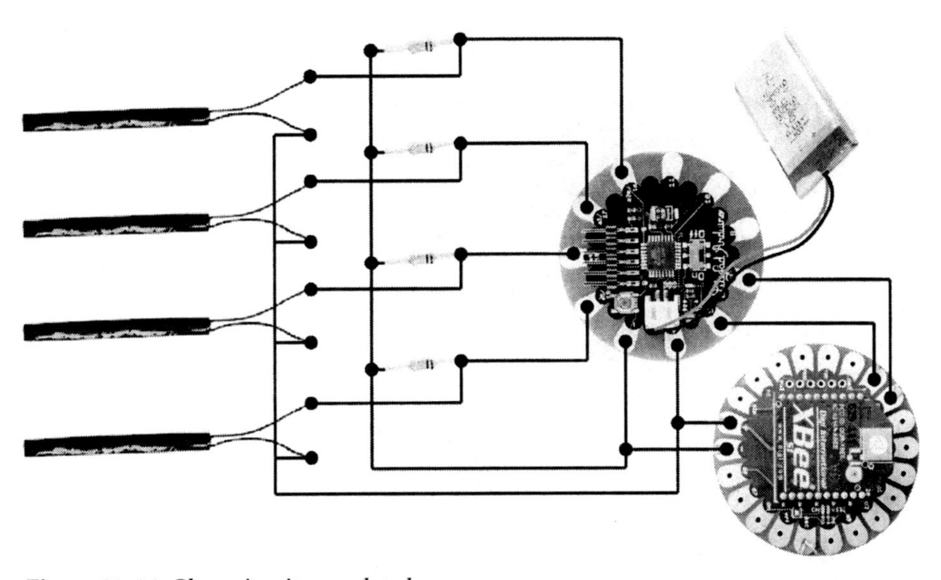

Figure 11-14. Glove circuit completed

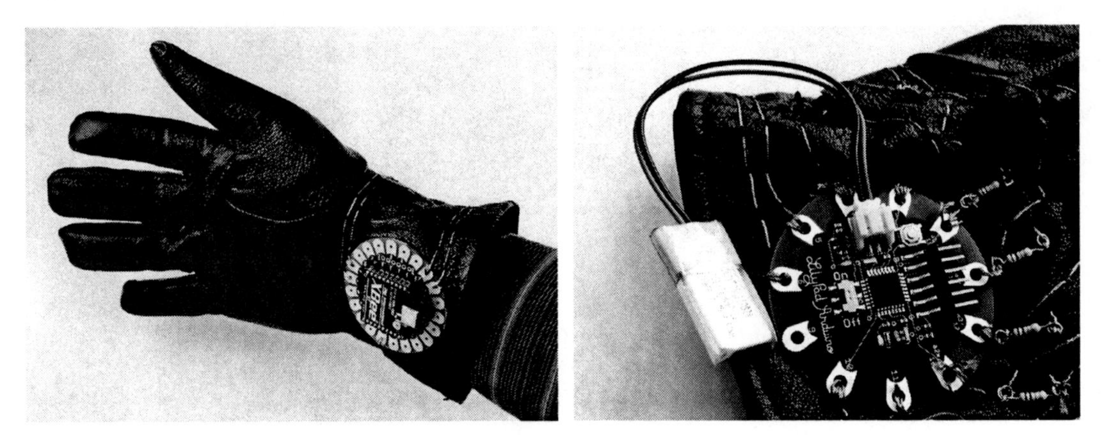

Figure 11-15. The LilyPad Xbee on the circuit (left) and the LiPo battery (right)

Implementing the Modeling Interface

Now that the physical device is ready and the code is uploaded to the Arduino LilyPad and sending serial data, it's time to write your 3D modeling interface program in Processing.

Let's define 3D modeling as the creation of 3D geometries with specialized software. These geometries are defined by surfaces and lines, which in turn are defined by points. You begin by defining points in space and then linking these points together with lines and surfaces, as you can see in Figure 11-16. You implement this by writing a very simple computer-aided design (CAD) software program (detailed in the "Geometric Classes" section) that you can control with your glove. This software is based on a main Processing routine and three helper classes that deal with particular functions and data structures.

The first class, `GloveInterface`, is derived from your previous Glove Data Receiver example and deals with the information coming from the glove via serial. The other three classes are the `Point`, `Line`, and `Shape` classes, which are the basic blocks of your geometrical constructions.

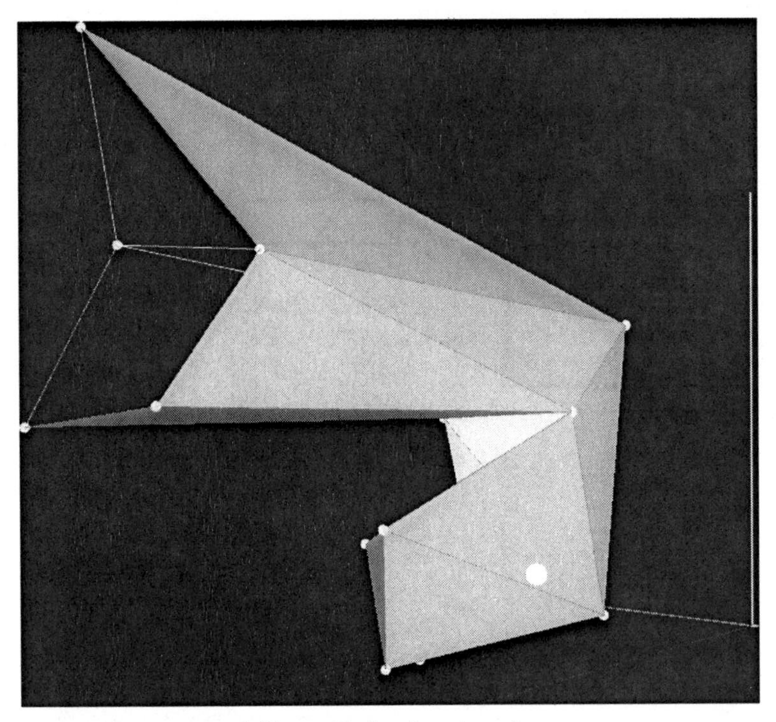

Figure 11-16. *3D modelling with the glove interface*

The GloveInterface Class

This class deals with the finger-bending information from the flex sensors. Due to the different sizes and forms of your fingers and the particularities of the different flex sensors, the ranges of the data coming from each one of the fingers will be slightly different. You want to control the incoming data with the same functions, so you need to map these values to standard ranges with a calibration function.

First, declare the GloveInterface class in its own tab, as you saw in Chapter 7. Then declare a File object to store the calibration information so you don't need to repeat the calibration process every time you run the program.

```
public class GloveInterface {
  File fingerCal;
```

You need three PVectors to handle the position of your hand in space, the position that you consider to be the origin of coordinates (which is determined by the position of the hand at the moment of recognition), and the position of the glove in your model space (pos).

```
 PVector pos = new PVector();
 PVector zeroPos = new PVector();
 PVector handPos = new PVector();
```

You also need an array to store the bending state of each finger as per the data from the Arduino (fingerState[]) and then after calibration (fingerStateCal[]). The calibration data is stored in the fingerCalibration[] nested array, which you initialize to predetermined values.

The integer currentFinger defines the finger that is "clicked" at any specific moment, which is determined using the setFingerValues() function. The default value (99) doesn't correspond to any of the four fingers you are following and it's identified as no finger clicked.

```
  float[] fingerState = new float[4];
  float[] fingerStateCal = new float[4];
  float[][] fingerCalibration = {{0,600},{0,600},{0,600},{0,600}};
  public int currentFinger = 99;
```

You only implement one constructor taking no arguments. Within this constructor, specify the file containing your calibration data. If the file already exists, get the data in the file using the function getCalibration().

```
  GloveInterface() {
    fingerCal = new File("data/fingerCal.txt");
    if (fingerCal.exists()) {
      getCalibration();
    }
  }
```

Setter Functions

You implement three setter functions: one to set the fingers' state, another to set the current glove position, and a third one to set the initial or zero position.

The function setFingerValues() is called from the main routine every time you get new data from Arduino via the serial interface. This function takes an array of strings as its first argument. The first step in the function is to convert each String in the array into a float and store them in the fingerState[] array. This code is very similar to the code from the Glove Data Receiver sketch, but it uses a try{} ... catch{} block to avoid the execution, stopping when incorrectly formatted strings are received.

```
  public void setFingerValues(String[] fingerStrings) {
      for (int i = 0; i < fingerState.length; i++) {
      try {
        float intVal = Float.valueOf(fingerStrings[i]);
        fingerState[i] = PApplet.lerp(fingerState[i], intVal, 0.2f);
      }
      catch (Exception e) {
      }
    }
```

After acquiring the fingers' states, use the fingerCalibration values to remap the incoming data from the calibration values to the desired range of 0-600. The way you calibrate the interface is by opening all your fingers and storing the "stretched" values into the zero index element of each array in fingerCalibration. Then you close your fingers and store the "maximum bending" state into the index 1 of each array in fingerCalibration. By mapping each of the incoming values from these ranges to the desired range (0-600), you make sure that a stretched finger means a value close to 0 and a fully bent finger casts a value close to 600. You analyze these calibration functions at the end of the class.

```
    for (int i = 0; i < fingerStateCal.length; i++) {
      fingerStateCal[i] = PApplet.map(fingerState[i], fingerCalibration[i][0],
fingerCalibration[i][1], 0, 600);
    }
```

Finally, you determine if one of the fingers is "clicked" by checking which one of the fingers presents a higher bending state and if its value is over a threshold (300). The current finger is stored in the integer currentFinger.

```
    float maxBending = 0;
    currentFinger = 5;
    for (int i = 0; i < fingerStateCal.length; i++) {
      if (fingerStateCal[i] > maxBending && fingerStateCal[i] > 300) {
        maxBending = fingerStateCal[i];
        currentFinger = i;
      }
    }
  }
```

The function setPosition() takes a PVector defining the current position of your hand in global coordinates, which is stored in the handPos PVector. Then you translate this value to your modeling coordinate system by subtracting the zeroPos PVector from it. The zeroPos PVector updates from the main routine by calling the setZeroPos() function.

```
  public void setPosition(PVector inputPos) {
    this.handPos = inputPos;
    this.pos = PVector.sub(inputPos, zeroPos);
  }

  public void setZeroPos(PVector handPos) {
    this.zeroPos = handPos;
  }
```

Display Functions

The draw() function simply draws a point on screen of the coordinates of your glove interface. The switch() statement changes the color of the point depending on the current finger.

```
  public void draw() {
    pushStyle();
    switch (currentFinger) {
    case 0:
      stroke(255, 0, 0);
      break;
    case 1:
      stroke(0, 255, 0);
      break;
```

```
    case 2:
      stroke(0, 0, 255);
      break;
    case 3:
      stroke(255, 255, 0);
      break;
    case 5:
      stroke(255, 255, 255);
      break;
    }
    strokeWeight(20);
    if(!record){
      point(pos.x, pos.y, pos.z);
    }
    popStyle();
  }
```

Note that you put the point() function within an if() statement that only draws if the Boolean record is not true. This is due to the dxf export library that you will be bringing in later; this Processing library can't export points so you need to avoid the printing of points when exporting. You do this every time you draw points in this program.

The drawData() function draws the same information on screen as the Glove Data Receiver sketch, letting you see the calibrated bending state of each finger.

```
  public void drawData() {
    for (int i = 0; i < fingerStateCal.length; i++) {
      rect(0, 50 + i * 50, (int) fingerStateCal[i], 20);
    }
  }
```

Calibrating the Interface

We have already explained how the calibration of the device works. The following three functions take care of setting the calibration data, storing it in the external file, and retrieving it on demand.

The function calibrateFinger() is called from the main Processing routine by pressing a key on the keyboard. It takes the finger to be calibrated (i) and the state being calibrated (j), which is 0 if calibrating the stretched state or 1 if calibrating the bent state. You first store the present non-calibrated state of the finger into the fingerCalibration[] array and then call the saveCalibration() function.

```
  public void calibrateFinger(int i, int j) {
    fingerCalibration[i][j] = fingerState[i];
    saveCalibration();
  }
```

saveCalibration() converts the fingerCalibration double array into a linear array of strings and uses the saveStrings() function to store all the values into the fingerCal file.

```
  private void saveCalibration() {
    String[] mySettings = new String[8];
    for (int k = 0; k < fingerCalibration.length; k++) {
      mySettings[k] = String.valueOf(fingerCalibration[k][0]);
      mySettings[k + 4] = String.valueOf(fingerCalibration[k][1]);
    }
    PApplet.saveStrings(fingerCal, mySettings);
  }
```

Lastly, the function getCalibration() performs the opposite process, using loadStrings() to retrieve the strings from the fingerCal file and storing these values into the fingerCalibration array after converting them into floats.

```
  private void getCalibration() {
    String[] mySettings = PApplet.loadStrings(fingerCal);
    for (int k = 0; k < fingerCalibration.length; k++) {
      fingerCalibration[k][0] = Float.valueOf(mySettings[k]);
      fingerCalibration[k][1] = Float.valueOf(mySettings[k + 4]);
    }
  }
}
```

Geometric Classes

In order for you to be able to draw in three dimensions with your new glove interface, you need to implement your own modeling software or link your Processing interface to existing CAD software. You're going to take the first approach by implementing a very, very simple and limited CAD program.

CAD, or computer-aided design, is the software that allows engineers, architects, and designers to draft with a computer. There are plenty of commercial CAD and 3-D modeling software packages—and a few good open source options. The very basic blocks of the geometrical constructions generated with modeling software are the notions of point, line, and surface. You are going to create a separate class for each of these basic geometric concepts, which you will then use to generate the geometry (Figure 11-17).

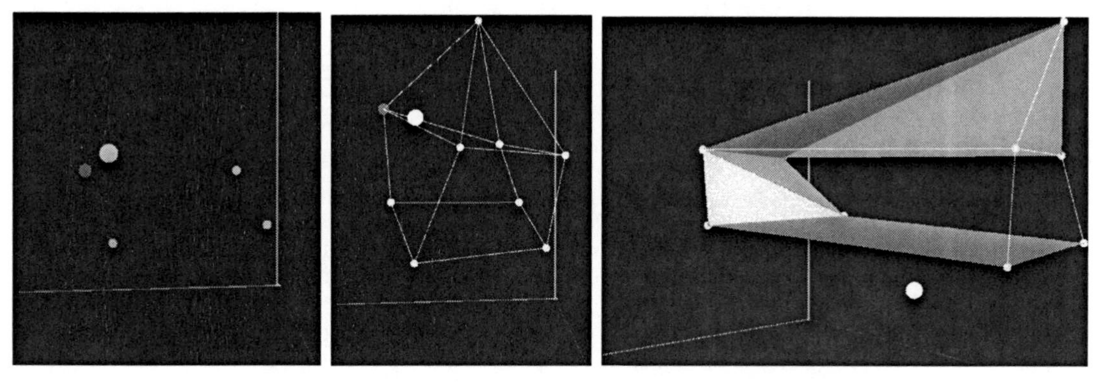

***Figure 11-17.** Points, lines, and surfaces defined by your program*

Point Class

As you build your geometry, a point is defined as a specific location in 3D space. This location is described by three coordinates: x, y, and z. For your purposes, this location can be treated as a vector (x,y,z), which means you could use the PVector class to define your points. Instead, you are going to write your own Point class to be able to add a couple more functions to your point objects.

Note that the class Point will have two fields: a PVector describing its position and a Boolean variable that is true if the element has been selected. You implement only one constructor taking a PVector as parameter that is used as the point's position.

```
public class Point {
```

```
  PVector pos;
  private boolean selected;

  Point(PVector pos) {
    this.pos = pos;
  }
```

The `draw()` function displays the point on screen. Note that you again put the `point()` function within an `if()` statement that only draws if the Boolean `record` is not true.

```
  void draw() {
    pushStyle();
    strokeWeight(10);
    if (selected) {
      stroke(255, 255, 0);
    }
    else {
      stroke(200);
    }
    if (!record){
     point(pos.x, pos.y, pos.z);
    }
    popStyle();
  }
```

Finally, implement three more functions that allow you to change the position of the point, select it, and unselect it (Figure 11-18).

```
  public void setPos(PVector newPos) {
    this.pos = newPos;
  }

  public void select() {
    this.selected = true;
  }
  public void unSelect() {
    this.selected = false;
  }
}
```

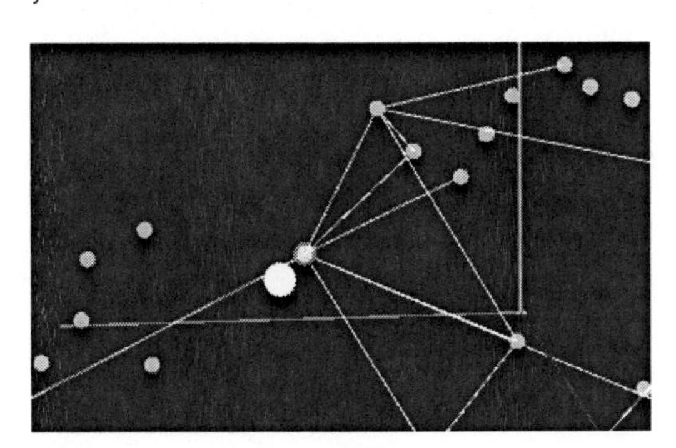

Figure 11-18. Selected point

Line Class

As you saw in the previous section, a line is a *one-dimensional* geometrical element. A line is by definition infinite, but in this case, whenever we talk about lines, we're actually referring to *line segments.*

A line segment is a portion of a line, and it can be defined by two points. You have probably heard *line* defined as "the shortest path between two points," which is pretty close to the sense the word *line* will have in your code.

Your Line class, then, has two Point objects as fields. These two points define the line and are used within the draw() function to display the line on screen by using Processing's line() function.

```
public class Line {
  Point p1;
  Point p2;

  Line(Point p1, Point p2) {
    this.p1 = p1;
    this.p2 = p2;
  }

  void draw() {
    pushStyle();
    stroke(255);
    line(p1.pos.x, p1.pos.y, p1.pos.z, p2.pos.x, p2.pos.y, p2.pos.z);
    popStyle();
  }
}
```

Shape Class

The Shape class represents triangular surfaces in your program. A surface is a *two-dimensional* geometrical element. Your Shape objects represent portions of planes, or flat surfaces, defined by three points (Figure 11-19). Triangular shapes are the basis of many polygonal modeling programs; for you, they constitute the only way in which your software will represent surfaces.

The Shape class uses an ArrayList of Point elements to store the points defining it. You could have used three points instead, but this way lets you extend the concept further and draw complex shapes consisting of more than three points. The draw() function makes use of Processing's beginShape(), vertex(), and endShape() functions to draw your shapes on screen. In the future, you might want to dynamically add points to this function in runtime, so add a function addPoint() for this purpose.

```
public class Shape {
  ArrayList<Point> points = new ArrayList<Point>();
  Shape( ArrayList<Point> points) {
    for (Point pt : points) {
      this.points.add(pt);
    }
  }

void draw() {
    pushStyle();
    fill(255);
    stroke(150);
    beginShape();
    for (int i = 0; i < points.size(); i++) {
```

```
      vertex(points.get(i).pos.x, points.get(i).pos.y, points.get(i).pos.z);
    }
    endShape(PConstants.CLOSE);
    popStyle();
  }

  void addPoint(Point newPoint) {
    points.add(newPoint);
  }
}
```

Figure 11-19. *Freeform three-dimensional geometry built with triangular shapes*

Modeling Interface Main Program

Now that you have your helper classes implemented, you can write the main routine. This code drives the program's workflow and makes use of the previous classes to achieve the ultimate goal of allowing you to make 3D geometries using your glove interface. Add more of Processing's core libraries so you can export the geometries you create to .dxf files and import them into other CAD software.

Imports and Fields

You need to import five libraries for this project: three core libraries (OpenGL, Serial, and DXF) and two external libraries (Simple-OpenNI and KinectOrbit).

```
import processing.opengl.*;
import processing.serial.*;
import processing.dxf.*;
import SimpleOpenNI.*;
```

```
import kinectOrbit.*;

KinectOrbit myOrbit;
SimpleOpenNI kinect;

Serial myPort; // Initialize the Serial Object
boolean serial = true; // Define if you're using serial communication
```

You are implementing hand recognition using NITE's XnVSessionManager, as you did in Chapter 5, so you must initialize the appropriate objects.

```
XnVSessionManager sessionManager;
XnVPointControl pointControl;
```

You then declare the GloveInterface object glove and four ArrayList collections defining the geometry's points, selected points, lines, and shapes. Use ArrayList collections because you don't know how many points, shapes, or lines the user will create; ArrayList collections provide you with the flexibility you need for the program. The integer thresholdDist defines how close the glove interface needs to be from a point for it to be selected.

```
private GloveInterface glove;
ArrayList<Point> points = new ArrayList<Point>();
ArrayList<Point> selPoints = new ArrayList<Point>();
ArrayList<Line> lines = new ArrayList<Line>();
ArrayList<Shape> shapes = new ArrayList<Shape>();
int thresholdDist = 50;
```

The boolean record serves to indicate when you want to save your geometry as a .dxf file. It is set to false by default, so you avoid creating a new file in the first loop.

```
boolean record = false;
```

Setup Function

The setup() function starts by defining the sketch size and renderer, and it then initializes the KinectOrbit object. You then set KinectOrbit to draw the orbiting gizmo on screen and the Coordinate System axes. You also use the function setCSScale() to set the size of the CS axes to 200 mm. (You can change these settings to any other configuration.)

```
public void setup() {
  size(800, 600, OPENGL);
  myOrbit = new KinectOrbit(this, 0, "kinect");
  myOrbit.drawGizmo(true);
  myOrbit.drawCS(true);
  myOrbit.setCSScale(200);
```

Now you initialize the Simple-OpenNI object and all the capabilities you need, including the sessionManager and pointControl objects.

```
  // Simple-openni
  kinect = new SimpleOpenNI(this);
  kinect.setMirror(false);
  kinect.enableDepth();
  kinect.enableGesture();
  kinect.enableHands();
```

```
  sessionManager = kinect.createSessionManager("Wave", "RaiseHand"); // setup NITE
  pointControl = new XnVPointControl();   // Setup NITE's Hand Point Control
  pointControl.RegisterPointCreate(this);
  pointControl.RegisterPointDestroy(this);
  pointControl.RegisterPointUpdate(this);
  sessionManager.AddListener(pointControl);
```

And finally, you initialize the serial communication and the GloveInterface object.

```
  if (serial) {
    String portName = Serial.list()[0]; // Get the first port
    myPort = new Serial(this, portName, 9600);
    myPort.bufferUntil('\n');
  }

  glove = new GloveInterface();  // Initialize the GloveInterface Object
}
```

Draw Function

As usual, you pack most of the processes into additional functions, so the draw() function presents a clear scheme of the program's workflow. The first step is to update the Simple-OpenNI object (kinect) and the sessionManager. Then you set the background to black.

```
public void draw() {
  kinect.update(); // Update Kinect data
  kinect.update(sessionManager); // update NITE
  background(0);
```

This program behaves differently depending on which finger is bent. You do this by means of a switch() statement with four case handlers. The four possible behaviors are moving an existing point, creating a new point, adding a line between two points, or adding a new shape (Figure 11-20).

```
  switch (glove.currentFinger) {
  case 0:
    movePoint();
    break;
  case 1:
    addPoint();
    break;
  case 2:
    addLine();
    break;
  case 3:
    addShape();
    break;
}
```

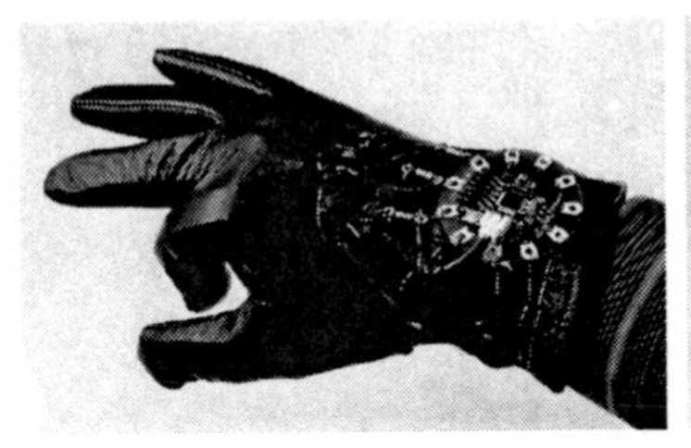 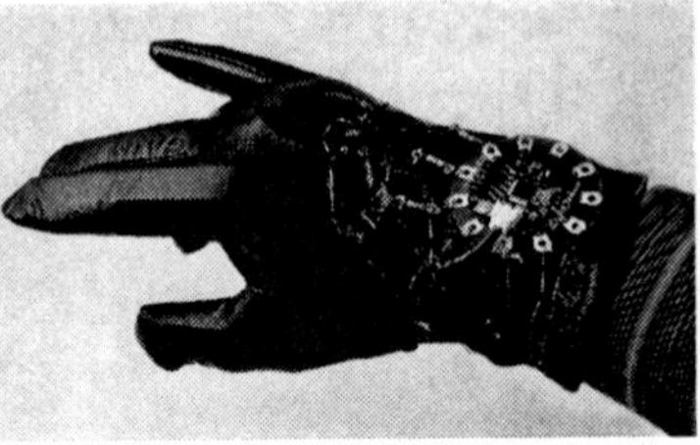 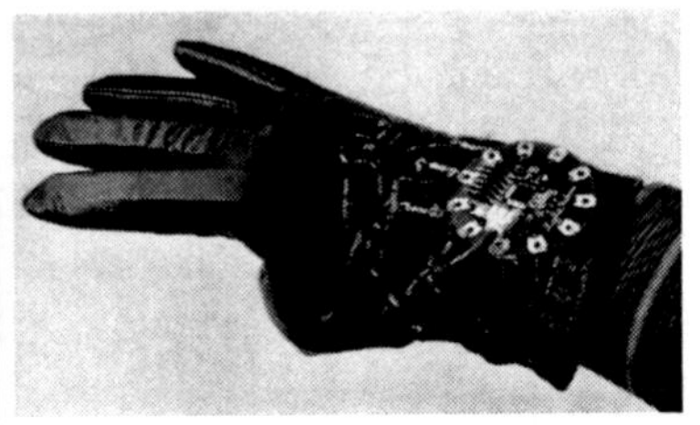

***Figure 11-20.** Triggering Actions with your Fingers and thumb*

After this switch() statement, all the geometrical operations have been performed, so you only need to draw the results on screen. You first push the KinectOrbit object so you can control your point of view of the geometry.

```
myOrbit.pushOrbit(this); // Start Orbiting
```

After pushing the KinectOrbit, check if the boolean record is true; if so, use the beginRaw() method to save the geometry into the file output.dxf. Then draw the glove and the points, lines, and shapes. The drawClosestPoint() function highlights the closest point to your glove interface if the distance is smaller than thresholdDist.

```
  if (record == true) {
    beginRaw(DXF, "output.dxf"); // Start recording to the file
  }

  glove.draw();
  drawClosestPoint();

  for (Point pt : points) {
    pt.draw();
  }
  for (Line ln : lines) {
    ln.draw();
  }
  for (Shape srf : shapes) {
    srf.draw();
  }
```

Before you pop the KinectOrbit, check again the state of the record variable. If it's true (and you are recording the .dxf file), you end the recording and set the record variable to false. After this, you pop the KinectOrbit and, if you want to see the finger data printed on screen, call the method drawData() from the glove object.

```
  if (record == true) {
    endRaw();
    record = false; // Stop recording to the file
  }
  myOrbit.popOrbit(this); // Stop Orbiting
  glove.drawData();
}
```

Additional Functions

The movePoint() function is called when you bend your thumb. It checks all the existing points; if their distance to your glove object is under thresholdDist, it changes the point's position to the position of your glove (which is your hand's position). In this way, you can actually move the points around with your hand, just as if they were standing around you in space.

You use a Java for-each loop to run through the ArrayList points. This is equivalent to using a standard for loop, simplifying this common form of iteration. You can read it as "For each Point object in the ArrayList points, do { }." The Point instance pt is set to a different Point object inside points in every loop.

```
void movePoint() {
  for (Point pt : points) {
    if (glove.pos.dist(pt.pos) < thresholdDist) {
      pt.setPos(glove.pos);
    }
  }
}
```

The addPoint() function creates a new point at the position of your glove. Bending your index finger triggers this function. To avoid the creation of multiple adjacent points, you won't create a point if there is another point closer than thresholdDist.

```
void addPoint() {
  Point tempPt = new Point(glove.pos.get());
  boolean tooClose = false;
  for (Point pt : points) {
    if (tempPt.pos.dist(pt.pos) < thresholdDist) {
      tooClose = true;
    }
  }
  if (!tooClose) {
    points.add(tempPt);
  }
}
```

Bending your middle finger calls the addLine() function. This function proceeds in two steps. First, it checks if there is a point closer than the defined threshold. If there is one and it's not already contained in the ArrayList selPoints, it selects the point and adds it to selPoints.

Then, if there is more than one point selected, it creates a new line between the two points in selPoints and adds it to the ArrayList lines. Finally, it unselects everything using the unSelectAll() function.

```
void addLine() {
  for (Point pt : points) {
    if (glove.pos.dist(pt.pos) < thresholdDist) {
      pt.select();
      if (!selPoints.contains(pt)) {
        selPoints.add(pt);
      }

      if (selPoints.size() > 1) {
        Line lineTemp = new Line(selPoints.get(0), selPoints.get(1));
        lines.add(lineTemp);
```

```
          unSelectAll();
        }
      }
    }
  }
```

The last geometry creation function, addShape(), is called when you bend your third finger. This function works in the same way as addLine(), but it waits until you have three selected points to create a new Shape object, adds it to the ArrayList shapes, and unselects all the points.

```
private void addShape() {
  for (Point pt : points) {
    if (glove.pos.dist(pt.pos) < thresholdDist) {
      pt.select();
      if (!selPoints.contains(pt)) {
        selPoints.add(pt);
      }
      if (selPoints.size() > 2) {
        Shape surfTemp = new Shape(selPoints);
        shapes.add(surfTemp);
        unSelectAll();
      }
    }
  }
}
```

The unSelectAll() function runs through all the points in your sketch and unselects them; then it clears the selPoints ArrayList.

```
void unSelectAll() {
  for (Point pt : points) {
    pt.unSelect();
  }
  selPoints.clear();
}
```

The previously used drawClosestPoint() function takes care of changing the onscreen appearance of the points that are close enough to the glove to be modified with the other functions.

```
void drawClosestPoint() {
  for (Point pt : points) {
    if (glove.pos.dist(pt.pos) < thresholdDist) {
      pushStyle();
      stroke(0, 150, 200);
      strokeWeight(15);
      point(pt.pos.x, pt.pos.y, pt.pos.z);
      popStyle();
    }
  }
}
```

Processing Callbacks

You use two Processing callback functions in this project. First, the serialEvent() function calls every time you receive a newline character from serial. This function acquires the string from the serial communication, splits it, and passes the resulting array of strings as a parameter to the glove object using the method setFingerValues(), which updates the finger-bending values.

```
public void serialEvent(Serial myPort) {
  String inString = myPort.readStringUntil('\n');
  if (inString != null) {
    String[] fingerStrings = inString.split(" ");
    if (fingerStrings.length == 4) {
      glove.setFingerValues(fingerStrings);
    }
  }
}
```

You use the keyPressed() Processing callback function to calibrate your glove interface. Numbers 1 to 4 set the stretched values for your fingers, and q, w, e, and r characters set the maximum bending values. Additionally, the character d triggers the saving of the geometry into the previously defined .dxf file.

```
public void keyPressed() {
  switch (key) {
  case '1':
    glove.calibrateFinger(0, 0);
    break;
  case '2':
    glove.calibrateFinger(1, 0);
    break;
  case '3':
    glove.calibrateFinger(2, 0);
    break;
  case '4':
    glove.calibrateFinger(3, 0);
    break;
  case 'q':
    glove.calibrateFinger(0, 1);
    break;
  case 'w':
    glove.calibrateFinger(1, 1);
    break;
  case 'e':
    glove.calibrateFinger(2, 1);
    break;
  case 'r':
    glove.calibrateFinger(3, 1);
    break;
  case 'd':
    record = true;
    break;
  }
}
```

Simple-OpenNI Callbacks

Finally, you add the Simple-OpenNI callback functions. Within onPointCreate(), you set the zero position of your glove. This has the effect of setting the origin of coordinates to the point in space where the waving gesture was recognized. This makes it easier for the user to be drawing at an acceptable distance from the origin.

```
public void onPointCreate(XnVHandPointContext pContext) {
  println("onPointCreate:");
  PVector handVec = new PVector(pContext.getPtPosition().getX(), pContext
    .getPtPosition().getY(), pContext.getPtPosition().getZ());
  glove.setZeroPos(handVec.get());
}

public void onPointDestroy(int nID) {
  println("PointDestroy: " + nID);
}
```

Within the onPointUpdate() function, you call the setPosition() method from your glove object to update the position of the glove image on screen to reflect your current hand position.

```
public void onPointUpdate(XnVHandPointContext pContext) {
  PVector handVec = new PVector(pContext.getPtPosition().getX(), pContext
    .getPtPosition().getY(), pContext.getPtPosition().getZ());
  glove.setPosition(handVec);
}
```

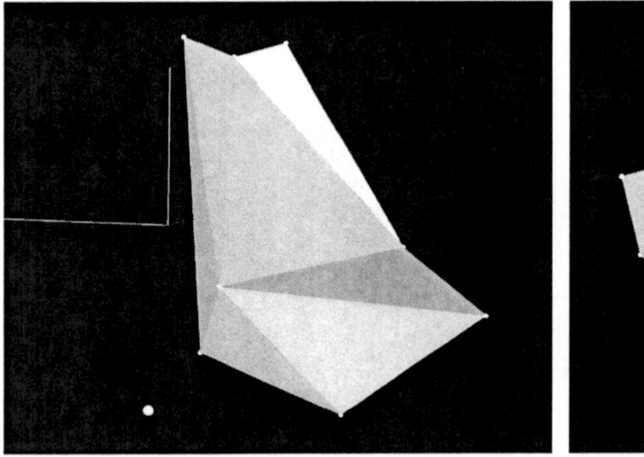
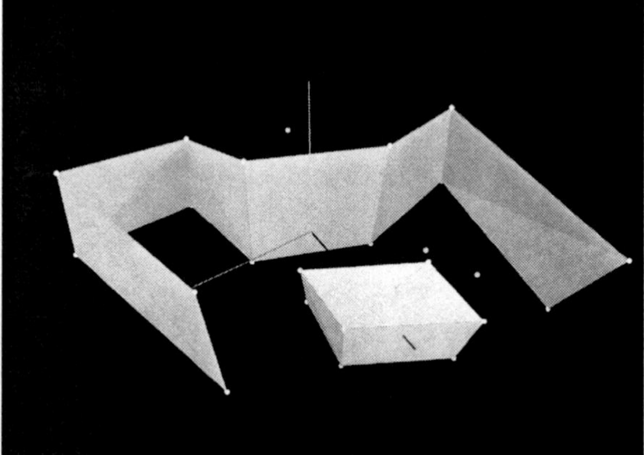

Figure 11-21. 3D geometries created with the interface

Run this program and start playing with the position of your hands and the bending of your fingers (Figure 11-21). Soon you will develop an intuitive understanding of the functioning of the program; after some time, you will be able to create some pretty complex 3D geometries without having touched your mouse.

When you are satisfied with your results, press the d key on your keyboard and you'll get a file called output.dxf in your Processing sketch folder. Note that .dxf is a file format created by Autodesk (the company behind AutoCAD) for importing and exporting data from AutoCAD to other programs.

You have chosen `.dxf` because there is a Processing core library that does all the exporting for you, but if you are curious enough to read the next chapter, you will learn how to implement your own routine to export `.ply` files, which are compatible with many other modeling programs.

Summary

This project led you through the building of a wearable, lightweight, wireless user interface designed to work in combination with Kinect's hand-tracking capabilities that ultimately allowed you to create 3D geometries in an intuitive way, without using any conventional human interface devices.

You built the interface on a glove, using the Arduino LilyPad board in combination with flex sensors to read the bending of your fingers and a couple of XBee radio modules to transmit the information to your computer wirelessly. You also learned how to use the Arduino SoftwareSerial library to use any pin on the Arduino for serial communication.

Finally, you implemented your own simplified CAD software to deal with the basic geometrical entities and data structures that constituted your constructions. This software is controllable by your glove interface.

This software and interface are just the bare bones of what could be the beginning of a new standard in human-computer interfaces. Thanks to the Kinect, you are no longer limited to the two dimensions of conventional devices. Now you can use the full mobility of your body as a way to input data into your applications.

CHAPTER 12

Turntable Scanner

by Enrique Ramos

If you're like most people, one of the first things that comes to mind when you see a Kinect point cloud, or any 3D scanner point cloud, is that it's only one-sided. There are a number of techniques that can be used to perform a 360-degree volumetric scan and then reconstruct the implicit surface into a three-dimensional mesh. One way is to use a turntable.

In this chapter, you will build a DIY turntable, connect it to your computer, and scan objects from every point of view. Then, you will learn how to process the scanned data and patch the different point clouds into one consistent point set. There will be a fair bit of trigonometry and vector math involved in this process, so you might want to keep a reference book handy.

You will then implement your own exporting routine to output files containing your point cloud data in a format readable by other software. Finally, you will learn how to use Meshlab to reconstruct the implicit surface from your point cloud so you can use or modify it in any modeling software, or even bring the modified volume back to the physical world with a 3D printer. Figure 12-1 shows the assembled system.

Figure 12-1. Assembled Kinect and turntable

The Theory

The ultimate goal of this project is to be able to scan an object from all 360 degrees around it to get the full point cloud defining its geometry. There are a number of techniques that you could use to achieve this.

First, you could use several Kinect cameras, all connected to the same computer and placed around the object you want to scan. There are a couple of downsides to this technique. First, you need to buy several Kinect devices, which is expensive. Then, every time you need to scan something you need to set up the Kinects, which takes a large amount of space.

Another option is to use SLAM and RANSAC-based reconstruction. These are advanced computational techniques that allow you to reconstruct a scene from different scans performed with a 3D-scanner without having any kind of information about the actual position of the camera. The software analyzes the different point clouds and matches them by detecting the presence of outliers (RANSAC stands for RANdom SAmple Consensus, SLAM for Simultaneous Localization and Mapping). Nicolas Burrus implemented a very successful scene reconstruction that is included in his software RGBdemo (`http://nicolas.burrus.name`), and Aaron Jeromin has adapted the RGBdemo code to create a homemade turntable scanner, the results of which you can see in Figure 12-2 and on its YouTube site at `http://www.youtube.com/user/ajeromin`.

But turntable scanners are nothing new, nor do they depend on the capabilities of the Kinect. Commercial 3D scanning companies offer turntables to be used with their 3D scanning cameras, like Mephisto from 4ddynamics. Structured light-based 3D scanners have been around for a while, and you can find a few DIY examples on the Internet, such as SpinScan (`http://www.thingiverse.com/thing:9972`).

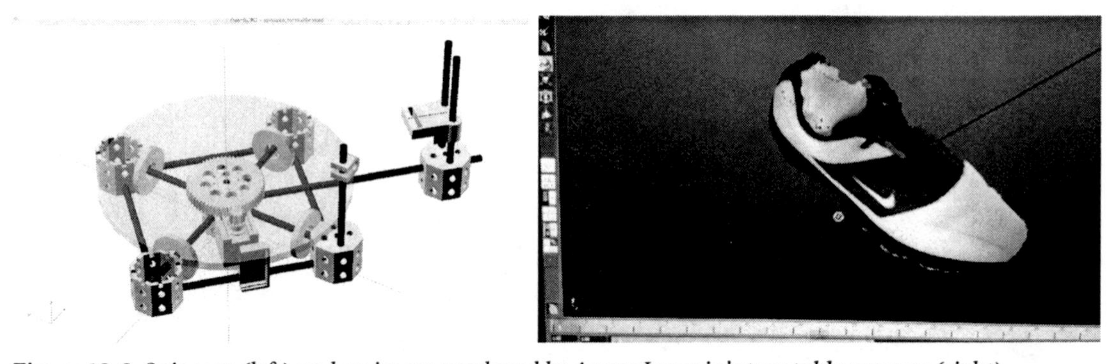

Figure 12-2. *Spinscan (left) and an image produced by Aaron Jeromin's turntable scanner (right)*

As clearly announced in the introduction, you're going to take the turntable approach. But you're not going to rely on any library or complicated algorithms for this project. You are going to build a simple but accurate turntable and use very basic geometrical rules to reconstruct a scene from several shots of the same geometry taken from different points of view. Then you will use another open source piece of software, Meshlab, to reconstruct the unstructured point cloud generated by your Processing sketch from the Kinect data.

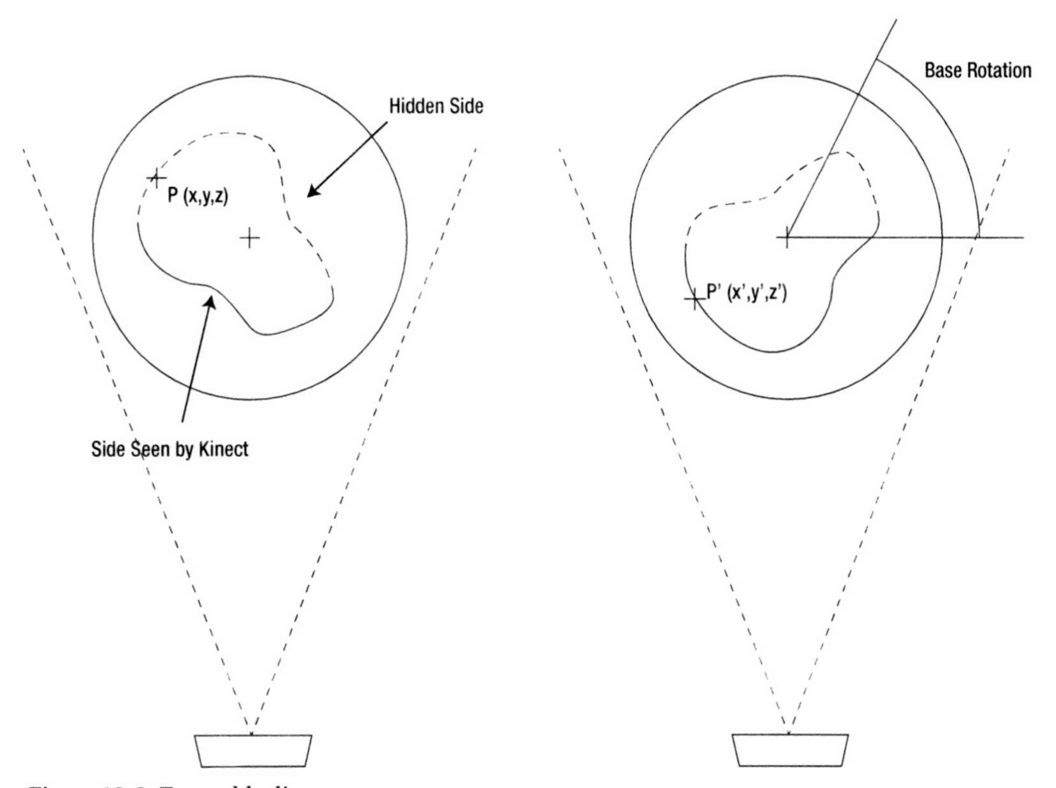

Figure 12-3. *Turntable diagram*

When you scan an object with the Kinect, there is always a visible side, the one that the Kinect can "see", and a hidden side that you can't capture with a single shot (Figure 12-3). If you mount your object on a rotating base, or turntable, you can rotate the object so that point P is on the visible side. You can then rotate the turntable to any angle and thus be able to see the object from every point of view.

But you want to reconstruct the whole object from different shots, so you need to have all points defining the object referred to the same coordinate system. In the original shot, point P was defined by the coordinates P(x,y,z); when you rotate the platform to see the point, the coordinates of P change, being now P(x',y',z'). This is only a problem if you don't know how much the platform has rotated. But if you know precisely the angle of the platform from the original shot, you can easily retrieve the original coordinates P(x,y,z) from the transformed ones P(x',y',z') using basic trigonometry. Let's call your angle α; you are rotating the point around the y-axis, so the transformation will be the following:

$$x = x' * \cos(\alpha) - z' * \sin(\alpha);$$
$$y = y'$$
$$z = x' * \sin(\alpha) + z' * \cos(\alpha);$$

By applying this simple transformation, you can scan from as many points of view as you want and then reconstruct the volume in a consistent coordinate system. You are going to build a turntable that will allow you to rotate the object 360 degrees, knowing at all times the precise angle at which you are rotating the object so you can retrieve the global coordinates of the scanned points (Figure 12-4). You are going to start by building the turntable, and then you will learn how to process the data.

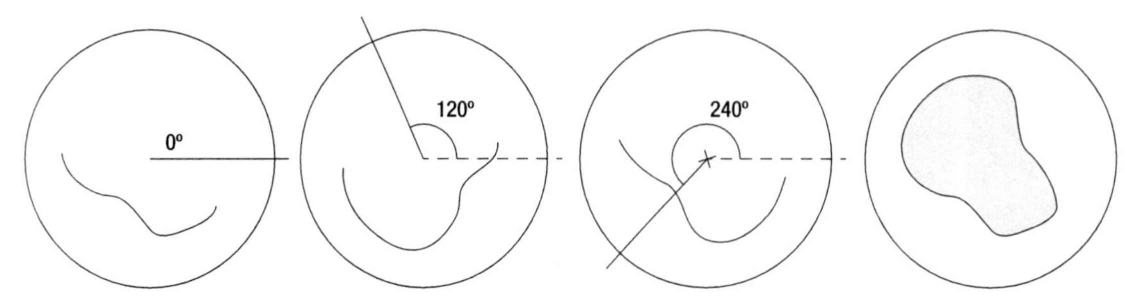

Figure 12-4. Reconstruction of an object from three different views.

This project can be built entirely from cheap parts you can find in hardware stores and hobbyist shops. Table 12-1 lists the requirements.

Table 12-1. The Parts List

Part	Description	Price
2 swivel discs	From the hardware store	$6.59 each
2 metal box shapes	We used some post bases we found in a DIY shop	$7.69 each
Aluminum angle 2.40m	From the hardware store	$12.95
Acrylic sheet	Thick enough to be stable	$17.59
Multi-turn potentiometer	DFRobot Rotation Sensor V2	$4.74
Continuous rotation servo	Parallax (Futaba)	$13.79
Servo bracket	Lynxmotion Aluminum	$11.95
Gears	We adapted the gear kit from Vex	$12.99
1 prototype shield	SparkFun DEV-07914, but any one will do.	$16.95

Building a Turntable

A turntable is a platform mounted on a swivel that can rotate around a center point. Yes, it's exactly like the turntable on any record player. We actually considered using an old record player for this project, but in the end it was cheaper to build our own.

There is something that you will need to bear in mind during the whole process of building the turntable: accuracy. The purpose of building such a device is to have a platform that can hold a reasonable weight without collapsing and that can rotate freely using a servo motor. The higher the precision of your physical device, the higher the consistency of the point clouds when you patch them together, so try to be thorough with your dimensions and tolerances!

We started this project by building a prototype of the turntable so we could test basic algorithms and principles. The prototype consisted of a cardboard box with a servo sticking out of its top face, as you can see in Figure 12-5. On the servo, we attached a circle of cardboard. That was it. This prototype was, obviously, not accurate enough: the servo could only rotate 180 degrees so we couldn't scan our object all around. However, it was very useful for understanding the basic principles and for developing the software that we will discuss in a later section.

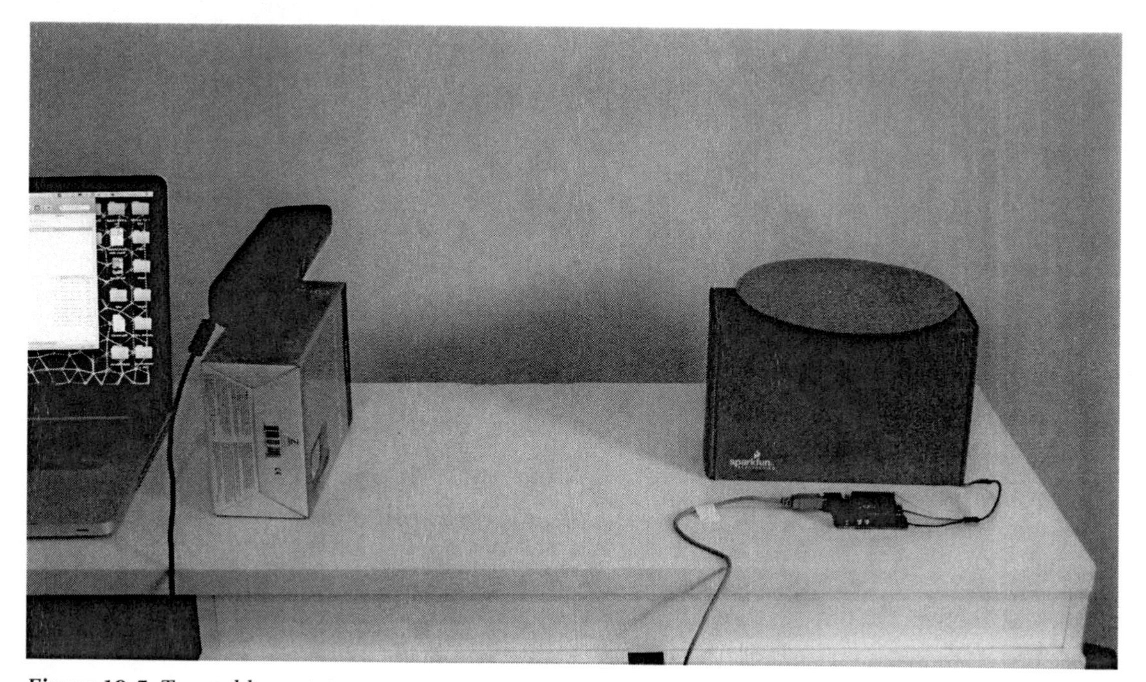

***Figure 12-5.** Turntable prototype*

The issue of the 180-degree rotation of standard servos was a key issue in the design of our final turntable. We could have used a continuous rotation servo or hacked a standard servo into rotating continuously, but then we would lose the feedback. In other words, we could drive the servo forward and backward, but we wouldn't know the precise position of the turntable. An option that immediately came to mind was a 360-degree servo. However, power was also an issue for the prototype, so we wanted to increase the torque of our motor with gears. The final solution was to use a continuous rotation servo and build our own feedback system with a multi-turn potentiometer. The two were connected with gears to the turntable, increasing the power of the servo seven times. Let's proceed step by step to the building of the turntable. Figure 12-6 shows all of the parts of the turntable.

***Figure 12-6.** Turntable parts*

Connecting the Gears

The gears you're using have a one-to-seven ratio, so the large gear rotates once every seven turns of the smaller gears. This is perfect for this project because your potentiometer can turn up to ten times, and the three gears fit perfectly onto the base. But they come from a gear kit from Vex Robotics, so the gears don't fit naturally with the Parallax servo or the potentiometer. The first thing you need to do is modify the gears so you can fit them tightly with the other two components (Figure 12-7).

For the potentiometer, drill a hole in the middle of the gear slightly smaller than the diameter of the potentiometer's shaft. This way you'll be able to push it in, and it will rotate the shaft without the need for gluing or attaching it permanently. Start with a small drill bit and go bigger step by step. Otherwise, you might drill a non-centered hole.

For the motor, you need to drill a hole only about a fourth of the depth of the gear (see Figure 12-7). The servo shaft is pretty short, and you're going to be attaching the gear on the other side of the metal base. Don't worry if the gear is slightly loose; you're going to secure it to the servo with a screw.

Figure 12-7. *Gear adapted to the servo*

Now you can attach the potentiometer and the servo to the base. First, check the position of the gears so you have an idea of where the holes need to be drilled for the potentiometer (see Figures 12-8 and 12-9).

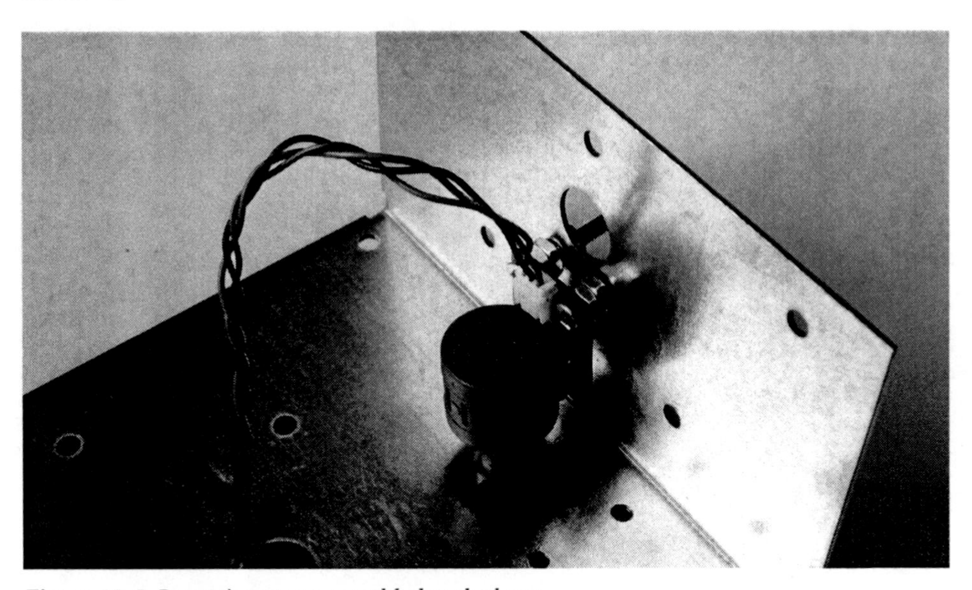

Figure 12-8. *Potentiometer asssembled to the base*

***Figure 12-9.** Gears assembled to the base*

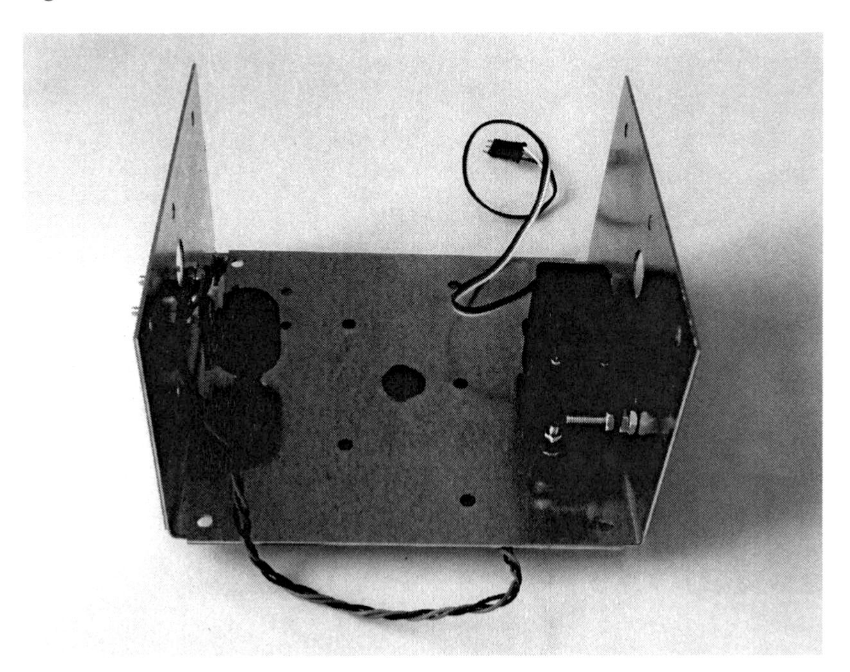

***Figure 12-10.** Potentiometer and servo assembled to the base*

The fixing of the servo and the potentiometer is highly dependent on the base element you have chosen. In our case, there was enough to space the servo bracket and the potentiometer with a couple of nuts. Note that we have attached the potentiometer and the servo to the sides of the base (Figure 12-10) instead of the top plate. This is to avoid any protruding bolts or nuts on the top plate that would get in the way of the swivel.

You should achieve a good match for the three gears so they rotate together without effort. The large gear will be attached to the acrylic sheet later. Make sure the gears are parallel to the base so there is no friction when the motor is in action.

The next step is to attach the swivel to the base (Figure 12-11), making sure that its center of rotation is as close to the center of rotation of the larger gear as you can possibly manage. This is very important because a badly centered swivel will not rotate when assembled!

Figure 12-11. *Swivel and gears*

The large gear won't be attached to the base. You attach it to the acrylic sheet that you previously cut in a disc shape 29cm in diameter. You fix the gear to the disc with a central screw, and you add another to avoid the differential rotation of the disc and the gear (Figure 12-12).

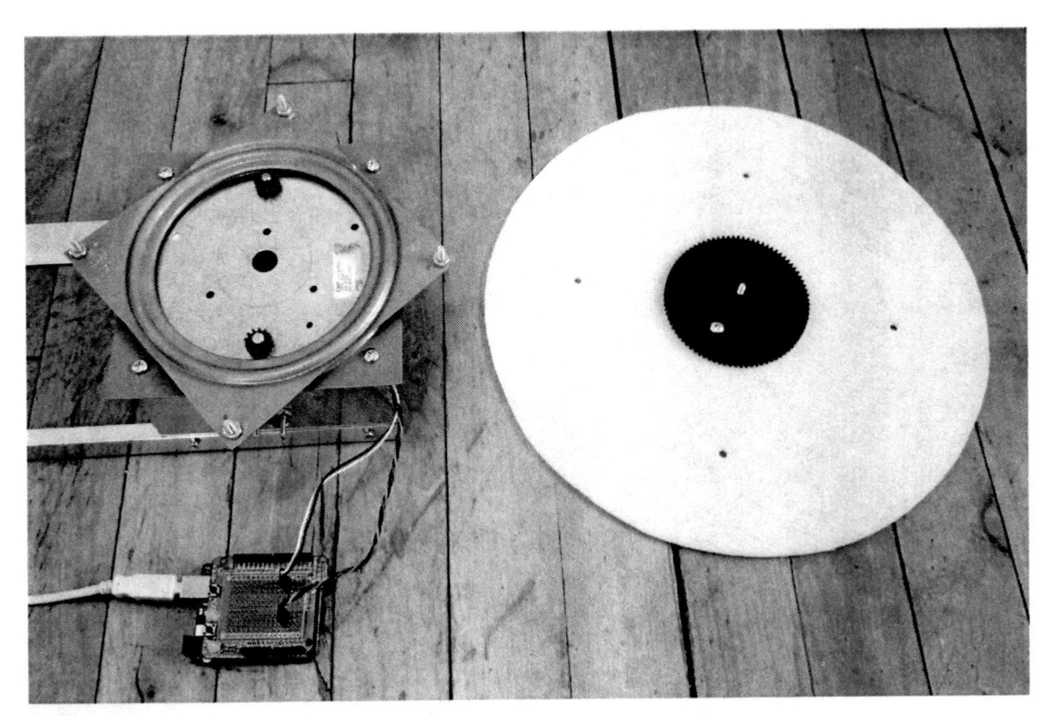

Figure 12-12. Disc and gear ready to be assembled

You could declare your turntable ready at this point, but you are going to attach a base for the Kinect to it so you don't have to calibrate the position of the Kinect every time you use the scanner.

The Kinect base is exactly like the one you used for the turntable, so both will be at the same height, and the Kinect will stand at pretty good position. First, cut the aluminum angle into two 1.2m pieces and drill three holes on each end to the same spacing of the holes on the bases (Figure 12-13).

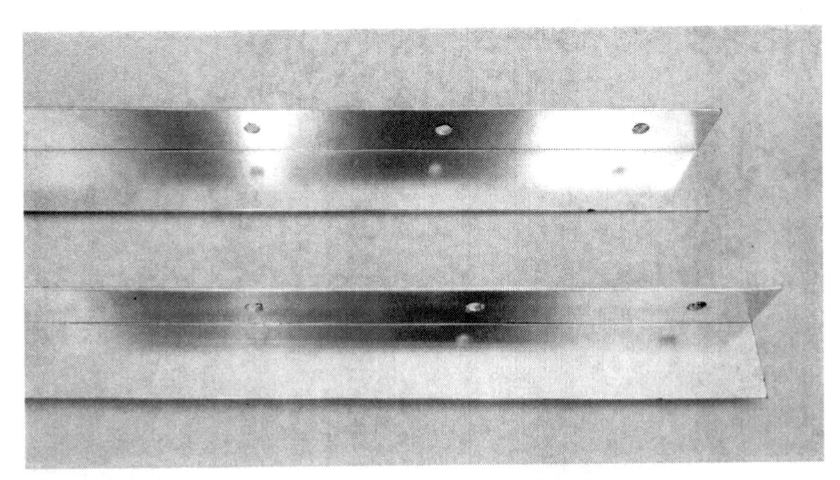

Figure 12-13. Bracing aluminum angles after drilling the holes

Figure 12-14. Assembled turntable

Figure 12-14 shows two bases separated by more than 1m, which is enough for the Kinect to be able to scan a medium-sized object. If you are thinking of scanning larger objects, play a little with your Kinect first to determine an ideal distance to the turntable. Once you have calibrated your Kinect and your turntable (check that the center of your turntable is coincident with the center of the scanning volume in your sketch), you can attach the Kinect to the base or simply draw a contour that will tell you where to place it every time. If you fancy a cooler finishing, you can countersink the screws and add a vinyl record to your turntable as a finishing surface (see Figure 12-15).

Figure 12-15. Assembled turntable

Building the Circuit

The turntable circuit is a very simple circuit with only a servo motor and a potentiometer. You're building it on a SparkFun Arduino prototype shield because you want the circuit to be usable for this and other projects. The circuit includes an analog reader on pin 0 and a servo, as shown in Figures 12-16 and 12-17. The circuit leaves two three-pin breakaway headers ready to for the servo and the potentiometer.

Figure 12-16. *Arduino prototype shield with breakaway headers*

Figure 12-17. *Arduino prototype shield (back)*

The circuit, as depicted in Figure 12-18, connects the three pins of the breakaway headers to 5v, ground, and pins analog 0 and digital 6.

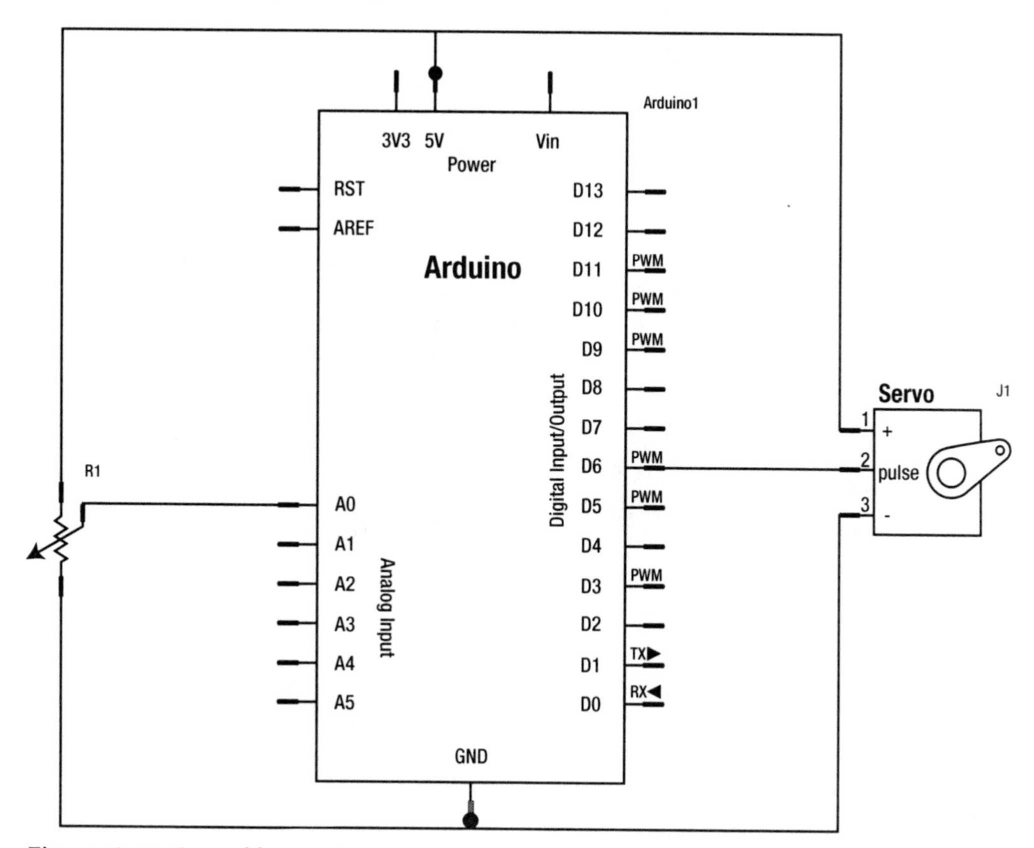

Figure 12-18. Turntable circuit

Arduino Code

You are going to implement an Arduino program that will drive the servo forward and backward according to signals coming from your main Processing program and send back the rotation of the potentiometer at every point in time. It will also inform the Processing sketch when it reaches the intended position.

The servo is attached to digital pin 6 and the potentiometer to analog pin 0, so create the necessary variables for those pins and the pulse for the servo. You also need a couple of long integers for the servo timer.

```
int servoPin = 6;
int potPin = 0;
int servoPulse = 1500;
long previousMillis = 20;
long interval = 20;
```

Figure 12-19. Testing the range of the potentiometer

After some intensive testing sessions, we worked out that 360 degrees on the turntable meant a range of 724 in the potentiometer values (Figure 12-19) . So use a start position of 100 and an end position of 824. In other words, whenever the potentiometer is at 100, you know your angle is zero; whenever the potentiometer reads 824, you know you have reached 360 degrees. The potentiometer values change in proportion to the angle, so you can always work out the angle from the values read from the potentiometer.

```
// Start and end values from potentiometer (0-360 degrees)
int startPos = 100;
int endPos = 824;
```

The continuous rotation servo doesn't work like a normal servo. You still have a range of 500-2500 microseconds for your pulses, but instead of the pulse meaning a specific angle, any pulse under 1500 microseconds makes the servo to turn forward, and any pulse over 1500 microseconds sends the servo turning backward. Create a couple of variables for the forward and backward pulses.

```
// Forward and backward pulses
int forward = 1000;
int backward = 2000;
```

Your Arduino sketch works in two states: `state=0` means stand and wait for orders and `state=1` means move toward the target angle. `PotAngle` is the variable storing the potentiometer values.

```
int state = 1;     // State of the motor
int targetAngle = startPos;  // Target Angle
int potAngle;      // Current angle of the potentiometer
```

In the `setup()` function, you only need to start the serial communication and define your servo pin as output.

```
void setup(){
  Serial.begin(9600);
  pinMode (servoPin, OUTPUT);
}
```

The main `loop()` function is completely engulfed within an "`if()`" statement that executes once every 20 milliseconds. This makes sure that you only talk to the servo within the acceptable range and that you don't send too much serial data and saturate the channel.

```
void loop() {
  unsigned long currentMillis = millis();
  if(currentMillis - previousMillis > interval) {
    previousMillis = currentMillis;
```

The first thing you do in the loop is read the value of the potentiometer and print it out to the serial port. After this, you perform a data validation check. If the read value is out of your range (100-824), you move the potentiometer in toward your range and you set the state to zero so it stops moving.

```
    potAngle = analogRead(0);  // Read the potentiometer
    Serial.println(potAngle);  // Send the pot value to Processing
    if(potAngle < startPos){  // If you have
      state = 0;
      updateServo(servoPin,forward);
      Serial.println("start");
    }
    if(potAngle > endPos){
      state = 0;
      updateServo(servoPin,backward);
      Serial.println("end");
    }
```

Now you need to check if you have any data in your serial buffer. You will implement a separate function for this, so here you simply have to call it. Finally, if your state is 1, you call the function `goTo(targetAngle)`, which moves the servo towards your goal.

```
    checkSerial();  // Check for values in the serial buffer
    if(state==1){
      goTo(targetAngle);
    }
  }
}
```

The function `goTo()` checks whether the current reading from the potentiometer is higher or lower than the target angle and updates the servo accordingly. The step of the servo is larger than one unit, so you have allowed a buffer zone of 10 units to avoid the oscillation around a target value. If you are within the accepted range from the target, you set your state to 0 and wait for orders and then you send a message through the serial channel saying that you have reached your intended position.

```
void goTo(int angle){
  if(potAngle-angle < -10) {
    updateServo(servoPin,forward);
  }
  else if(potAngle - angle > 10){
    updateServo(servoPin,backward);
  }
  else {
    Serial.println("arrived");
    state=0;
  }
}
```

The orders you are waiting for will come via serial communication. In the checkSerial() function, you look into the serial buffer, and if there is at least two data, you read the first one. If this happens to be the trigger character 'S', you check the next one, which you interpret as the new target angle for the turntable. You map this value from 0-255 to the whole range of the potentiometer, and you set your state to 1, which allows the turntable to move toward this new value.

```
void checkSerial(){
  if (Serial.available() > 1) { // If data is available to read,
    char trigger = Serial.read();
    if(trigger == 'S'){
      state = 1;
      int newAngle = Serial.read();
      targetAngle = (int)map(newAngle,0,255,startPos,endPos);
    }
  }
}
```

The updateServo() function is the same as that used throughout the book. You write a high signal to the servo pin for the amount of microseconds that you have passed as an argument.

```
void updateServo (int pin, int pulse){
  digitalWrite(pin, HIGH);
  delayMicroseconds(pulse);
  digitalWrite(pin, LOW);
}
```

This simple code allows you to control the rotation of the turntable from any serial-enabled software (In your case, Processing). On top of that, this Arduino code is sending out messages with the current angle of the turntable and whether the target position has been reached. Let's see how to use this data and your turntable to scan 3D objects.

Processing Code

You are going to write a slightly complex piece of code that will communicate with the Arduino board and interpret the incoming data to reconstruct a three-dimensional model of the objects standing on the turntable. We covered the geometrical principles behind the patching of the different scans at the beginning of the chapter, but we will expand on this in the pages that follow.

We looked for a complex and colorful shape for this example and, because we are writing this in November, the first thing we found was a little figure of Santa in a dollar store (Figure 12-20). We will be using this figure for the rest of the chapter. If you can find one too, you'll have a complete, full-color model of your favorite Christmas gift-bringer inside of your computer by the end of the chapter, as per Figure 12-21. Let's start coding!

Figure 12-20. *Our example model*

Figure 12-21. *Our example model in scanning space*

Variable Declaration

The imports are the usual ones: Serial, OpenGL, Simple-OpenNI, and KinectOrbit. You define a Boolean variable called serial that initializes the serial ports only if it is true. This is useful if you want to test stuff with the serial cable disconnected and you don't want to get those nasty errors. If you want to play with the sketch without connecting the Arduino, change it to false. The String turnTableAngle holds the string coming from Arduino containing the current angle of the turntable.

```
import processing.serial.*;
import processing.opengl.*;
import SimpleOpenNI.*;
import kinectOrbit.*;
// Initialize Orbit and simple-openni Objects
KinectOrbit myOrbit;
SimpleOpenNI kinect;
// Serial Parameters
Serial myPort; // Initialize the Serial Object
boolean serial = true; // Define if you're using serial communication
String turnTableAngle = "0"; // Variable for string coming from Arduino
```

You initialize four ArrayLists. The first two, ScanPoints and scanColors, contain the current scanned points and their colors. This means the points being scanned currently by the Kinect and contained within your scanning volume. The ArrayLists objectPoints and objectColors contain all the points that configure your scanned objects patched together. These are the points that will be exported in .ply format.

```
// Initialize the ArrayLists for the pointClouds and the colors associated
ArrayList<PVector> scanPoints = new ArrayList<PVector>(); // PointCloud
ArrayList<PVector> scanColors = new ArrayList<PVector>(); // Object Colors
ArrayList<PVector> objectPoints = new ArrayList<PVector>(); // PointCloud
ArrayList<PVector> objectColors = new ArrayList<PVector>(); // Object Colors
```

The following variables define the "model space," the volume that contains your objects. The float baseHeight is the height of your base from your Kinect coordinate system. In our case, it was placed 67mm under the Kinect camera. modelWidth and modelHeight are the dimensions of the model space. Anything that falls within this volume will be scanned.

```
// Scanning Space Variables
float baseHeight = -67; // Height of the Model's base
float modelWidth = 400;
float modelHeight = 400;
PVector axis = new PVector(0, baseHeight, 1050);
```

You need to define some additional variables. The scanLines parameter defines how many rows of the Kinect depth image you use at every scan (Figure 12-22). Think of a traditional structured 3D scanner. A single line is scanned at every angle. This parameter can be set to low values if you have a pretty symmetrical geometry around its center. Low scanLines values need to be used with a high number of scans, defined by the size of shotNumber[]. Play with these two parameters until you achieve a smooth result in your point cloud. In our example, we used a high number of lines, 200, and only three shots or patches.

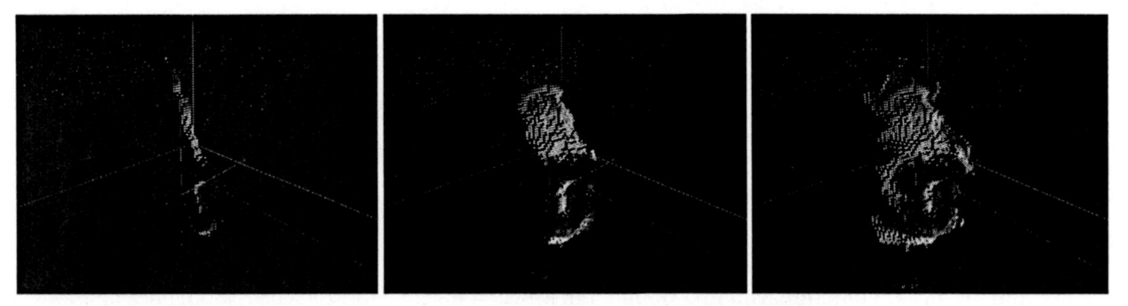

Figure 12-22. *The scanLines parameter set to 5 (left), 25 (center), and 65 (right)*

The variable scanRes defines the number of pixels you take from your scan. One is the higher resolution. If you chose a higher number than 1, you will skip a number of pixels in every scan. The Boolean variables and currentShot integer are used to drive the flow of the sketch depending on the incoming data.

```
// Scan Parameters
int scanLines = 200;
int scanRes = 1;
boolean scanning;
boolean arrived;
float[] shotNumber = new float[3];
int currentShot = 0;
```

Setup and Draw Functions

Within the setup() function, you initialize all your KinectOrbit, Simple-OpenNI, and serial objects. You also need to work out the angle corresponding to every shot you want to take. The idea is to divide the 365 degrees of the circle, or 2*PI radians, into the number of shots specified by the size of the shotNumber[] array. Later, you send these angles to the Arduino and you take a 3D shot from every one of them. Lastly, you move the turntable to the start position, or angle zero, so it's ready to scan.

```
public void setup() {
  size(800, 600, OPENGL);
  // Orbit
  myOrbit = new KinectOrbit(this, 0, "kinect");
  myOrbit.drawCS(true);
  myOrbit.drawGizmo(true);
  myOrbit.setCSScale(200);
  myOrbit.drawGround(true);
  // Simple-openni
  kinect = new SimpleOpenNI(this);
  kinect.setMirror(false);
  kinect.enableDepth();
  kinect.enableRGB();
  kinect.alternativeViewPointDepthToImage();

  // Serial Communication
  if (serial) {
    String portName = Serial.list()[0]; // Get the first port
```

```
    myPort = new Serial(this, portName, 9600);
    // don't generate a serialEvent() unless you get a newline
    // character:
    myPort.bufferUntil('\n');
  }

  for (int i = 0; i < shotNumber.length; i++) {
    shotNumber[i] = i * (2 * PI) / shotNumber.length;
  }
  if(serial) { moveTable(0); }
}
```

Within the `draw()` function, you first update the Kinect data and start the Orbit loop. Pack all the different commands into functions so the workflow is easier to understand.

```
public void draw() {
  kinect.update(); // Update Kinect data
  background(0);

  myOrbit.pushOrbit(this); // Start Orbiting
```

First, you draw the global point cloud with the `drawPointCloud()` function, which takes an integer as parameter defining the resolution of the point cloud. As you're not very interested in the global point cloud, you speed things up by only drawing one every five points.

```
  drawPointCloud(5);
```

Then you update the object points, passing the `scanLines` and `scanRes` defined previously as parameters. As you will see later, this function is in charge of transforming the points to the global coordinate system.

```
  updateObject(scanLines, scanRes);
```

You have two Boolean values defining if you are in the scanning process and whether you have reached the next scanning position. If both are `true`, you take a shot with the function `scan()`.

```
  if (arrived && scanning) { scan(); }
```

The last step is to draw the objects, the bounding box, and the camera frustum. And of course, close the Kinect Orbit loop.

```
  drawObjects();
  drawBoundingBox(); // Draw Box Around Scanned Objects
  kinect.drawCamFrustum(); // Draw the Kinect cam
  myOrbit.popOrbit(this); // Stop Orbiting
}
```

Additional Functions

The `drawPointCloud()` function is similar to the one previously used to visualize the raw point cloud from the Kinect. You take the depth map, bring it to real-world coordinates, and display every point on screen. Skip the number of points defined by the `steps` parameter. The dim, background points on Figure 12-23 are the result of this function.

```
void drawPointCloud(int steps) {
  // draw the 3D point depth map
  int index;
  PVector realWorldPoint;
```

```
  stroke(255);

  for (int y = 0; y < kinect.depthHeight(); y += steps) {
    for (int x = 0; x < kinect.depthWidth(); x += steps) {
      index = x + y * kinect.depthWidth();
      realWorldPoint = kinect.depthMapRealWorld()[index];
      stroke(150);
      point(realWorldPoint.x, realWorldPoint.y, realWorldPoint.z);
    }
  }
}
```

***Figure 12-23.** Scanned figure in the bounding box*

The `drawObjects()` function displays on screen the point cloud defining the object being scanned. This function has two parts: first, it displays the points stored in the `objectPoints` ArrayList, which are the object points patched together. Then it displays the points stored in the `scanPoints` ArrayList, which are the points being currently scanned. All these points are displayed in their real color and with `strokeWeight = 4`, so they stand out visually, as you can see on Figure 12-23.

```
void drawObjects() {
  pushStyle();
  strokeWeight(4);

  for (int i = 1; i < objectPoints.size(); i++) {
    stroke(objectColors.get(i).x, objectColors.get(i).y, objectColors.get(i).z);
```

```
    point(objectPoints.get(i).x, objectPoints.get(i).y, objectPoints.get(i).z + axis.z);
  }

  for (int i = 1; i < scanPoints.size(); i++) {
    stroke(scanColors.get(i).x, scanColors.get(i).y, scanColors.get(i).z);
    point(scanPoints.get(i).x, scanPoints.get(i).y, scanPoints.get(i).z + axis.z);
  }

  popStyle();

}
```

The drawBoundingBox() function displays the limits of the scanning volume. Anything contained within this volume is scanned.

```
void drawBoundingBox() {
  stroke(255, 0, 0);
  line(axis.x, axis.y, axis.z, axis.x, axis.y + 100, axis.z);
  noFill();
  pushMatrix();
  translate(axis.x, axis.x + baseHeight + modelHeight / 2, axis.z);
  box(modelWidth, modelHeight, modelWidth);
  popMatrix();
}
```

The scan() function is in charge of storing the current point cloud into the object point cloud. You need to avoid multiplying the number of points unnecessarily, so you implement a loop in which you check the distance from each new point to every point already stored in the objectPoints ArrayList. If the distance from the new point to any of the previously stored points is under a certain threshold (1mm in our example), you won't add this point to the object point cloud. Otherwise, you add the point and the corresponding color to the object ArrayLists.

```
void scan() {
  for (PVector v : scanPoints) {
    boolean newPoint = true;
    for (PVector w : objectPoints) {
      if (v.dist(w) < 1)
        newPoint = false;
    }

    if (newPoint) {
      objectPoints.add(v.get());
      int index = scanPoints.indexOf(v);
      objectColors.add(scanColors.get(index).get());
    }
  }
```

After taking the shot and adding it to the object point cloud, check if the shot taken was the last of the list. If the shot is not the last one, you tell the turntable to move to the next position and you set the "arrived" Boolean to false. This prevents the sketch from taking any other shot before you get an "arrived" signal from Arduino indicating you have reached the target position. If the shot happens to be the last one, you set the scanning Boolean to false, declaring the end of the scanning process. Figure 12-24 shows the three steps in this scanning process.

```
  if (currentShot < shotNumber.length-1) {
    currentShot++;
```

```
      moveTable(shotNumber[currentShot]);
      println("new angle = " + shotNumber[currentShot]);
      println(currentShot);
      println(shotNumber);

    }
    else {
      scanning = false;
    }
    arrived = false;
  }
```

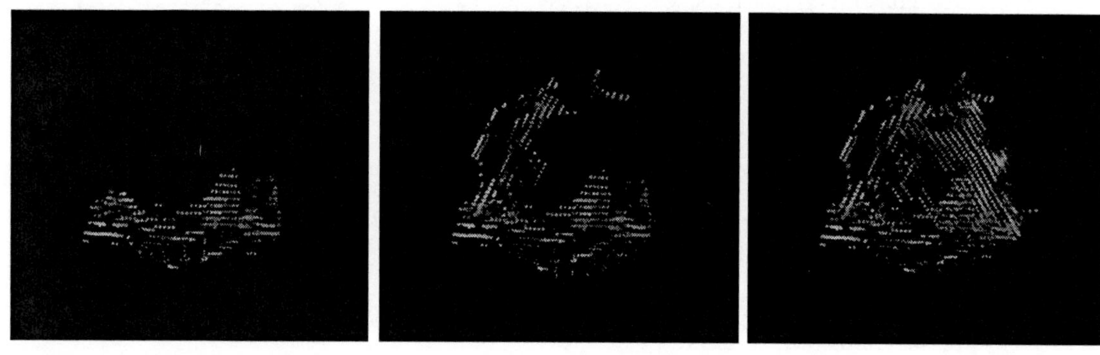

Figure 12-24. *Three-step scan*

The `updateObject()` function is where the computation of the world coordinates of the scanned points takes place. You declare an integer to store the vertex index and a PVector to store the real coordinates of the current point. Then you clear the ArrayLists of your scanned points so you can update them from scratch.

```
void updateObject(int scanWidth, int step) {
  int index;
  PVector realWorldPoint;
  scanPoints.clear();
  scanColors.clear();
```

You need to know the current angle of the turntable in order to compute the global coordinates of the points. You work out this angle by mapping the integer value of your `turnTableAngle` string from its range (100-824) to the 365-degree range in radians (0-2*PI). Remember that `turnTableAngle` is the string coming from Arduino, so it changes every time you rotate your turntable.

```
  float angle = map(Integer.valueOf(turnTableAngle), 100, 824, 2 * PI, 0);
```

The next lines draw a line at the base of the bounding box to indicate the rotation of the turntable.

```
  pushMatrix();
  translate(axis.x, axis.y, axis.z);
  rotateY(angle);
  line(0, 0, 100, 0);
  popMatrix();
```

Now, you run in a nested loop through your depth map pixels, extracting the real-world coordinates of every point and its color.

```
  int xMin = (int) (kinect.depthWidth() / 2 - scanWidth / 2);  int xMax = (int)
(kinect.depthWidth() / 2 + scanWidth / 2);
  for (int y = 0; y < kinect.depthHeight(); y += step) {
    for (int x = xMin; x < xMax; x += step) {
      index = x + (y * kinect.depthWidth());
      realWorldPoint = kinect.depthMapRealWorld()[index];
      color pointCol = kinect.rgbImage().pixels[index];
```

If the current point is contained within the defined scanning volume or bounding box (this is what the scary-looking "`if()`" statements check), you create the PVector `rotatedPoint` to store the global coordinates of the point.

```
      if (realWorldPoint.y < modelHeight + baseHeight && realWorldPoint.y > baseHeight) {
        if (abs(realWorldPoint.x - axis.x) < modelWidth / 2) {  // Check x
          if (realWorldPoint.z < axis.z + modelWidth / 2 && realWorldPoint.z > axis.z -
modelWidth / 2) {  // Check z

            PVector rotatedPoint;
```

The coordinate system transformation process happens in two steps. First, you need to get the coordinates of the vector from the center of the turntable. The axis vector defined the coordinates of the center of the turntable from the Kinect coordinate system, so you only need to subtract the coordinates of the axis vector to the real-world point coordinates to get the transformed vector. Then you need to rotate the point around the y-axis by the current angle of your turntable. You use the function `vecRotY()`to do this transformation (Figure 12-25).

```
            realWorldPoint.z -= axis.z;
            realWorldPoint.x -= axis.x;
            rotatedPoint = vecRotY(realWorldPoint, angle);
```

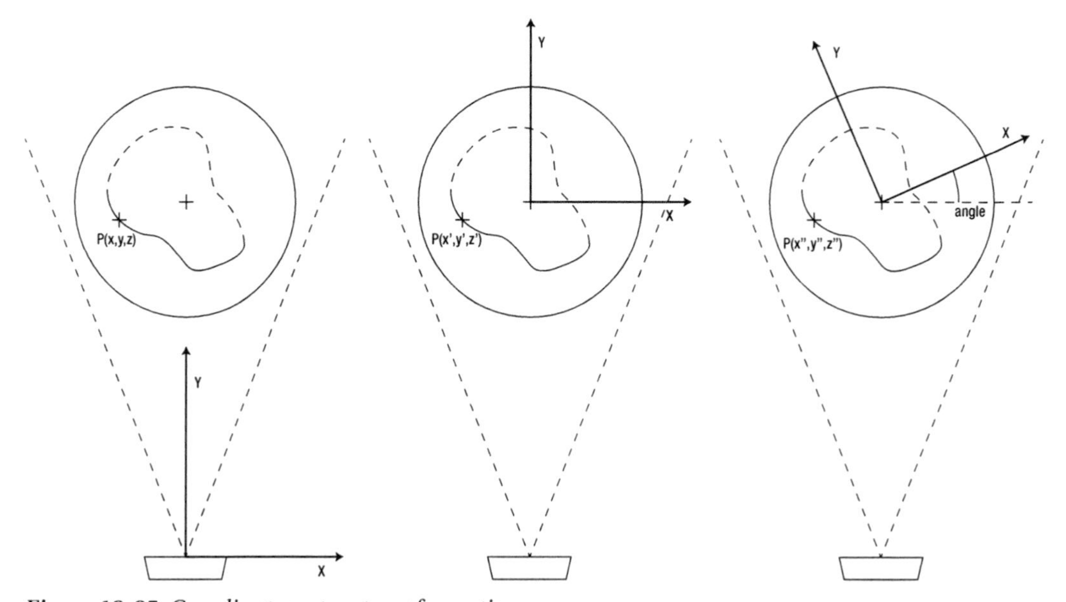

***Figure 12-25.** Coordinate system transformations*

Now your rotatedPoint should contain the real-world coordinates of the scanned point, so you can add it to the scanPoints ArrayList. You also want to add its color to the scanColors ArrayList so you can retrieve it later. Finally, you close all the curly brackets: three for the if() statements, two for the for() loops, and the last one to close the function updateObject() function.

```
          scanPoints.add(rotatedPoint.get());
          scanColors.add(new PVector(red(pointCol), green(pointCol), blue(pointCol)));
        }
      }
    }
  }
 }
}
```

The vecRotY() function returns the PVector resulting from the operation of rotating of the input PVector by the input angle.

```
PVector vecRotY(PVector vecIn, float phi) {
  // Rotate the vector around the y-axis
  PVector rotatedVec = new PVector();
  rotatedVec.x = vecIn.x * cos(phi) - vecIn.z * sin(phi);
  rotatedVec.z = vecIn.x * sin(phi) + vecIn.z * cos(phi);
  rotatedVec.y = vecIn.y;
  return rotatedVec;
}
```

You use the function moveTable(float angle) every time you want to send a new position to your turntable. The input angle is in degrees (0-360). This function sends a trigger character 'S' through the serial channel to indicate that the communication has started. Then you send the input angle remapped to one byte of information (0-255). It prints out the new angle for information.

```
void moveTable(float angle) {
  myPort.write('S');
  myPort.write((int) map(angle, 0, 2*PI, 0, 255));
  println("new angle = " +  angle);
}
```

All the preceding functions are called from the draw() loop or from other functions. You also need to include two additional functions that you won't be calling explicitly but will be called by Processing in due time. The first one is the serialEvent() function. It is called every time you get a new line character in your serial buffer (you specified this in the setup()function). If you receive a string, you trim off any white spaces and you check the string. If you receive a "start" or "end" message, you just display it on the console. If you receive an "arrived" message, you also set your Boolean arrived to true so you know that you have reached the target position.

If the message is not one of the preceding ones, that means you should be receiving the current angle of the turntable, so you update your turnTableAngle string to the incoming string.

```
public void serialEvent(Serial myPort) {
  // get the ASCII string:
  String inString = myPort.readStringUntil('\n');
  if (inString != null) {
    // trim off any whitespace:
    inString = trim(inString);
    if (inString.equals("end")) {
      println("end");
    }
```

```
      else if (inString.equals("start")) {
        println("start");
      }
      else if (inString.equals("arrived")) {
        arrived = true;
        println("arrived");
      }
      else {
        turnTableAngle = inString;
      }
    }
  }
```

The keyPressed() function is called whenever you press a key on your keyboard. You assign several actions to specific keys, as described in the following code.

```
public void keyPressed() {
  switch (key) {
  case 'r': // Send the turntable to start position
    moveTable(0);
    scanning = false;
    break;
  case 's': // Start scanning
    objectPoints.clear();
    objectColors.clear();
    currentShot = 0;
    scanning = true;
    arrived = false;
    moveTable(0);
    break;
  case 'c': // Clear the object points
    objectPoints.clear();
    objectColors.clear();
    break;
  case 'e':      // Export the object points
    exportPly('0');
    break;
  case 'm': // Move the turntable to the x mouse position
    moveTable(map(mouseX, 0, width, 0, 360));
    scanning = false;
    break;
  case '+':  // Increment the number of scanned lines
    scanLines++;
    println(scanLines);
    break;
  case '-':  // Decrease the number of scanned lines
    scanLines--;
    println(scanLines);
    break;
  }
}
```

Note that you implement mouse actions for moving to the start position; also, if you press the [S]key, you start the scan from the start position. The turntable is first sent to the zero position and then you start scanning.

The M key helps you control the turntable. Pressing it sends the turntable to the angle defined by the x-coordinate of your mouse on the Processing frame (remapped from 0-`width` to 0-360). For example, pressing M while your mouse is on the left edge of the frame sends the turntable to 0 degrees, and pressing it while the mouse is on the right edge sends it to 360 degrees.

The key E calls the function `exportPly` to export the point cloud. This leads to the next step. At the moment, you have being able to scan a figure all around and have on your screen a full point cloud defining its shape (Figure 12-26) . But remember you wanted to convert this point cloud to a mesh so you can bring it to 3D modeling software or 3D print it. You are going to do this with another open source software package, so you need to be able to export a file recognizable by this software. You will use the polygon file format, or `.ply`, for this exporting routine, which you will implement yourself. You'll see how in the next section.

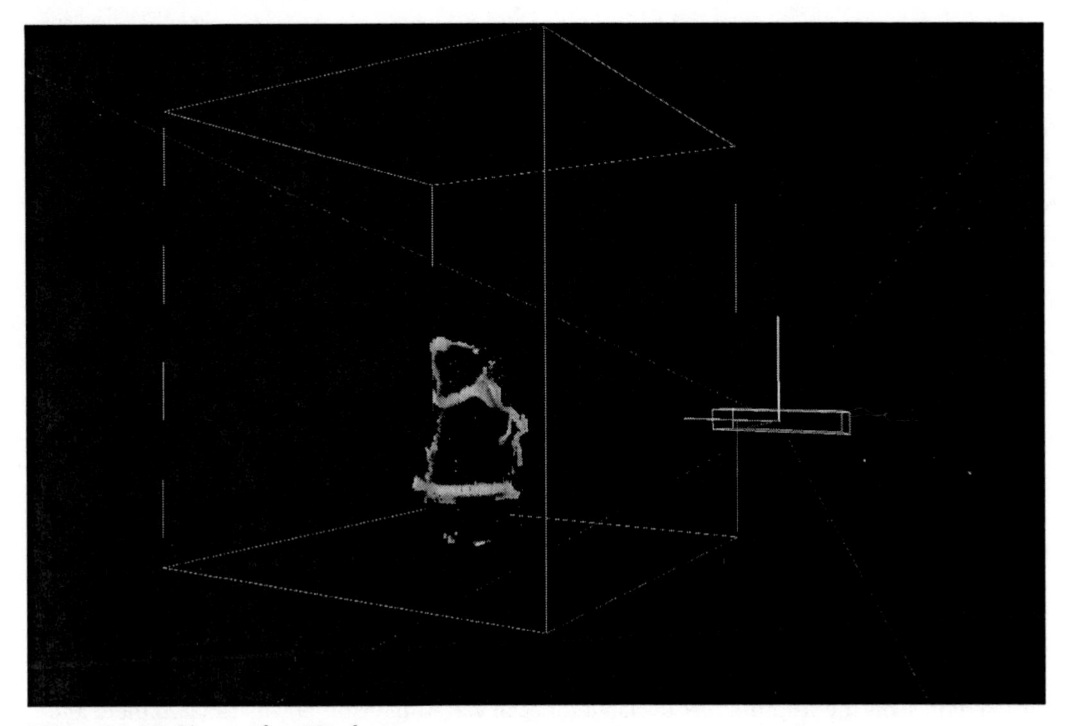

Figure 12-26. *Figure after 360-degree scan*

Exporting the Point Cloud

As stated in the introduction, you are going to mesh the point cloud using an open source software package called Meshlab, which can import unstructured point clouds in two formats, `.ply` and `obj`. There are several Processing libraries that allow you to export these formats, but exporting a `.ply` file is so simple that you are going to implement your own file exporter so you can control every datum you send out.

The `.ply` extension corresponds to polygon file format, or stanford triangle format. This format can be stored in ASCII and binary files. You will use ASCII so you can actually write your data directly to an external file and then read it from Meshlab.

The .ply format starts with a header formed by several lines. It starts by declaring "I am a ply file" with the word "ply" written on the first line. Then it needs a line indicating the type of .ply file. You indicate that you are using ASCII and the version 1.0. You then insert a comment indicating that you are looking at your Processing-exported .ply file. Note that comments start with the word "comment"—pretty literal, eh?

■ **Note** The next lines are not Processing code, but the lines that will be written on your .ply file!

```
ply
format ascii 1.0
comment This is your Processing ply file
```

These three first lines are constant in every file you export. You could add a comment that changes with the name of the file or any other information. If you want to add a comment, just make sure your line starts with the word "comment".

The next lines, still within the header, declare the number of vertices, and the properties you are exporting for each of them. In your case, you are interested in the coordinates and color, so you declare six properties: x, y, z, red, green, and blue. You now close the header.

```
element vertex 17840
property float x
property float y
property float z
property uchar red
property uchar green
property uchar blue
end_header
```

After closing the header, you have to write the values of all the properties stated for each vertex, so you get a long list of numbers like this:

```
0.0032535514 0.1112856 0.017800406 125 1 1
0.005102699 0.1112856 0.017655682 127 1 81
-0.0022937502 0.10943084 0.018234566 130 1 1
.........
```

Each line represents the six properties declared for each vertex in the order you declared them and separated by a white space. Let's have a look at the code you need to implement to effectively producing .ply files containing the object point cloud that you previously generated.

■ **Note** In this case, you are only exporting points, but .ply files also allow you to export edges and faces. If you wanted to export edges and faces, you would need to add some lines to your header defining the number of edges and faces in your model and then a list of faces and a list of edges. You can find information on the .ply format at http://paulbourke.net/dataformats/ply.

The exportPly Function

First, you declare a `PrintWriter` object. This class prints formatted representations of objects to a text-output stream. Then you declare the name of the output file and you call the method `createWriter()`, passing file name in the data path as a parameter.

```
void exportPly(char key) {
  PrintWriter output;
  String viewPointFileName;
  viewPointFileName = "myPoints" + key + ".ply";
  output = createWriter(dataPath(viewPointFileName));
```

Now you need to print out all the header lines. The number of vertices is extracted from the size of the `objectPoints` ArrayList.

```
  output.println("ply");
  output.println("format ascii 1.0");
  output.println("comment This is your Processing ply file");
  output.println("element vertex " + (objectPoints.size()-1));
  output.println("property float x");
  output.println("property float y");
  output.println("property float z");
  output.println("property uchar red");
  output.println("property uchar green");
  output.println("property uchar blue");
  output.println("end_header");
```

After closing the header, you run through the whole `objectPoints` ArrayList, and you print out its `x`, `y`, and `z` coordinates and the integers representing their color parameters. You want the output to be in meters, so you scale down the point coordinates by dividing the values by 1000.

```
  for (int i = 0; i < objectPoints.size() - 1; i++) {
    output.println((objectPoints.get(i).x / 1000) + " "
      + (objectPoints.get(i).y / 1000) + " "
      + (objectPoints.get(i).z / 1000) + " "
      + (int) objectColors.get(i).x + " "
      + (int) objectColors.get(i).y + " "
      + (int) objectColors.get(i).z);
  }
```

When you are done with your point cloud, you flush the `PrintWriter` object and you close it. Now you should have a neat `.ply` file in the data path of your Processing file!

```
  output.flush(); // Write the remaining data
  output.close(); // Finish the file
}
```

You have now gone through the whole process of scanning a three-dimensional object and exporting the resulting point cloud to a `.ply` file. You will next learn how to use this point cloud to generate a surface using Meshlab.

Surface Reconstruction in Meshlab

Surface reconstruction from unorganized point sets is a complicated process requiring even more complicated algorithms. There is a lot of research out there on how this process can be performed, and luckily for you, there are several open source libraries like CGAL (`http://cgal.org`) and open source software like Meshlab (`http://meshlab.sourceforge.net`) that you can use to perform these operations without being an accomplished mathematician.

Meshlab is an open source system for the processing and editing of 3D triangular meshes. It was born in 2005 with the intention of helping with the processing of the unstructured models arising in 3D scanning. That is pretty much the state you are in: you have 3D scanned a figure and need to process the point cloud and reconstruct the implicit surface.

Meshlab is available for Windows, Mac OSX and Linux, so whatever your OS, you should now go to the web site and download the software. You'll start using it right now!

If you have already installed Meshlab, `.ply` files should be automatically associated to it, so you can double-click your file and it will open in Meshlab. If this is not the case, open Meshlab and import your file by clicking the File menu, and then Import Mesh.

If your file was exported correctly and the number of vertices corresponds to the declared number of vertices, when you open the file you should see on screen a beautifully colored point cloud closely resembling your scanned object, like the one shown in Figure 12-27.

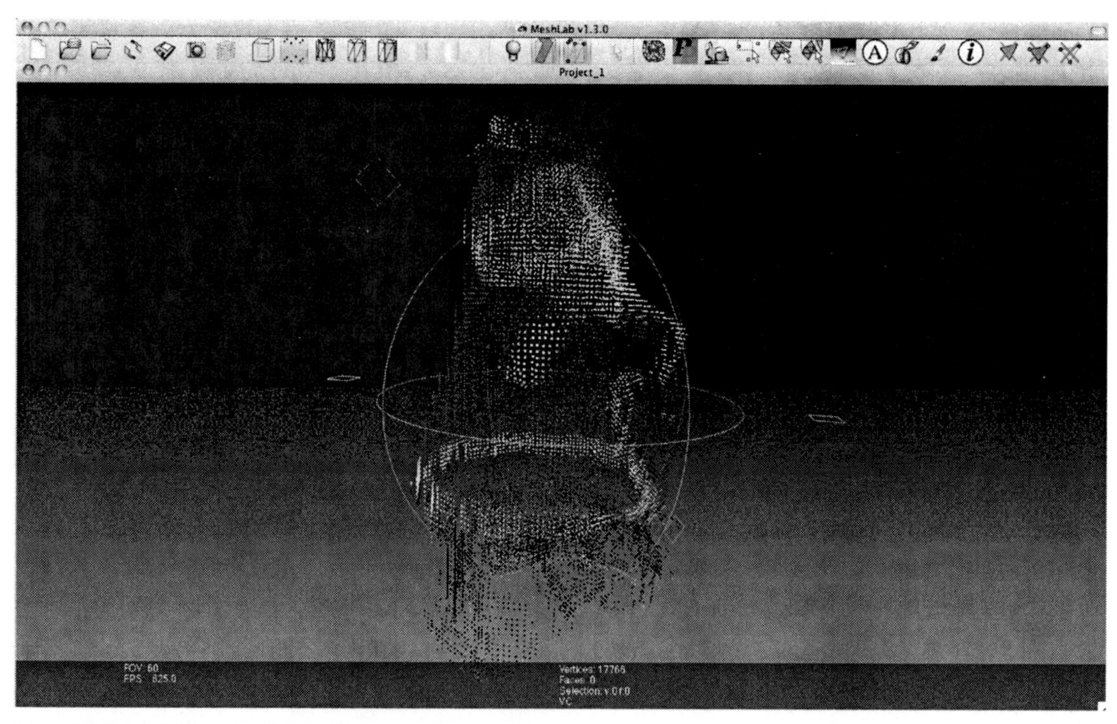

Figure 12-27. Point cloud .ply file opened in Meshlab

If your file contains inconsistencies (like a different number of points declared and printed), Meshlab can throw an "Unexpected end of file" error at you. If you click OK, you won't see anything on the screen. Don't panic! The points are still there, but you have to activate the Points view on the top bar, and then you will see the point cloud. It won't be colored, though. You can then go to the Render menu and choose Colors/Per Vertex. After this, you should see an image similar to the previous one.

Now orbit around your point cloud and identify any unwanted points. Sometimes Kinect detects some points in strange places, and it helps the reconstruction process to get rid of those. Sometimes the Kinect is not perfectly calibrated with the turntable, so you will have slight inconsistencies in the patched edges. To delete the unwanted points, select the Select Vertexes tool from the top menu bar. You can select several points together by holding the Control or Command key down. After selecting, you can click Delete Selected Vertices on the right of the top menu bar (Figure 12-28).

***Figure 12-28.** Delete unwanted points*

You are now ready to start the surface reconstruction process. First, you need to compute the normal vector on your points, which is the vector perpendicular to the surface of the object at each point, pointing out of the object. This provides the reconstruction algorithm information about the topology of your object. Go to the Filters/Point Set menu and choose "Compute normal for point sets." This should bring up a menu. Choose 16 as the number of neighbors and click Apply. The process shouldn't take more than a couple of seconds. It may seem like nothing changed, but if you go to the Render menu and tick Show Vertex Normals, you should now see the point cloud with all the normals in light blue.

Go to the Filters/Point Set menu again and click Surface Reconstruction: Poisson. In the menu, there are several parameters that we won't have the time to discuss in this book. These parameters have an influence on how precise the reconstruction is, how many neighboring points are taking into account, and so on. Chose 16 as Octree Depth and 7 as Solver Divide; depending on your object and the quality of the scan, the optimum parameters can greatly change, so try to find yours.

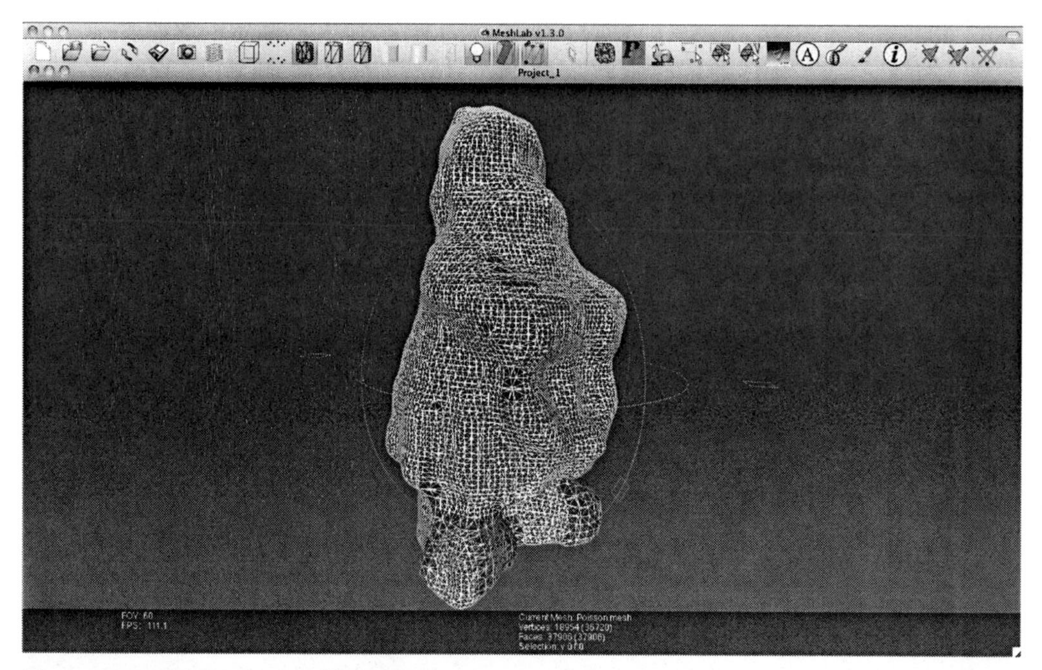

Figure 12-29. Mesh generated by Poisson reconstruction

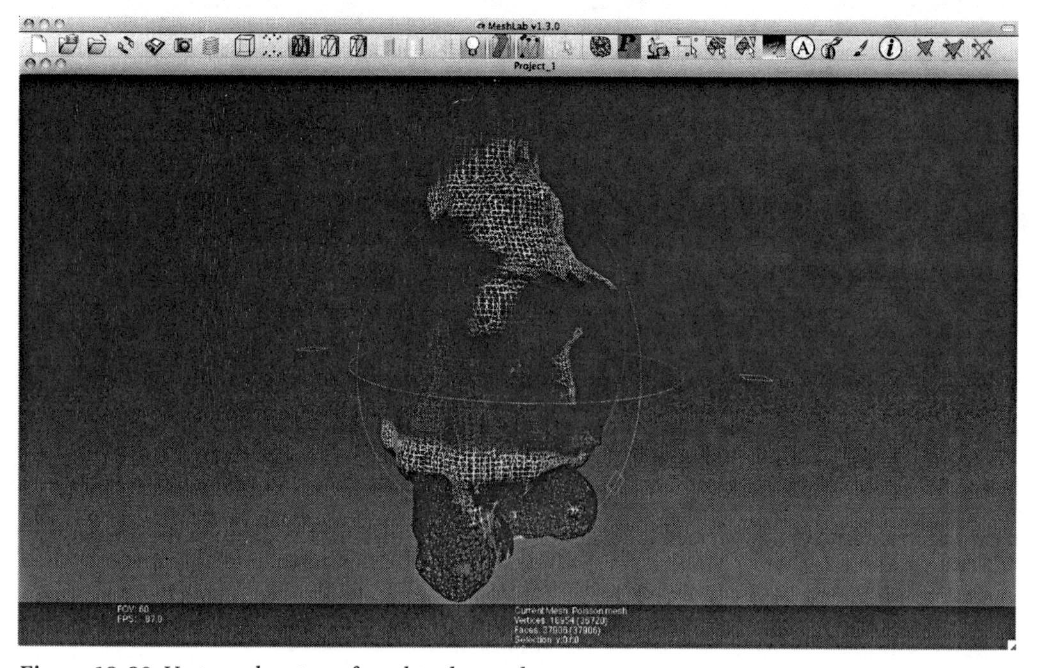

Figure 12-30. Vertex colors transferred to the mesh

After a couple of painfully long minutes, you should get a pretty neat surface like the one in Figure 12-29. The last step is transferring the colors of the vertices to your mesh, so you can export the fully colored mesh to any other software. Go to Filters/Sampling and choose Vertex Attributes Transfer. Make sure that the Source Mesh is your point cloud, and the Target Mesh is your Poisson Mesh. Tick only Transfer Color and click Apply. The mesh should show now a pretty similar range of colors to the original figure (Figure 12-30), and you are done!!

If your point cloud was not what you would call "neat," sometimes the Poisson reconstruction will throw something much like a potato shape slightly smaller than your point cloud. Don't be discouraged; Meshlab can fix it! In the previous step, instead of transferring the colors only, transfer the geometry as well. You will get then a nicely edged shape like the one in the Figure 12-31.

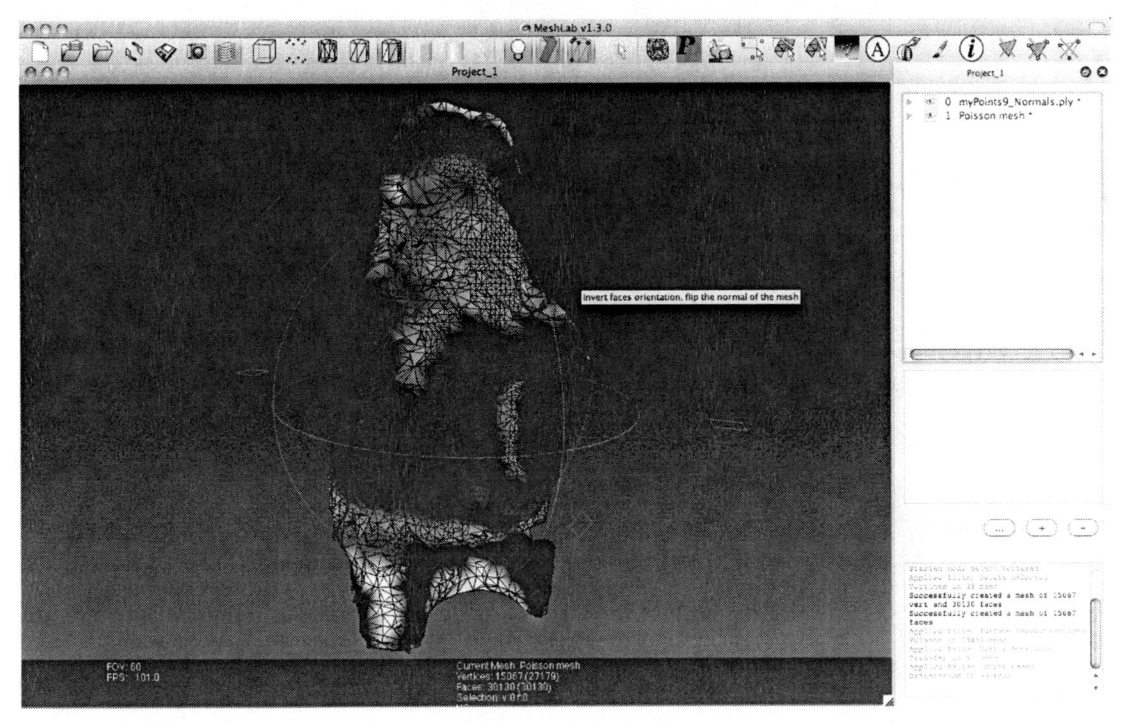

***Figure 12-31.** Vertex geometry and color transferred to the mesh*

Go to "Filters/Smoothing, Fairing, and Deformation" and click HC Laplacian Smooth. Apply the filter as many times as needed to obtain a reasonably smooth surface. With a little luck, the result will be reasonably good, considering you started from a pretty noisy point cloud.

One last tip on this process: sometimes you'll get an error on the importation process, the computed normal will be inverted, and the figure will look pretty dark. Again, you can use Meshlab to invert your face normals. Go to "Filters/Normals, Curvature, and Orientation" and click Invert Faces Orientation. This should make your mesh appear brighter. Check that the light toggle button is on. Sometimes the switch is off and you might get the impression that your normals are inverted; if you have just forgotten to turn on the light, do it on the top menu bar!

You have generated a mesh that can be exported to a vast number of file formats compatible with CAD packages, 3D printers and other 3D modeling software. Go to File/Export Mesh As and choose the file format you need. We exported it as a `.ply` file with colors and imported it into Blender (see Figure 12-32), a great open source modeling software freely available from `www.blender.org`.

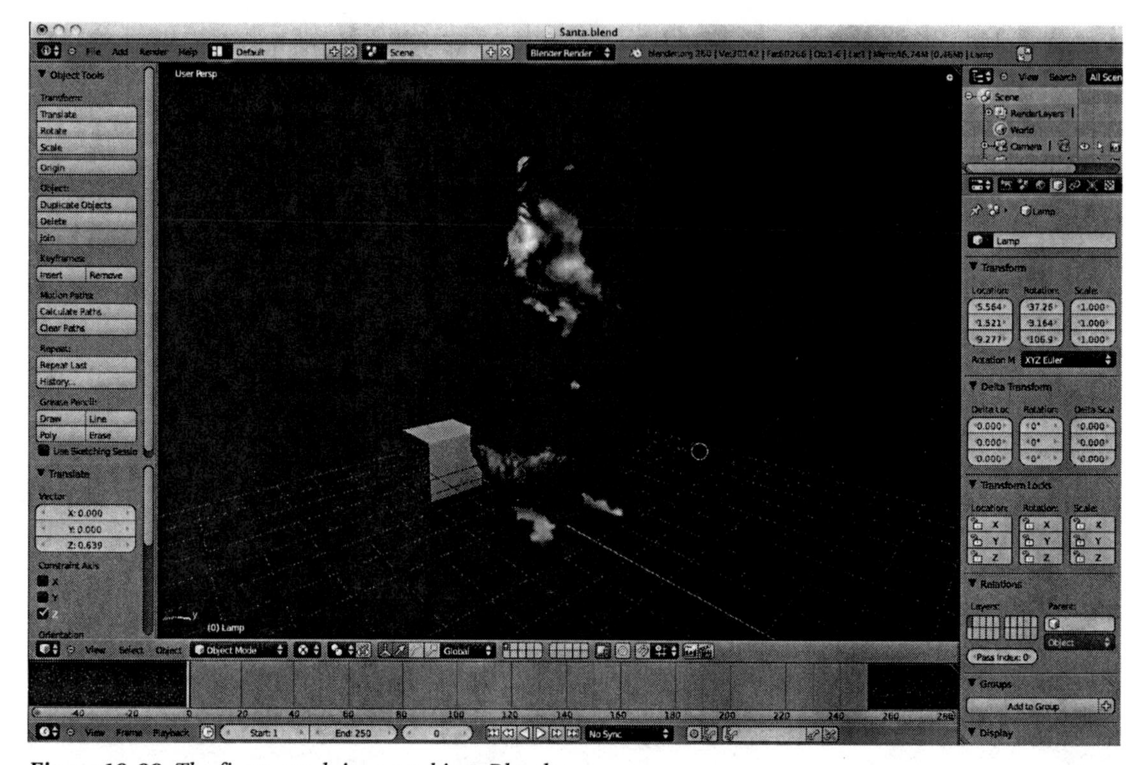

Figure 12-32. The figure mesh imported into Blender

If your goal is to replicate your object with a 3D printer, there are plenty of services online that will take your STL model, print it (even in color), and send it to you by mail. But, being a tinkerer, you might be willing to go for the DIY option and build yourself a RepRap. This is a pretty astonishing concept of self-replicating 3D-printer, which means that RepRap is partially made from plastic components that you can print using another RepRap. It is certainly the cheapest way to have your own 3D printer, so have a look at `http://reprap.org/` for plenty of information on where to buy the parts and how to build it.

Summary

In this chapter you learned several techniques for working with raw point clouds. You haven't made use of any of the NITE capabilities, so actually this project could be done using any of the available Kinect libraries, as long as you have access to the Kinect point data.

You used gears, a servo, and a multi-turn potentiometer to build a precise turntable that can be controlled from Processing and that provides real-time feedback of its rotational state. You acquired 3D data with your Kinect and patched the data together using geometrical transforms to form a 360-degree scan of an object. You even built your own file-exporting protocol for your point cloud and learned how to process the point cloud and perform a Poisson surface reconstruction using Meshlab. The outcome of this process is an accurate model of your 3D object that you can export to any 3D modeling software package.

The possibilities of 360-degree object scanning for architects, character artists, and people working in 3D in general are endless. Having this tool means that for less than $200 (including your Kinect!) you have a device that can take any three-dimensional shape and bring its geometry and colors into your computer to be used by any application. It's your task now to imagine what to do with it!

CHAPTER 13

Kinect-Controlled Delta Robot

by Enrique Ramos and Ciriaco Castro

This last chapter of the book will guide you through the process of building and programming your own Kinect-controlled delta robot (Figure 13-1). You will learn everything from the inverse kinematics of this specific kind of robot to the programming of the hand tracking routines and the process of translating the data from the virtual simulation into electrical pulses for the operation of the physical robot.

Delta robots, or parallel robots, are not only great for industrial use, where they perform a variety of tasks at incredibly high speeds, but they are also beautiful to watch and relatively simple to build. They rely on the use of parallelograms to translate the rotational movement of three independent motors into a precise 3D position of the robot's effector. You are building a basic gripper in this project, but the effector could actually be any manipulator you can think of. Imagine this effector changed into a magnetic gripper, a laser welding/cutting tip, or a nozzle for 3D printing. After you understand the concepts in this project, you will build a platform extendable to any kind of robotic application.

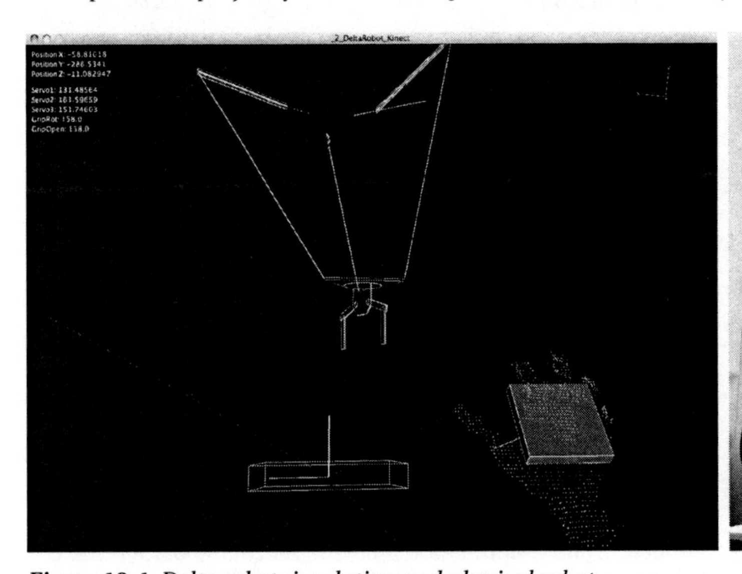

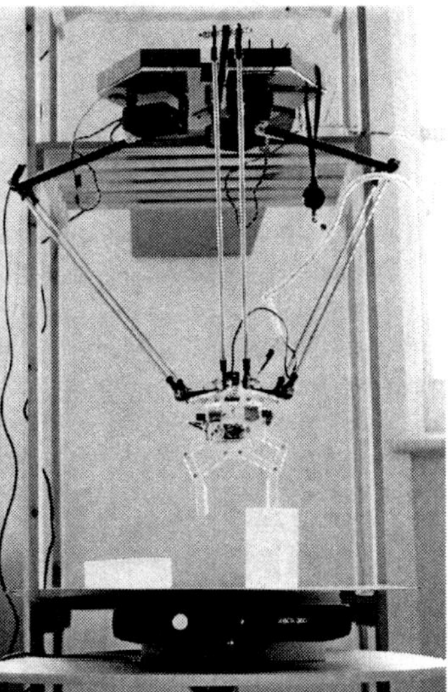

***Figure 13-1.** Delta robot simulation and physical robot*

You will need quite a few components and materials for the development of this project, but they are all widely available and reasonably priced (see Table 13-1). You will find most components in your local hardware store; the motors are standard servos you can find at SparkFun and similar retailers. The ball joints are perhaps the most specific pieces for this project, but there are a number of online shops where you can find them.

Table 13-1. *The Parts List*

Part	Description	Price
6 aluminum rods	M6 diameter, 450 mm length	$10.82
12 ball joints	M6 diameter. We found them at reidsupply	$3.86 each
1 aluminum pipe	M8 diameter, to be used as a spacer	$2.25
3 aluminum square bars	240 mm length, 20 mm side	$8.24
3 aluminum square bars	60 mm length, 20 mm side	$2.12
1 steel corrugated rod	M6 diameter, to be cut in 60 mm length pieces	$5.49
Perspex board	Clear acrylic board 3 x 420 x 720 mm	$15.80
L brackets	2 x 20 mm side, 6 x 50 mm side. From hardware store	$4.05
Timber board	MDF 1200 x 350 x 20 mm	$5.85
Power supply	We found it at Amazon.	$34.99
3 servo motors	Hitec HS-805BB	$39.99 each
2 servo motors	Hitec HS-311	$7.99 each
Extension lead cables	1 for each servo and 1 for LED lights	$1.75 each
Strip board	94 x 53 mm copper	$1.05
2 LEDs	Basic LED 5 mm	$0.35 each
1 resistor 220ohm	You can buy them in packs.	$0.25 each
23 breakaway headers-straight	You should have a bunch of these already somewhere in your lab or bedroom.	$1.50

About This Project

This project was born at the Adaptive Architecture and Computation MSc program at the Bartlett, UCL. We created the first version of the Kinect-driven delta robot together with Miriam Dall'Igna under installation artist Ruairi Glynn. After finishing the project, we had the opportunity to develop a new, much larger version of the installation called "Motive Colloquies" at the prestigious Centre Pompidou in

Paris, together with Ruairi Glynn and the London School of Speech and Drama. You can view a first version of the project at `http://vimeo.com/20594424` and the performance at Pompidou Centre at `http://vimeo.com/25792567`.

The Delta Robot

What exactly is a delta robot? A delta robot is a type of parallel robot invented by Reymond Clavel in the 1980s. The basic idea behind it is the use of parallelograms in the design of the robot's legs in order to keep the orientation of an end platform, or effector, that has three degrees of translational freedom but no rotation. The robot consists of only three motors that control the rotation of the three legs, which transform these rotations into a specific pose determined by the connectivity and the nature of the joints of the structure.

There are many industrial applications for this type of robot, mainly due to their low inertia, which allows high accelerations. This makes delta robots some of the fastest robots ever built; they are especially useful for picking and packaging processes. Run a search on the Internet for "delta robot" and you will be surprised by the variety of applications and the amazing speeds these robots can achieve.

Building a Delta Robot

Let's start putting things together so you can get your fantastic delta robot working. As mentioned, your delta robot has three limbs (even though there are some four-legged members of the family) attached to a fix base on one end and to the robot's effector on the other end. The fix base accommodates three servos and the Arduino board; it is the only static part of the robot. The legs constitute a kinematic chain that translates the servos' rotational states into specific spatial positions of the effector, which is composed of a rotating base and a gripper, controlled by two small servos.

The Legs

In order to describe the robot's limbs, we're going to use an analogy of a human leg. Each leg is composed of "thigh" (one strong piece that stands the torque coming from the motor), "shin" (two parallel pieces that are connected to the effector and the thigh), that are connected through a "knee" consisting of ball joints. These legs connect to the effector on an "ankle" joint and to the base on the "hip," which is the joint controlled by the servo motors.

For the thigh part, we selected hollow square sections due to their higher resilience to bending moment and torsion. Before starting to build the first limb, you need to drill small holes on these bars in order to connect them to the servos. These holes should be the same diameter as the bolt that you use to connect this bar to the servos. It's important to be really accurate because this union needs to be as strong as possible if you want to achieve a good level of accuracy (Figure 13-2).

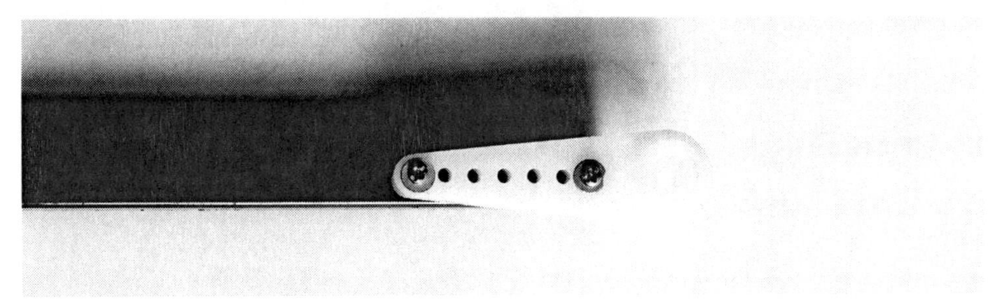

Figure 13-2. *Thigh connected to the servo horn*

You need to measure the position of the holes according to the plastic servo horn. Once you have drilled these holes (you need a minimum of two), you can join the square bar to the servo leg using a couple of bolts. On the other end, drill one hole in the perpendicular plane. This time it's bigger: the same diameter as the corrugated rod that you are going to pass through (6 mm diameter). The next job is to prepare the shin's aluminum rods. You have to thread them using a diestock in order to screw them into the ball joint (Figure 13-3). If you don't have the right tool, you can use standard steel corrugated rods instead of the aluminum ones. Note that this option may require that you use shorter legs or more powerful servos, because of the increase in weight.

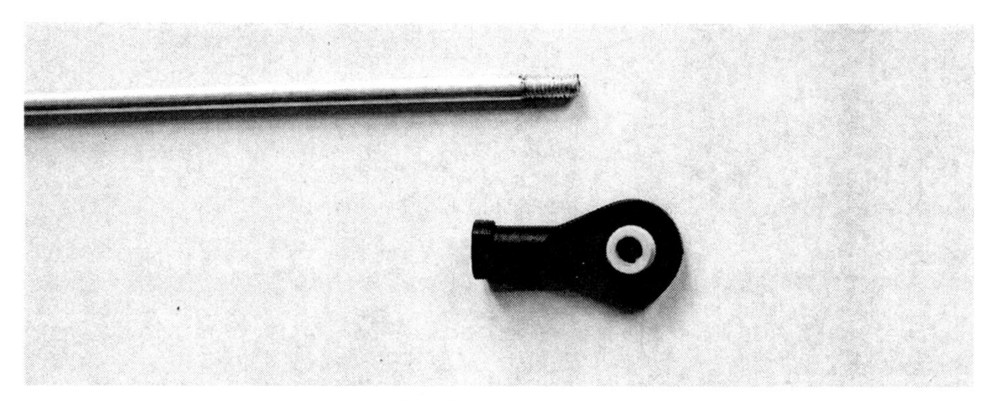

Figure 13-3. *Threaded aluminum rod and ball joint ready for assembly*

Once you put together the aluminum rod and the ball joint on each end, repeat the process with another aluminum rod, making sure they are exactly the same length.

Next, you join all these elements together and finish your first leg. The use of ball joints is key for this project. The leg's joints should allow full rotational freedom for the correct movement of the robot. Pass the corrugated rod trough the ball joint, the square section bar, and the second ball joint, as shown in Figure 13-4.

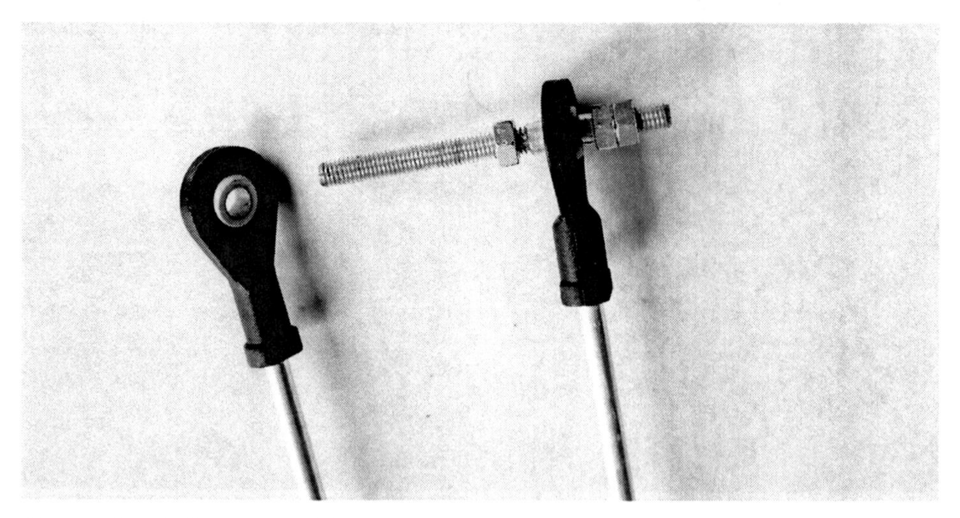

Figure 13-4. *Ball joint assembly*

In order to allow the free rotation of the ball joint, use a series of spacers. First, you screw in a couple of nuts on both sides of the square section bar in order to fix it with the rod, and then you fit a short section of the aluminum tube (5 mm will do); this allows for ball joint movement without clashing with the corrugated rod. These spacers are fitted before and after the ball joint. Finally, put a nylon nut on the ends (Figure 13-5). If you don't have nylon nuts around, you can use two nuts, locking them together tightly in order fix the articulation.

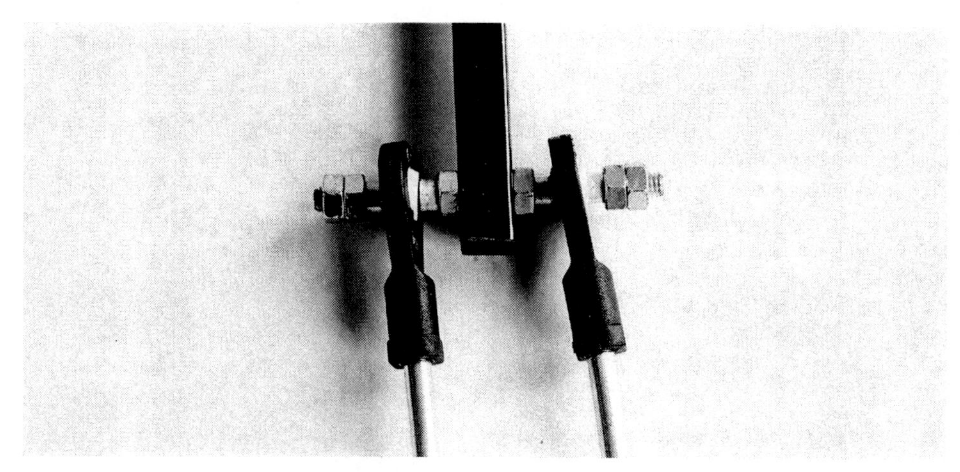

Figure 13-5. *Leg articulation finished*

You now have one leg ready. Just repeat the same process two times (Figure 13-6), paying attention to the position of the holes in the square section bars (dimensions should be consistent on the three legs!) and the length of the aluminum rods. These two factors play a key role in the final accuracy of the robot's movements. Once you have your legs ready, move on to building the base.

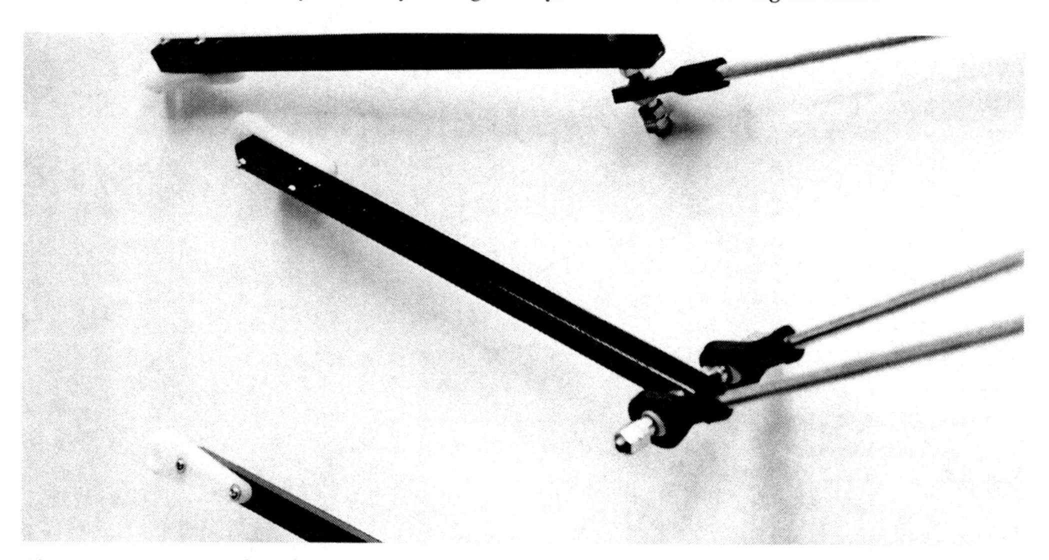

Figure 13-6. *Legs ready to be assembled to the base and effector*

The Base

The base holds the servo motors and is the piece connecting the whole robot to the cantilevered timber plank acting as a support. If you are planning to build your robot differently, just pay attention to the clearance for the legs. When the robot moves, the legs come higher than the base, so you need to make room for them to move freely.

You are going to fabricate the base out of Perspex, but any other rigid material will work. We built our base using a laser cutter, so we got a pretty neat and accurate shape and holes. If you want to laser-cut your base, there are some good online services where you can send the vector file (`dwg`, `ai` or `pdf`) and they will ship the finished product. Otherwise, you can use the files from the book's web site to trace over and cut your base with a good old scalpel.

The base shape is a triangle (Figure 13-7) on which you place each leg's axis aligned to the middle point of each side. Use a triangle of 440 mm per side. The servos are held via L brackets so the rotation happens on the right plane. You subtract a rectangular shape in each side in order to allow the thigh to lift over the plane of the base. You also chop the triangle vertices to save some Perspex and have a smaller base (Figure 13-8).

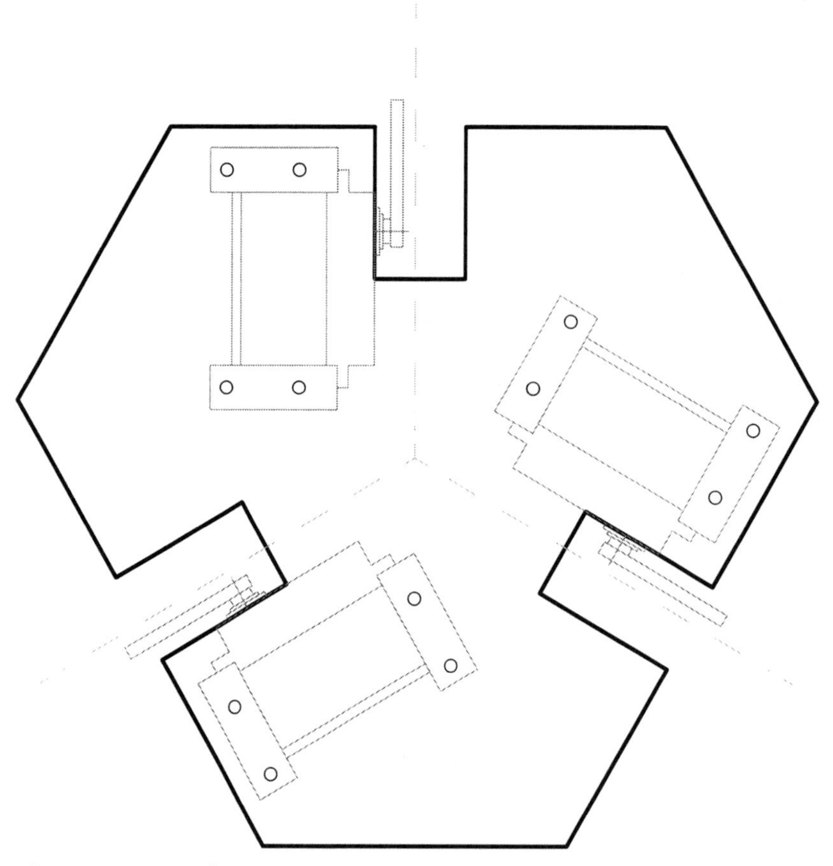

Figure 13-7. Base shape

Figure 13-8. *Laser-cut base and servo with an L bracket*

Once the piece is cut, you just have to screw the servo motors to the brackets, fixing their position; pay special attention to the distance that they are fixed from the border, which must be equal (Figure 13-9). The next step is to screw the brackets that hold the servos to the base.

Figure 13-9. *Assembled base*

The Effector

The effector is a triangular piece that moves parallel to the plane of the base (Figure 13-10). You build it out of a triangular piece of Perspex with 135mm sides. The aluminum rods are joined to this piece.

Use the same type of joint that you saw in the union between thigh and shin. Start by fixing the small square bars to the triangle. Draw a series of holes in the medians of the triangle and a rectangular shape where you insert a new servo motor for the gripper. Don't forget to also draw the holes for fixing the servos to this piece! Once the piece is cut, drill some small holes into the square bars, making them match up with the ones in the effector. These square bars stick out of the triangular shape in order to have enough room to connect them to the legs. Before fixing them with some screws and nuts, drill a hole in one of the extremes (in the perpendicular plane) to build the same type of joint previously described. This hole must be big enough to let an M6 rod pass through it.

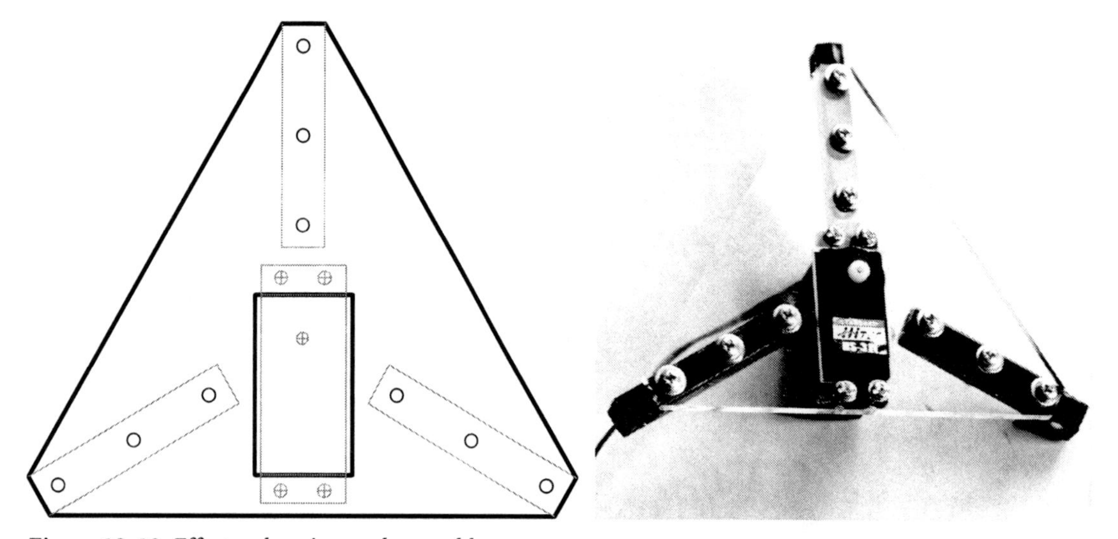

Figure 13-10. *Effector drawing and assembly*

Once all the holes are drilled, fix the square bars to the Perspex piece, conforming to the effector. Now join it to the legs. Remember to use the same type of node you used previously: square bar in the center, nut fixing it on both sides, spacer, ball joint, spacer, nylon nut, or double nut.

Now that the three legs are joined to the effector, you can join them to the base! Just fix the servo leg to the servomotor and screw them in so they don't come off (believe me, they do). And that's it! Your delta robot is nearly finished. But no robot is complete without an effector end. The robot needs a tool to act in the world; you're going to build a gripper, a robotic hand that will catch anything within its reach.

The Gripper

The gripper is used by the robot to pick up objects. It's a really simple mechanism that can open and close using just one servo. You will laser-cut the Perspex pieces and assemble them into the gripper mechanism. The gripper shown in Figure 13-11 is a modified version of a design from `http://jjshortcut.wordpress.com`, which is a modification on a design from `http://www.diniro.net` (one beautiful example of Internet sharing!). We altered the design to account for the thickness of the material, the specific servos, and the effector size; we also added a circular base.

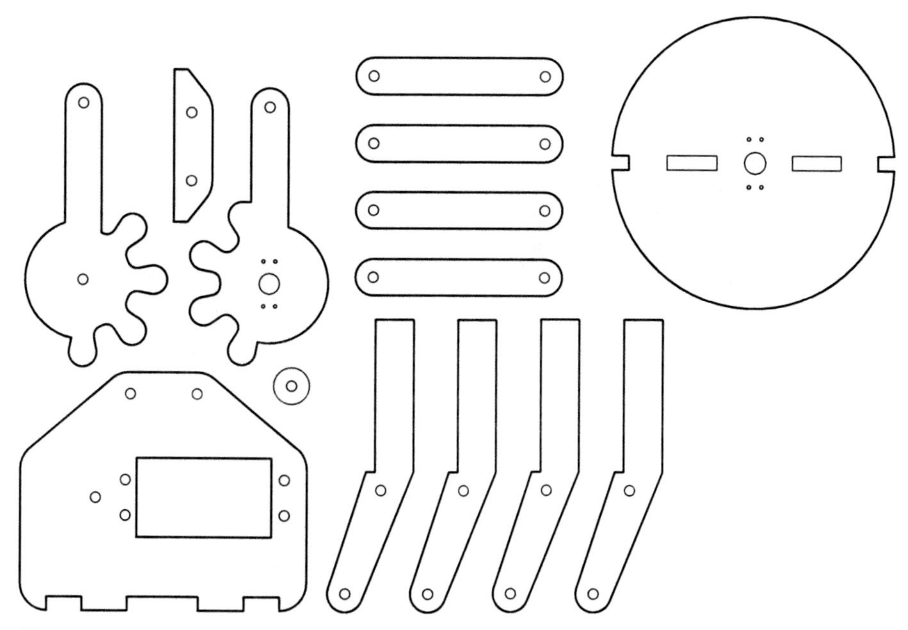

Figure 13-11. Gripper parts

Note that all the holes have been marked in order to make the assembly easier and also to allow the rectangular shape to host a servo. The rotation of the servo is transmitted to the gears; this closes the gripper. The gripper is assembled to a base that allows rotation in the perpendicular plane using another servomotor.

You start by assembling the first gear with the servo horn, then assemble the first servo to the big piece, and continue with the rest of pieces, using bolts and nylon nuts or double nuts. Once you have the complete mechanism, test it by moving it with your hand. It should have a pretty smooth movement (Figure 13-12). Secure all bolts but don't over tight them. Add some rubber adhesives in order to have a more stable grip (Perspex doesn't have enough friction, so objects just slip away!).

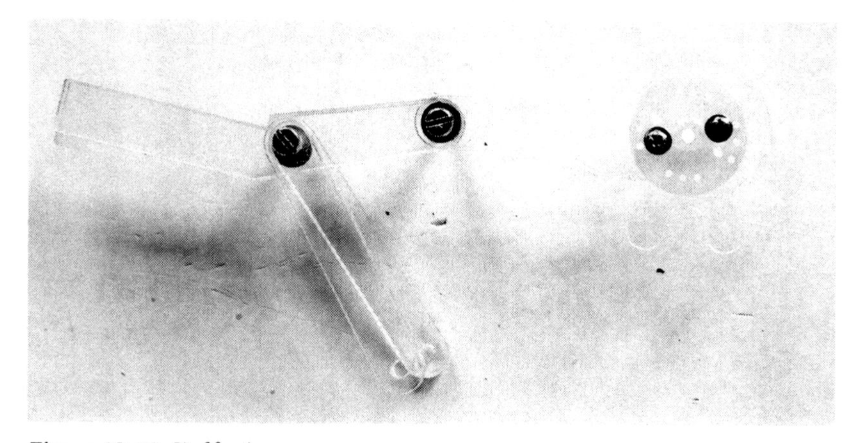

Figure 13-12. Half gripper

Figure 13-13. *Motor assembled on the gripper*

Once your gripper is assembled, connect it to the effector. Remember that you left enough space to fix one servo to the effector (Figure 13-13). Then adjust a servo horn to the gripper circular base. After that, fix it to the servo and screw it. Lastly, assemble the gripper with the circular base. To avoid too much vibration, use two mini L brackets, one on each side. The final piece should look like the one in Figure 13-14.

Figure 13-14. *Assembled gripper*

As a final touch, we added a couple of red LEDs on both sides of the gripper simply as an effect (Figure 13-15). The quaint face-like effect it conveys is quite theatrical when the robot is in action.

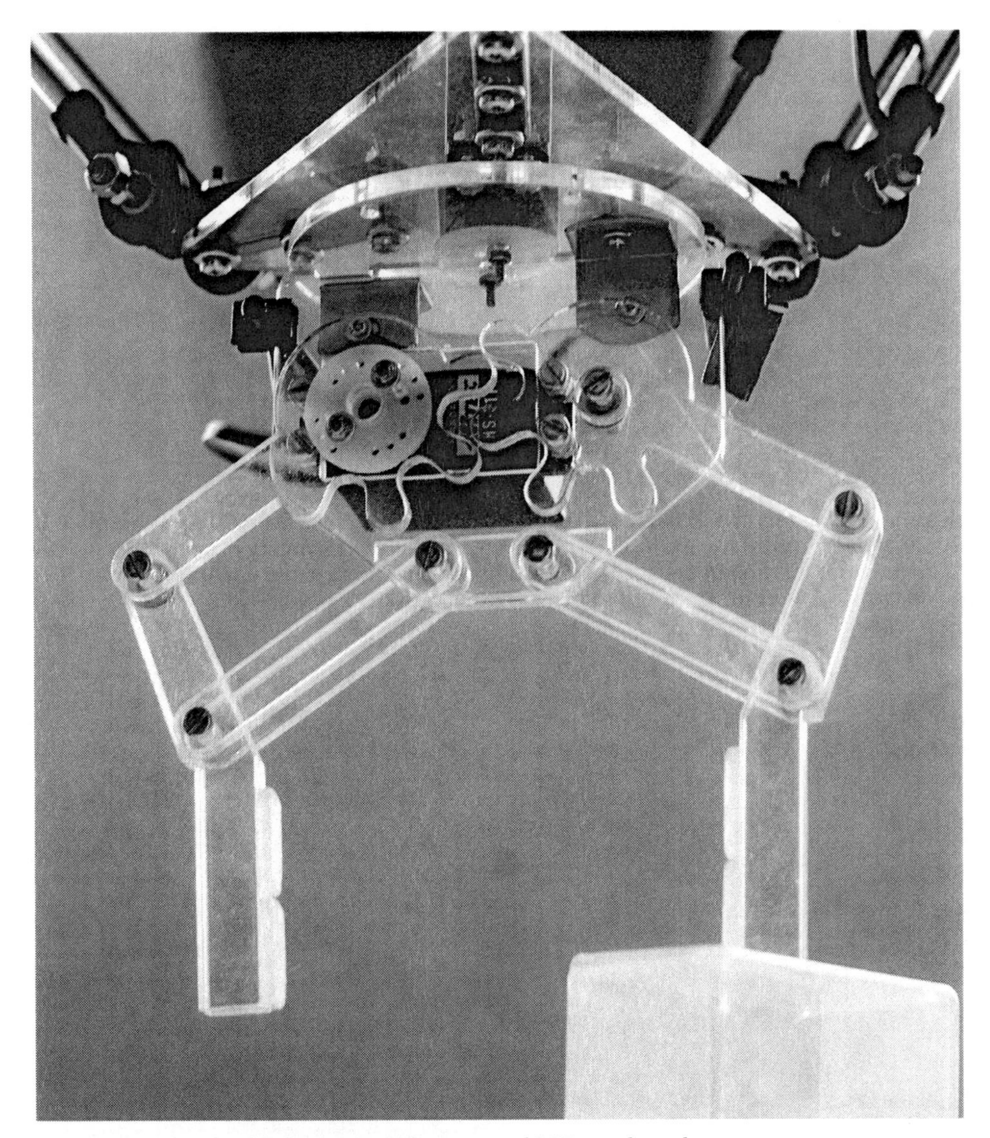

***Figure 13-15.** Assembled gripper with servo and LEDs on board*

Hanging the Robot

You cantilever the robot using an MDF board that you previously cut in the same shape as the robot's base (Figure 13-16), allowing the legs to go over the base plane. It is very important that you allow this movement. If you attach the robot directly to a flat surface (say you screw the base to your ceiling), the legs won't be able to go over the base plane and the robot's movements will be severely limited.

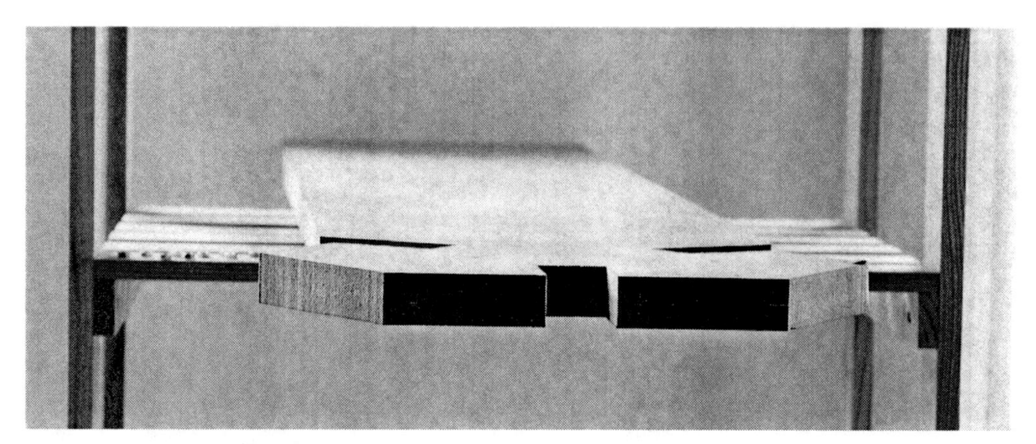

Figure 13-16. *Cantilevered MDF board*

Mark the holes and use long screws to fix this MDF board to the base. Once everything is assembled, use a timber shelve as a platform and use a couple of clamps to fix the robot hanging in space (Figure 13-17). It's important to measure this height because you need to add a working plane if you want to have a surface where you can leave objects for the robot to pick (Figure 13-18).

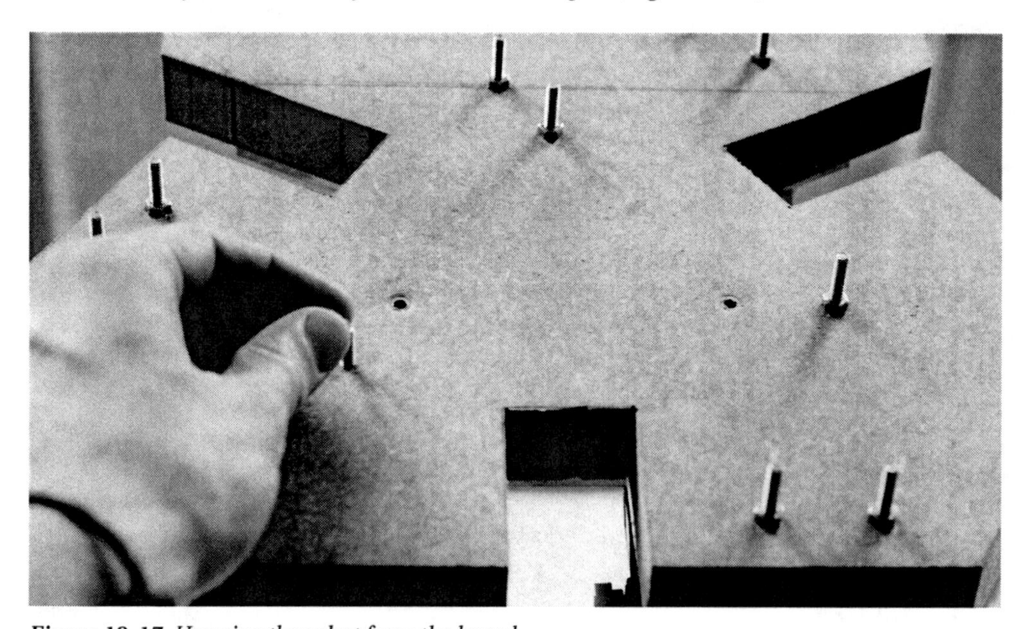

Figure 13-17. *Hanging the robot from the board*

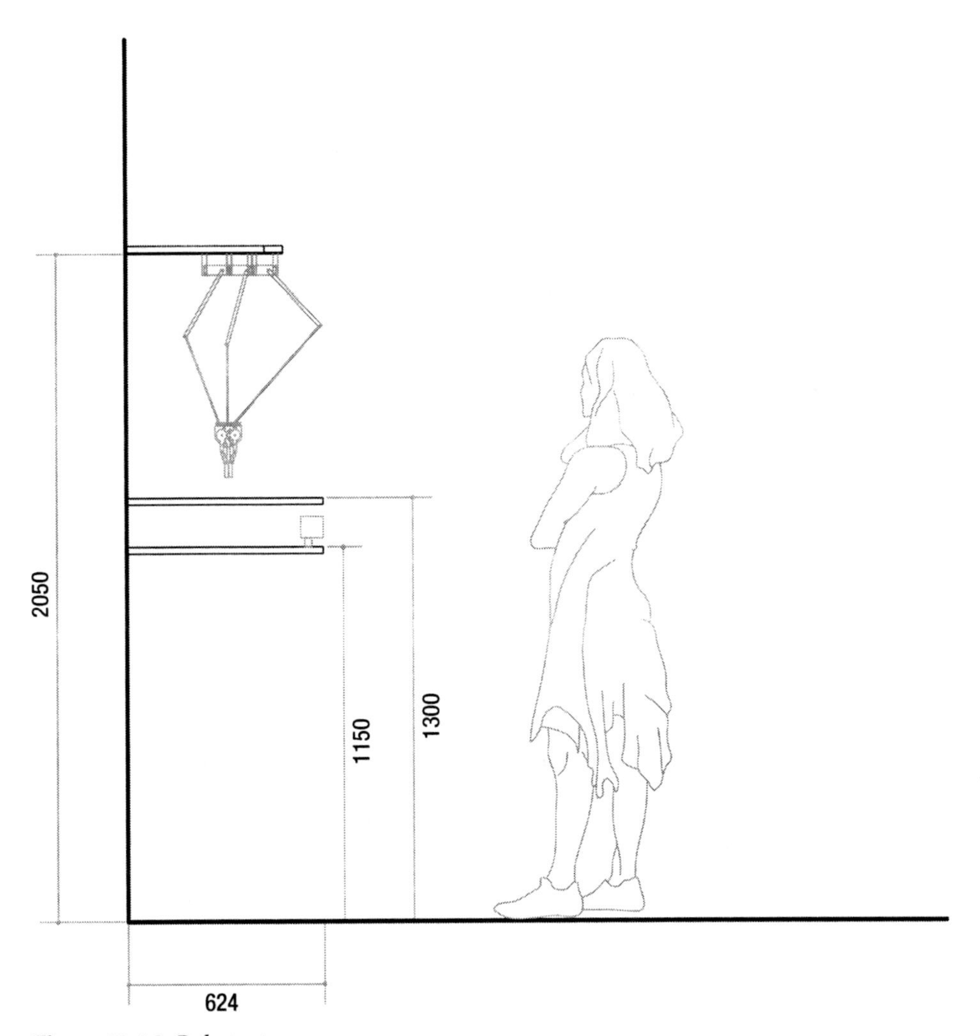

Figure 13-18. Robot setup

You can use a normal table as a working plane. You place the Kinect under the working plane—at the right height to detect your movements but also far enough from the robot as you never know what these creatures are capable of!

Building the Circuit

Now it's time to deal with the wiring. You have two LEDs and two servos in the gripper plus three big servos for the legs, and you need to connect them to your main circuit. Connect some extension lead cables from the gripper to connect the LEDs and the servos, running it from one leg through cable ties. Another option is to use some cable tidy (a plastic spiral that collects all cables inside), shown in Figure 13-19.

Figure 13-19. Cable tidy

Then you connect shorter extension cables for the three servos on the legs. Note that you need an external 5V power supply for this project; the 5V power supply coming from the Arduino is not stable enough for the power requirements of all servos working at the same time.

The circuit contains five servos (three for the legs and two for the gripper), two LEDs with their resistors, your Arduino, and an external power supply. Arduinos pins 9, 10, and 11 control three servos for the legs. Pins 5 and 6 control the gripper's servos and pin 3 controls the LEDs, as per Figure 13-20.

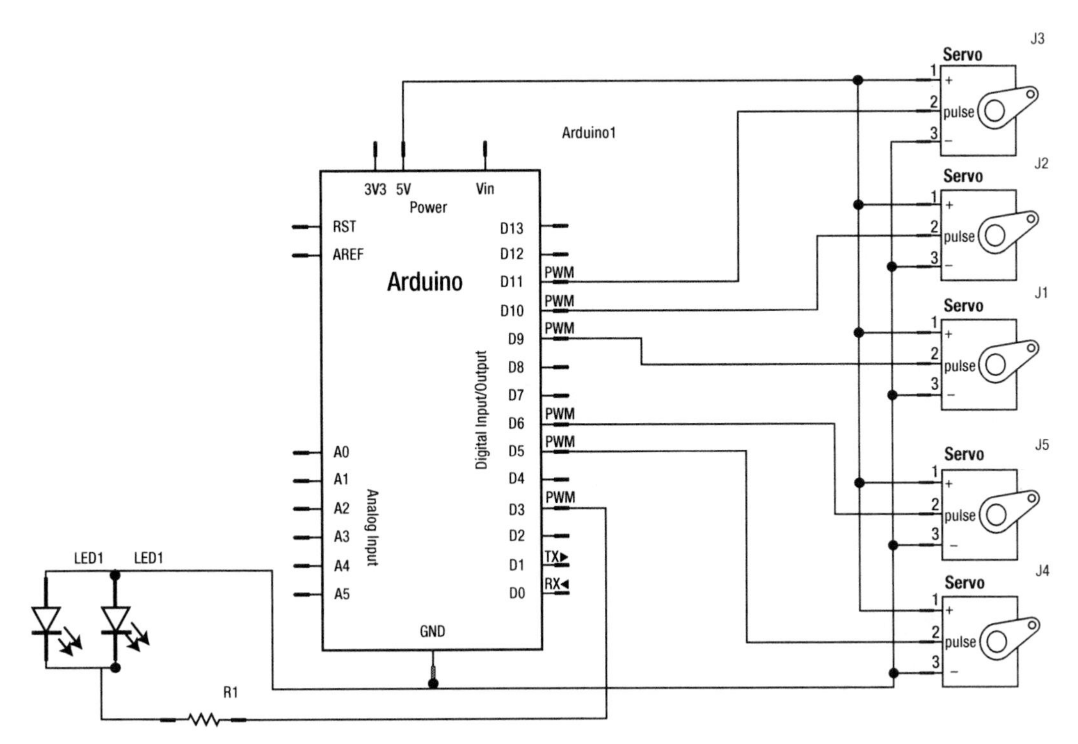

Figure 13-20. Delta robot circuit

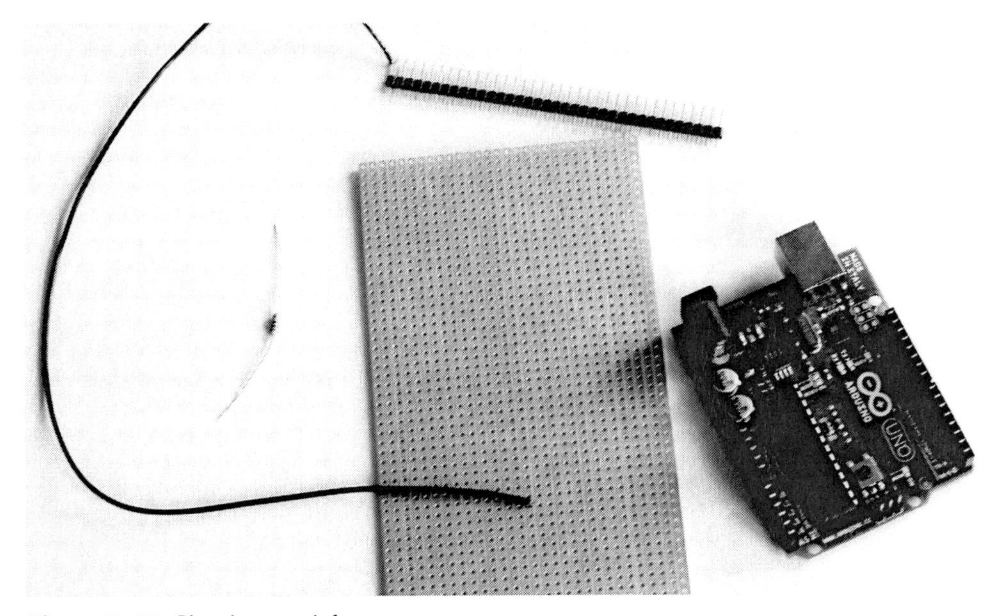

Figure 13-21. Circuit materials

To build the circuit, you are going to use a strip board (Figure 13-21). Start by measuring it and soldering the breakaway headers that connect to your Arduino board (Figure 13-22). It's important that you have a good look at the back of the strip board and confirm the direction of the copper strips before starting to solder! If you want to give it more stability, you can add more headers than necessary. The important ones are ground and pins 9, 10, 11, 5, 6, and 3.

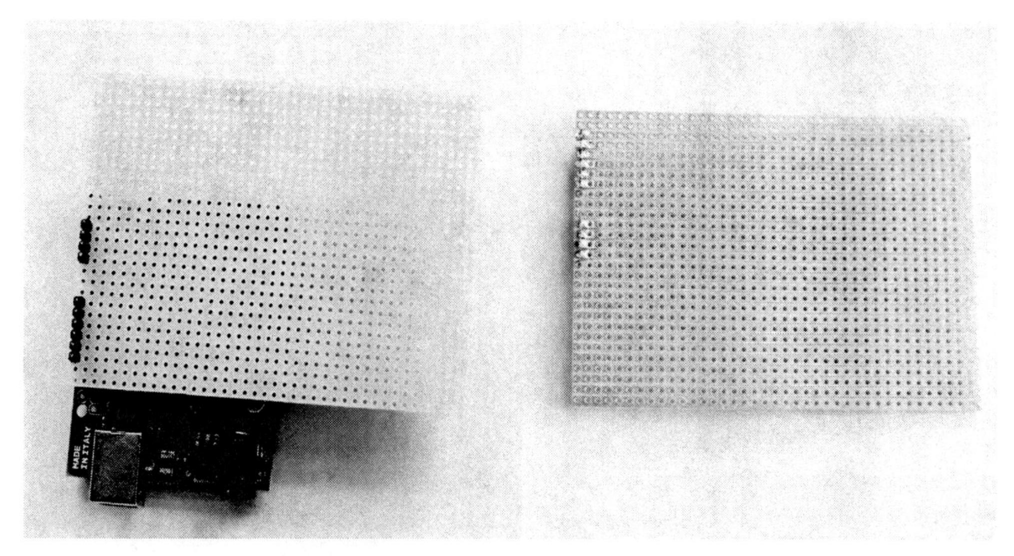

Figure 13-22. *Strip board with soldered headers*

After soldering, double check that they are placed correctly by plugging the shield to your Arduino board, making sure that it comes in easily. You may have to bend some of the headers with a nipper to get the shield properly connected.

The next step is to solder strips of three headers that are plugged to the female lining from the servos. Put them in the same column, leaving a distance between them; leave three rows in between. On each pin you get power (red wire), ground (black wire), and signal (yellow wire), as you saw in the previous chapters when we introduced servo motors. All servos are going to share ground and power, which come from an external power supply; leave the third pin connected independently to the corresponding Arduino pin (Figure 13-23).

Figure 13-23. *Headers for the servos*

After soldering five of them (one for each servo), solder a strip of two for later connecting the LED. In this case, you share just the ground (common for all the circuit) so it is shifted respective to the previous column; in other words, one pin is in the same column as the ground and the other completely independent in another strip.

Now turn over the strip board and scratch some lines into the copper so you disconnect pins that shouldn't be connected (servos' signal). You can use a scalpel to remove the copper that connects the line. Then check that the connection is completely broken by using a multimeter.

The next step is to connect the ground to the Arduino ground pin (ground is common in the circuit). After that, connect each signal pin to the corresponding pin on the Arduino. Now all your servos are connected to the Arduino. For the LED, connect a resistor between the strip that is connected with pin 3 on the Arduino board and the power free pin that you left before for the LED (Figure 13-24).

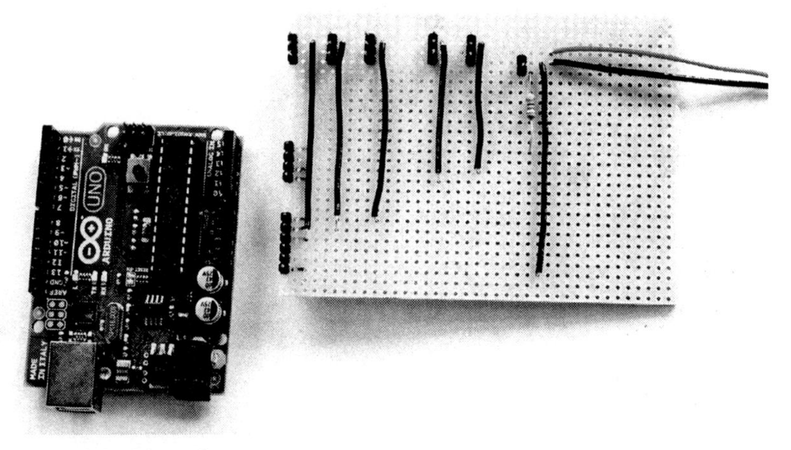

Figure 13-24. *Finished board*

The last step is to solder two cables in the strip's power and ground for the servos. These two cables are connected to an external power source. You can use a PC power supply for a smooth 5V power out. The only trick is to hack it so it works without being connected to a computer motherboard.

You will use the so-called paperclip trick. Make sure that the power is off and the power supply is disconnected (Figure 13-25); find the 20/24-pin connector that is used to connect the power supply to the motherboard. In the 20-pin block, find the green wire; there is only one and beside it are two black (ground cables). On the other end, connect the green wire with any of the grounds using a small piece of wire; it doesn't matter which ground you choose as both will work.

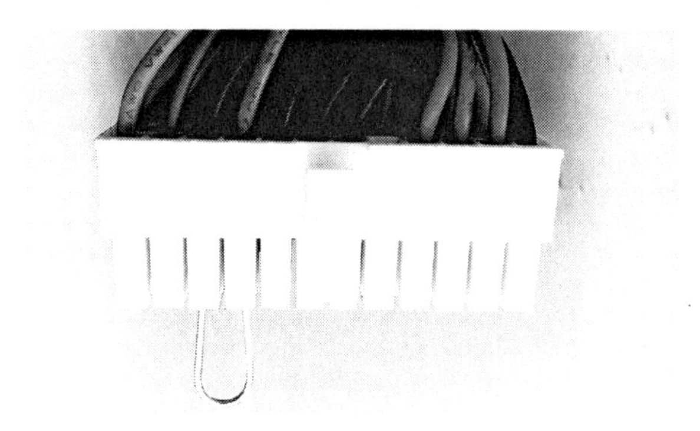

***Figure 13-25.** The paperclip trick*

And that's it! For safety, wrap some electrical tape around this connection so it's not exposed. Now you can pick any thin cable (5V) and connect red (power) and black (ground) to your circuit

Everything is ready now! Plug your homemade shield to Arduino, plug all the servos in the correspondent pins (making sure that they are the correct ones, and that ground, power, and signals are correct!). Connect your LED cable (again making sure that you correctly connect ground and power!), connect the power and ground to your external power supply (Figure 13-26), and you're ready to bring in the code!

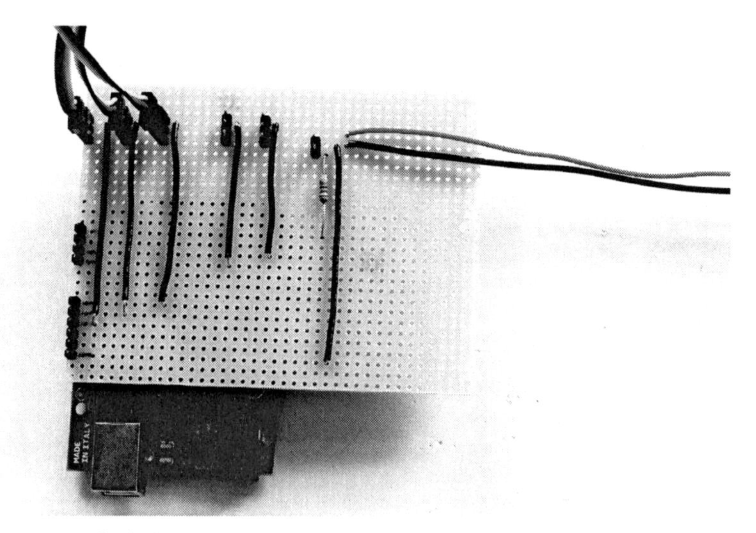

***Figure 13-26.** Finished circuit*

Delta Robot Simulation Software

You now have a (theoretically) working delta robot hanging from a support, and you have built a circuit attached to an Arduino board to control the robot. It's time to give it a brain! This role is fulfilled by your Processing sketch, which will tell the three servos to go to certain positions in order to move the effector to a precise point in space. You will also send the necessary data to determine the rotation and the state of the gripper. You will, of course send all this information to Arduino through serial communication; your board is in charge of transmitting the right messages to the right devices at the right times.

You are going to proceed step by step. First, you are going to develop the delta robot classes and test them with a simple example. When you are sure the simulation is working properly, you'll add all the Kinect and Serial methods, and then you will implement the Arduino program that will translate the serial data into pulses that the servos can understand. But first, we need to clarify an important concept.

Inverse Kinematics

When you work with robots, you often need to make use of inverse kinematics. But what on Earth is that? When you have a jointed, flexible object, or *kinematic chain* (like your delta robot, or any other articulated robot), there are two problems that you usually need to solve when studying its movements.

- If I move all the motors or actuators in the structure to certain states, what is the resulting pose of the robot?
- If I want to achieve a specific pose, what are the required states of my actuators in order to achieve it?

The first question can be answered by *forward kinematics*, the second one by *inverse kinematics*. In your case, what you want is to be able to drive the robot's effector to specific points in space so it can follow your hand; this is within the realm of *inverse kinematics*. You have a pretty straightforward structure, driven by only three servos (the other two are unrelated to the robot's position in space), but you need to find out how the three independent spatial coordinates of x, y, and z that define the desired position of the effector translate back through the kinematic chain to the three rotations of the servos. This is achieved by analyzing the connection and geometry of the robot and the constraints this geometry imposes on the robot's kinematics.

The goal of the inverse kinematics of a leg is to find the servo rotation necessary to achieve a specific effector position in space. Let's assume for a second you are looking at a simpler, two-dimensional problem. If you need to find out the servo angle for the diagram in Figure 13-27, the answer is obtained through planar geometry. The code that gives you the answer is the following:

```
float c = dist(posTemp.x + effectorSize, posTemp.y, baseSize, 0);
float alpha = acos((-a2 + thigh * thigh + c * c) / (2 * thigh * c));
float beta = -atan2(posTemp.y, posTemp.x);
servoAngle = alpha - beta;
```

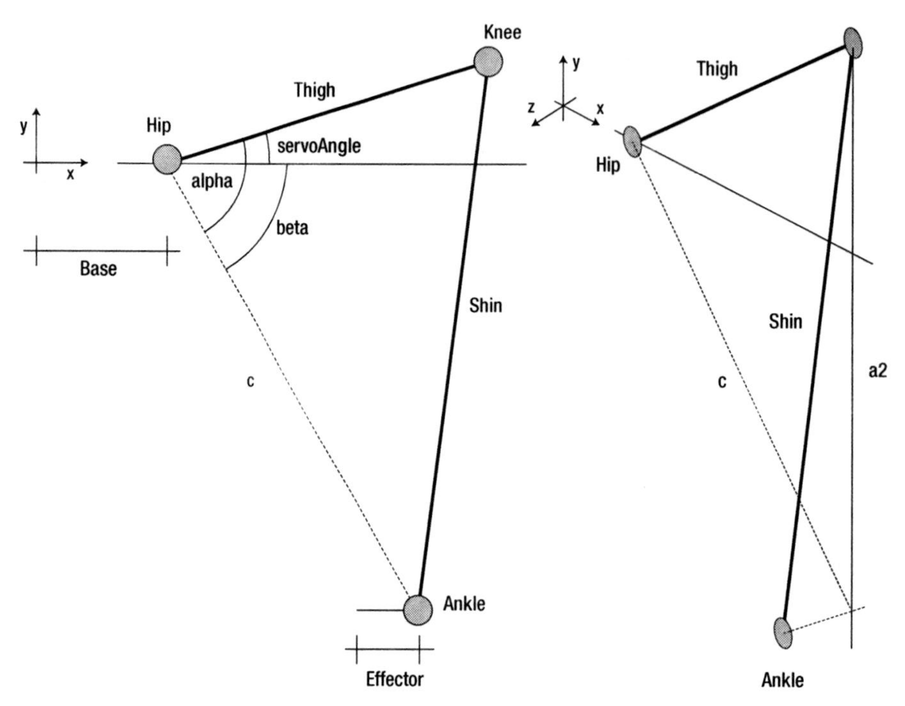

Figure 13-27. Delta robot's inverse kinematics in 2D and 3D

But you are not in a two-dimensional world, so you need to resolve the problem for a three-dimensional structure. If you look closely at the Figure 13-27, you can see that the problem can be reduced to a two-dimensional problem in which the dimension of the shin is reduced to its projection on the plane of the servo rotation. This projection, called a2 in the code, is obtained by simple triangle theory.

```
float a2 = shin * shin - posTemp.z * posTemp.z;
```

Once a2 is known, you only need to substitute it for the shin in the previous equations and you get the answer to the three-dimensional problem.

```
float c = dist(posTemp.x + effectorSize, posTemp.y, baseSize, 0);
float alpha = acos((-a2 + thigh * thigh + c * c) / (2 * thigh * c));
float beta = -atan2(posTemp.y, posTemp.x);
servoAngle = alpha - beta;
```

■ **Note** This has been a very brief introduction to delta robot's inverse kinematics, but there are plenty of resources on the Internet where you can find more information on this topic. Another advantage of delta robots is the extent to which they have been studied and the availability of good online information about them.

Now you need to structure your delta robot classes.

DeltaRobot Class

The delta robot simulation consists of the DeltaRobot class (Figure 13-28), containing the main parameters and routines, and a DeltaLeg class, where you deal with the inverse kinematics and the drawing functions for each leg. Let's start with the DeltaRobot class. You need a series of fields defining the position of the effector at each moment in time, and a zeroPos PVector that establishes the position from which you define the movements of your effector.

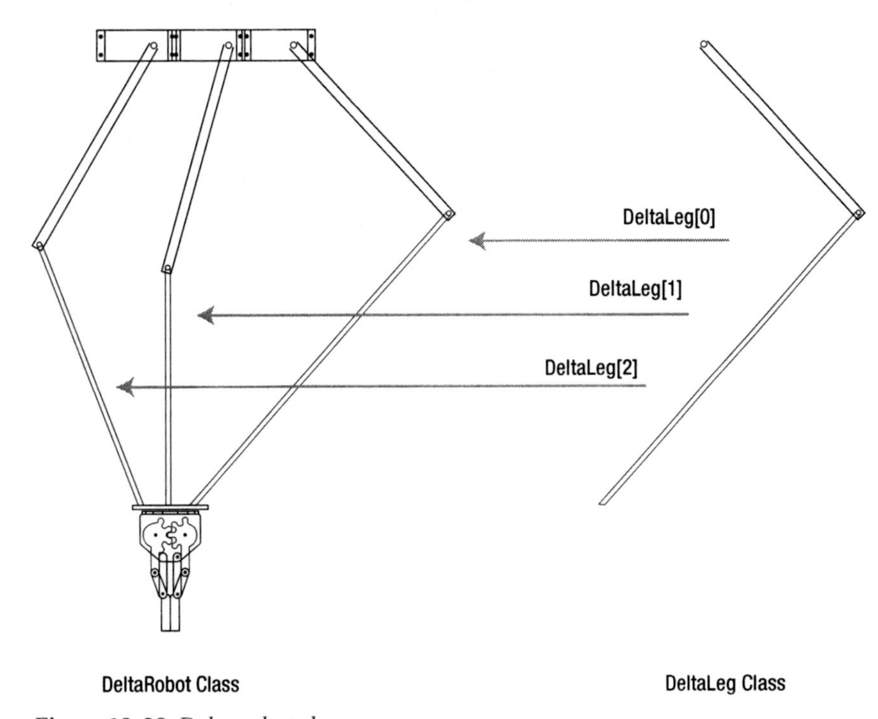

Figure 13-28. *Delta robot classes*

You define a variable called numLegs that defines the number of legs of your robot. Yes, you can actually increase the number of legs to 4, 5, or any higher number. The fact that there is no 23-legged parallel robot out there is because any high number of legs would just be a waste of material (three being the minimum necessary number of legs for stability). Having said that, one of the fastest commercial delta robots, the Adept Quattro, is based on a four-leg architecture, so feel free to play with the parameter. The results are certainly curious (Figure 13-29).

```
class DeltaRobot {
   PVector posVec = new PVector(); // Position of the Effector
   PVector zeroPos;
   int numLegs = 3; // Number of legs
```

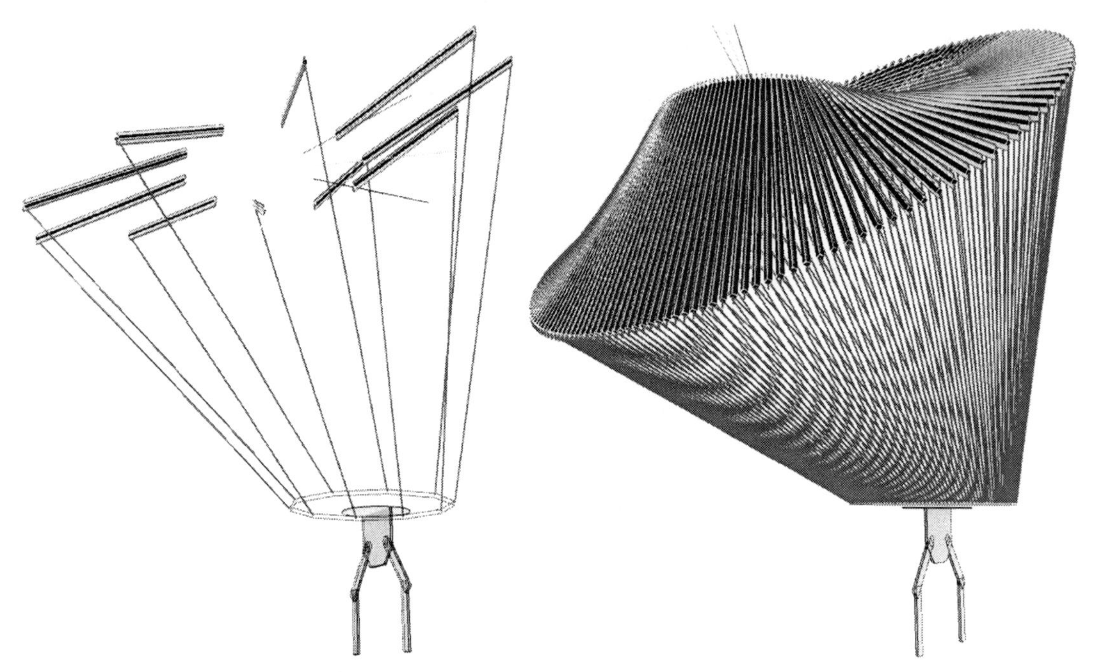

Figure 13-29. Two distant (and pretty bizarre) cousins of your delta robot, with 5 and 150 legs

You then set an array called DeltaLeg to contain the three (or any other number of) legs of the robot, and an array of floats for the angles of the joints driven by the servos (the upper joints). You also declare the variables that set the dimensions of your specific delta robot, and three variables defining the maximum span of your robot in the three axes: robotSpanX, robotSpanZ, and maxH). You need to respect these distances if you don't want to send unreachable locations to your robot, as this would lead to the legs disappearing in the case of the simulation, or worst, a damaged robot if you are driving the physical robot.

```
DeltaLeg[] leg = new DeltaLeg[numLegs]; // Create an array of deltaLegs
float[] servoAngles = new float[numLegs]; // Store the angles of each leg
float thigh, shin, baseSize, effectorSize; // Delta-Robot dimensions
float gripRot = 100;
float gripWidth = 100;
// Maximum dimensions of the Robot space
float robotSpanX = 500;
float robotSpanZ = 500;
float maxH;
```

You have come to the main constructor of the DeltaRobot class. It takes as parameters the lengths of the hip and ankle elements, and the size of the base and the effector. Note that when you refer to the base and effector sizes, you mean the distance between the center of the base/effector and the axis of the joint attached to the element (Figure 13-30).

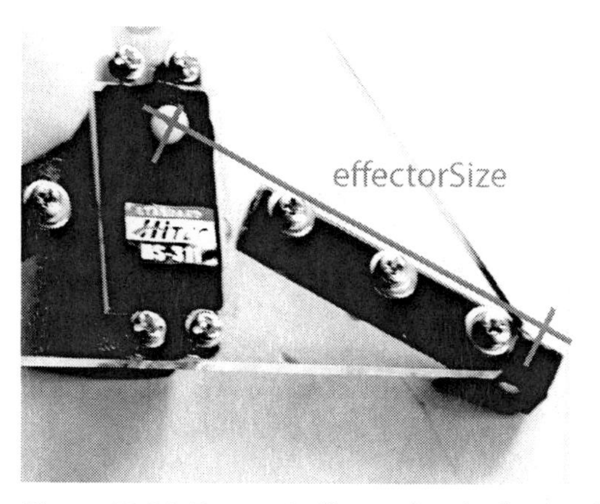

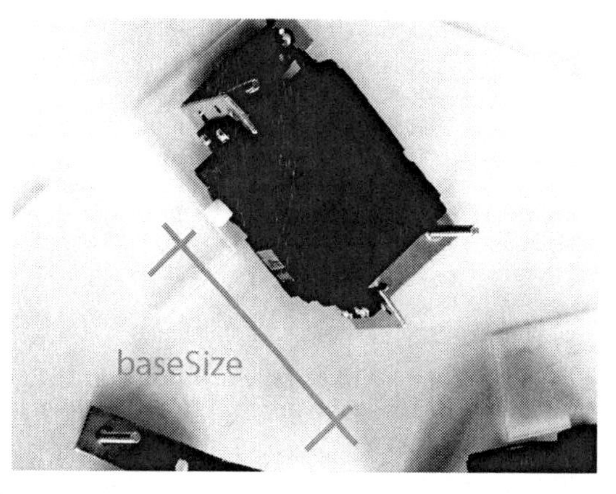

***Figure 13-30.** Base and effector sizes in the sketch*

```
DeltaRobot(float thigh, float shin, float baseSize, float effectorSize) {
    // Set the variables
    this.thigh = thigh;
    this.shin = shin;
    this.baseSize = baseSize;
    this.effectorSize = effectorSize;
    this.maxH = -(thigh + shin) * 0.9f; // Avoid the effector going out of range
    zeroPos = new PVector(0, maxH / 2, 0);
```

The DeltaLegs need to be initialized to the right dimensions, those that you extract from your robot's parameters. Each leg is initialized to a certain angle, one that you find by dividing the full circle (2*Pi radians) by the number of legs. In our sketch, we added PI/6 because we wanted to center the legs from the Kinect point of view.

```
for (int i = 0; i < numLegs; i++) {
      float legAngle = (i * 2 * PI / numLegs) + PI / 6;
      leg[i] = new DeltaLeg(i, legAngle, thigh, shin,  baseSize, effectorSize);
    }
  }
```

Once all your variables are initialized, you add some methods to your main delta robot class. The public methods you call from the main sketch are moveTo() and draw().

The moveTo() method takes a PVector as a parameter. You use this method to tell the robot to go to a specific position from your main sketch. You consider this PVector to be the relative position of the robot's effector to the point specified as zero position. The resulting position of your robot's effector is the result of adding the incoming PVector to the zero position.

```
public void moveTo(PVector newPos) {
    posVec.set(PVector.add(newPos, zeroPos));
```

Within this method, and before passing the new position to each leg, you need to make sure that the position is a "legal" position, which means a position physically reachable by the robot. The position is legal if it is contained within a parallelepiped of the dimensions defined by the variables robotSpanX, robotSpanZ, and maxH. If one of your coordinates is out of this virtual volume, trim the value to the maximum value of the movement on that axis.

```
    float xMax = robotSpanX * 0.5f;
    float xMin = -robotSpanX * 0.5f;
```

```
    float zMax = robotSpanZ * 0.5f;
    float zMin = -robotSpanZ * 0.5f;
    float yMax = -200;
    float yMin = 2*maxH+200;

    if (posVec.x > xMax) posVec.x = xMax;
    if (posVec.x < xMin) posVec.x = xMin;
    if (posVec.y > yMax) posVec.y = yMax;
    if (posVec.y < yMin) posVec.y = yMin;
    if (posVec.z > zMax) posVec.z = zMax;
    if (posVec.z < zMin) posVec.z = zMin;
```

Finally, move each leg to its new position and inquire what angle is necessary to move your servos to in order to achieve it. The servo angle is stored in an array for later.

```
    for (int i = 0; i < numLegs; i++) {
      leg[i].moveTo(posVec); // Move the legs to the new position
      servoAngles[i] = leg[i].servoAngle; // Get the servo angles
    }
  }
```

By this step, you have all the data you need to drive the robot (the three servo angles), so you can now implement a serial communication protocol and start driving your robot straight away. But let's develop the class a little further to include a whole visualization interface for your robot. You will ultimately have a virtual model of the delta robot that behaves exactly like the physical model and that you can visualize on screen. We think this is important for the project because it's a good idea to make sure everything is working before you output the values to the servos, but the whole project would work perfectly without the visual side. We also consider this simulation to be of great help in understanding the behavior of the machine, and it will later prove very useful when you bring the Kinect input along.

The public method `draw()`is called from your main routine when you want to print your delta robot on screen. This method changes the main matrix to a different position so the delta robot is shown at a certain height instead of at the origin of coordinates, because later this origin is the Kinect device. Then you call each leg's own `draw()` function and the `drawEffector()` function that draws the gripper.

```
  public void draw() {
    stroke(50);
    pushMatrix();
    translate(0, -maxH, 0);

    for (int i = 0; i < numLegs; i++) {
      leg[i].draw();
    }

    drawEffector();
    popMatrix();
  }
```

The `drawEffector()` function displays the effector and the gripper you have attached to it. We won't go into much detail on this because it is mainly an exercise of changing transformation matrices and drawing elements so you end up having a simplified model of your gripper that can be rotated, open, and closed at will (Figure 13-31).

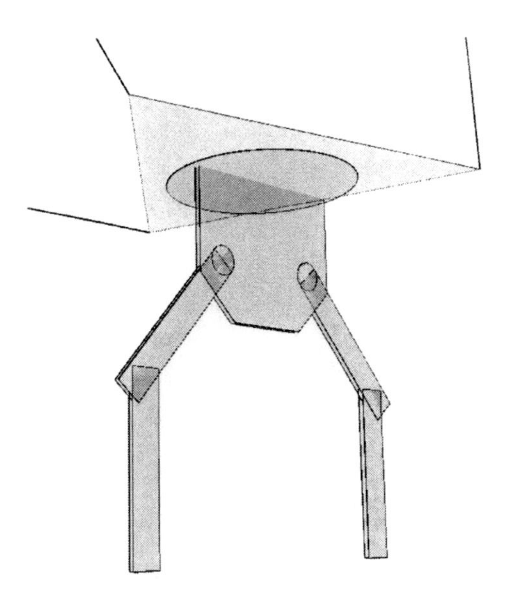

Figure 13-31. *Effector and gripper simulation*

```
void drawEffector() {
 // Draw the Effector Structure
 stroke(150);
 fill(150, 50);
   beginShape();
   for (int i = 0; i < numLegs; i++) {
     vertex(leg[i].ankleVec.x, leg[i].ankleVec.y,
     leg[i].ankleVec.z);
   }
   endShape(CLOSE);
 // Draw Gripper
 stroke(200, 200, 255);
 fill(200, 50);

 // Translate your Coordinate System to the effector position
 pushMatrix();
 translate(posVec.x, posVec.y - 5, posVec.z);
 rotateX(-PI/2);     // Rotate The CS, so you can drwa
 ellipse(0, 0, effectorSize / 1.2f, effectorSize / 1.2f);
 rotate(map(gripRot, 35, 180, -PI / 2, PI / 2));

 for (int j = -1; j < 2; j += 2) {
   translate(0, 2 * j, 0);
   beginShape();
   vertex(-30, 0, 0);
   vertex(30, 0, 0);
   vertex(30, 0, -35);
   vertex(15, 0, -50);
   vertex(-15, 0, -50);
   vertex(-30, 0, -35);
```

```
        endShape(CLOSE);

        for (int i = -1; i < 2; i += 2) {
          pushMatrix();
          translate(i * 20, 0, -30);
          rotateX(PI / 2);
          ellipse(0, 0, 10, 10);
          rotate(i * map(gripWidth, 50, 150, 0, PI / 2.2f));
          rect(-5, -60, 10, 60);
          translate(0, -50, 0);
          rotate(-i * map(gripWidth, 50, 150, 0, PI / 2.2f));
          rect(-5, -60, 10, 60);
          popMatrix();
        }
      }
      popMatrix();
    }
```

There is one more function to add, but you only use in the next example. The updateGrip() method can be called from the main sketch whenever you want to change the rotation and width of the gripper.

```
  public void updateGrip(float gripRot, float gripWidth) {
      this.gripRot = gripRot;
      this.gripWidth = gripWidth;
  }
}
```

deltaLeg Class

Your main DeltaRobot class is complete now, and you have built the necessary interfaces to drive the robot from your main program, but the most important part of the code is yet to be implemented: working out the inverse kinematics of the robot and getting the servo angles necessary to achieve the desired pose.

In the deltaRobot class, when moving the robot to a new position, you use a "black box" that gives you the servo angles. This black box is the deltaLeg class. This class has a series of variables defining the id of the leg, position of the effector, angle of the servo, angle of the leg, and the real-world coordinates of the leg's joints, which is used for visualization.

```
class DeltaLeg {
  int id; // id of the leg
  PVector posVec = new PVector(); // Effector Position
  float servoAngle; // Anlgle between the servo and the XZ plane
  float legAngle; // Y rotation angle of the leg

  // Universal position of the joints
  PVector hipVec, kneeVec, ankleVec;

  float thigh, shin, baseSize, effectorSize; // Sizes of the robot's elements
   DeltaLeg(int id, float legAngle, float thigh, float shin, float base, float effector) {
    this.id = id;
    this.legAngle = legAngle;
    this.baseSize = base;
    this.effectorSize = effector;
    this.thigh = thigh;
```

```
    this.shin = shin;
  }
```

The following function, moveTo, functionis actually where the magic happens. This short function converts the input position vector into a servo rotation. This is the equivalent of saying that this is the function dealing with the robot's inverse kinematics. (We ran through the delta robot's inverse kinematics previously, so we just repeat the code here.)

```
  void moveTo(PVector thisPos) {
    posVec.set(thisPos);
    PVector posTemp = vecRotY(thisPos, -legAngle);

    // find projection of a on the z=0 plane, squared
    float a2 = shin * shin - posTemp.z * posTemp.z;

    // calculate c with respect to base offset
    float c = dist(posTemp.x + effectorSize, posTemp.y, baseSize, 0);
    float alpha = (float) Math.acos((-a2 + thigh * thigh + c * c) / (2 * thigh * c));
    float beta = -(float) Math.atan2(posTemp.y, posTemp.x);
    servoAngle = alpha - beta;
    getWorldCoordinates();
  }
```

The function getWorldCoordinates() function updates the PVectors defining the joints of the leg in world coordinates, so it can draw them on screen. Use the helper functions vecRotY() and vecRotZ() to perform the vector rotations.

```
   void getWorldCoordinates () {
    // Unrotated Vectors of articulations
    hipVec = vecRotY(new PVector(baseSize, 0, 0), legAngle);
    kneeVec = vecRotZ(new PVector(thigh, 0, 0), servoAngle);
    kneeVec = vecRotY(kneeVec, legAngle);
    ankleVec = new PVector(posVec.x + (effectorSize * (float) Math.cos(legAngle)), posVec.y,
    posVec.z - 5 + (effectorSize * (float) Math.sin(legAngle)));
  }

   PVector vecRotY(PVector vecIn, float phi) {
    // Rotates a vector around the universal y-axis
    PVector rotatedVec = new PVector();
    rotatedVec.x = vecIn.x * cos(phi) - vecIn.z * sin(phi);
    rotatedVec.z = vecIn.x * sin(phi) + vecIn.z * cos(phi);
    rotatedVec.y = vecIn.y;
    return rotatedVec;
  }

   PVector vecRotZ(PVector vecIn, float phi) {
    // Rotates a vector around the universal z-axis
    PVector rotatedVec = new PVector();
    rotatedVec.x = vecIn.x * cos(phi) - vecIn.y * sin(phi);
    rotatedVec.y = vecIn.x * sin(phi) + vecIn.y * cos(phi);
    rotatedVec.z = vecIn.z;
    return rotatedVec;
  }
```

The public draw() function functionis called from the deltaRobot's draw() function and takes care of displaying the geometry of the leg on screen. Once again, we won't go into much detail here, but you can follow the comments in the code.

```
public void draw() {
    // Draw three lines to indicate the plane of each leg
    pushMatrix();
    translate(0, 0, 0);
    rotateY(-legAngle);
    translate(baseSize, 0, 0);
    if (id == 0) stroke(255, 0, 0);
    if (id == 1) stroke(0, 255, 0);
    if (id == 2) stroke(0, 0, 255);
    line(-baseSize / 2, 0, 0, 3 / 2 * baseSize, 0, 0);
    popMatrix();

    // Draw the Ankle Element
    stroke(150);
    strokeWeight(2);
    line(kneeVec.x, kneeVec.y, kneeVec.z, ankleVec.x, ankleVec.y,
    ankleVec.z);
    stroke(150, 140, 140);
    fill(50);
    beginShape();
    vertex(hipVec.x, hipVec.y + 5, hipVec.z);
    vertex(hipVec.x, hipVec.y - 5, hipVec.z);
    vertex(kneeVec.x, kneeVec.y - 5, kneeVec.z);
    vertex(kneeVec.x, kneeVec.y + 5, kneeVec.z);
    endShape(PConstants.CLOSE);
    strokeWeight(1);

    // Draw the Hip Element
    stroke(0);
    fill(255);

    // Align the z axis to the direction of the bar
    PVector dirVec = PVector.sub(kneeVec, hipVec);
    PVector centVec = PVector.add(hipVec, PVector.mult(dirVec, 0.5f));
    PVector new_dir = dirVec.get();
    PVector new_up = new PVector(0.0f, 0.0f, 1.0f);
    new_up.normalize();
    PVector crss = dirVec.cross(new_up);
    float theAngle = PVector.angleBetween(new_dir, new_up);
    crss.normalize();

    pushMatrix();
    translate(centVec.x, centVec.y, centVec.z);
    rotate(-theAngle, crss.x, crss.y, crss.z);
    // rotate(servoAngle);
    box(dirVec.mag() / 50, dirVec.mag() / 50, dirVec.mag());
    popMatrix();
  }
}
```

Driving the Delta Robot Simulation with the Mouse

You have wrapped all functions you need for driving and visualizing a delta robot in two classes that you can add to any Processing sketch, so let's run a first test.

You are going to implement the simplest of applications and drive a virtual model of a delta robot with your mouse position. Import OpenGL and your Kinect Orbit library; then initialize a `deltaRobot` object to play with.

```
import processing.opengl.*;
import kinectOrbit.KinectOrbit;

import SimpleOpenNI.*;

// Initialize Orbit and simple-openni Objects
KinectOrbit myOrbit;
SimpleOpenNI kinect;

// Delta Robot
DeltaRobot dRobot;
PVector motionVec;

public void setup() {
  size(1200, 900, OPENGL);
  smooth();

  // Orbit
  myOrbit = new KinectOrbit(this, 0, "kinect");
  myOrbit.setCSScale(100);

  // Initialize the Delta Robot to the real dimensions
  dRobot = new DeltaRobot(250, 430, 90, 80);
}
```

Now the only thing you need to do is, within the orbit loop (so you can rotate around), create a motion PVector with your mouse coordinates, move the delta robot to that point in space, and draw it on screen to see the result. Done!

```
public void draw() {
  background(0);
  myOrbit.pushOrbit(this); // Start Orbiting
  motionVec = new PVector(width/2-mouseX, 0, height/2-mouseY);// Set the motion vector
  dRobot.moveTo(motionVec); // Move the robot to the relative motion vector
  dRobot.draw();  // Draw the delta robot in the current view.
    myOrbit.popOrbit(this); // Stop Orbiting
}
```

Run the sketch. You should get a neat image of a delta robot on screen (Figure 13-32). If you move your mouse over the sketch, the robot should move around accordingly.

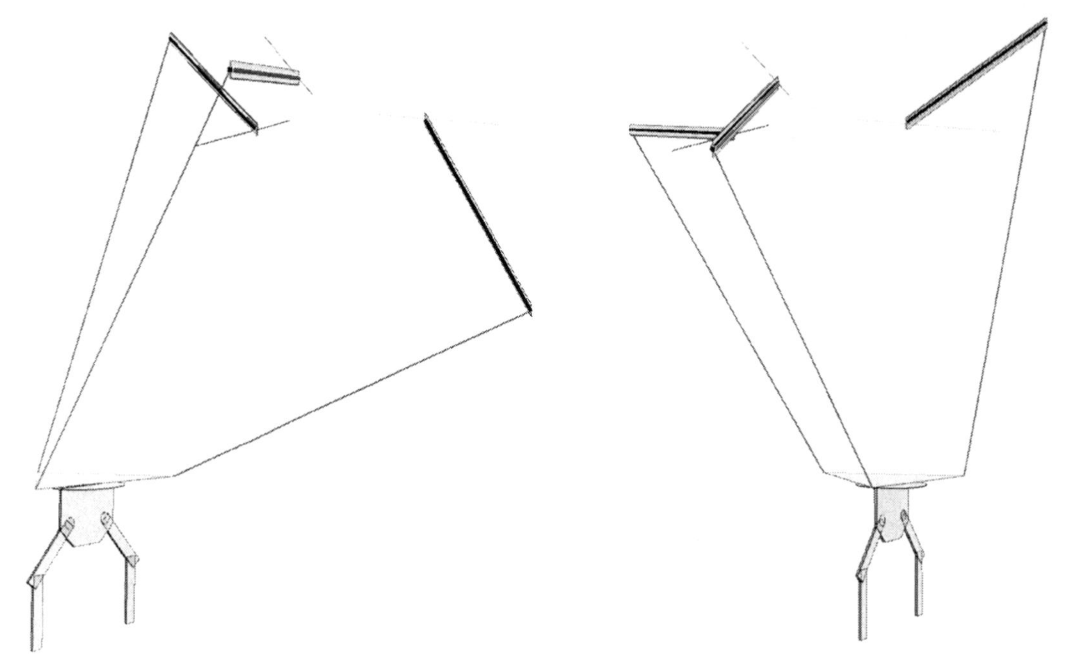

Figure 13-32. *Mouse-driven delta robot simulation*

Spend some time with your new virtual toy and be delighted by its movements. When we first saw a delta robot in action, we were amazed by the richness of the movements that emerged from such a simple structure. When you have made friends with this quaint creature, you can move on to the next step: driving it with your bare hands.

Kinect-Controlled Delta Robot

You are going to reuse the two classes you already created to develop a more complex application. The goal is to introduce Kinect and NITE's hand tracking capabilities to drive the delta robot simulation, and then implement your own routine to drive the gripper by tilting, opening, and closing your hand (Figure 13-33). Then you will add a serial communication routine to send the servo state to Arduino so you can plug the software straight into your physical robot and drive it with your hand.

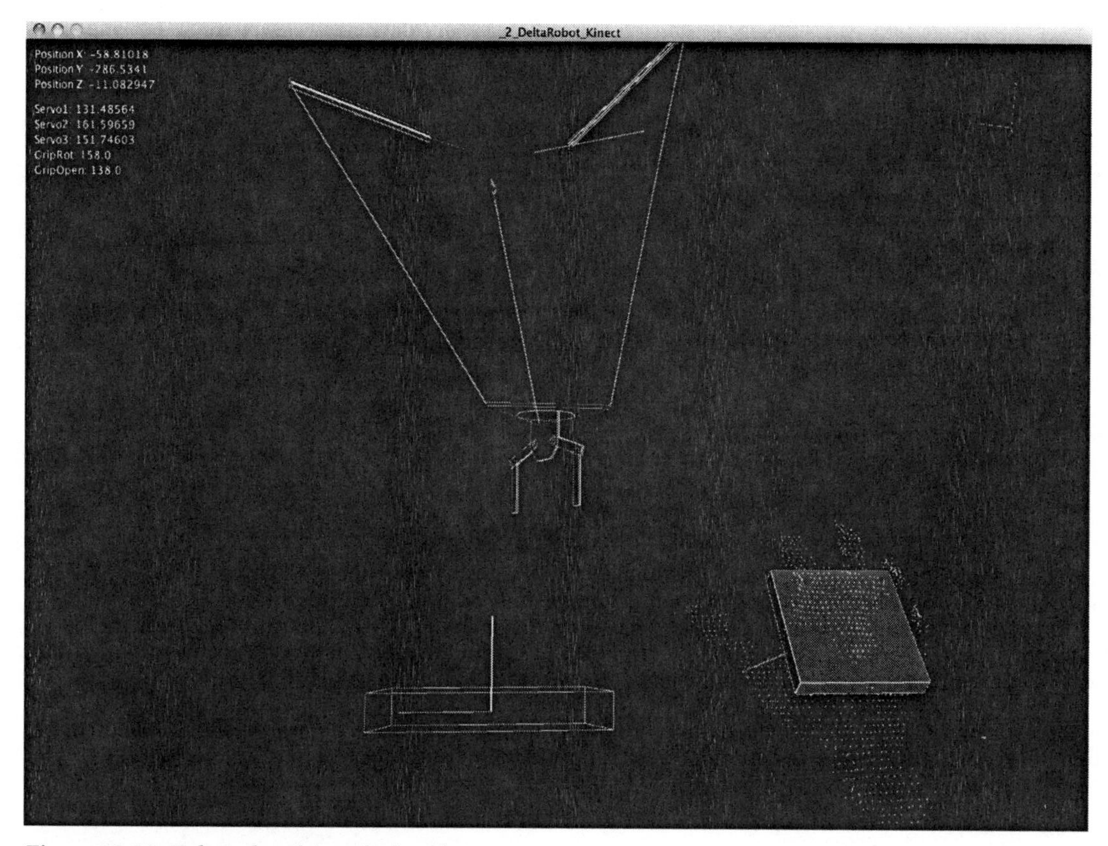

Figure 13-33. Delta robot driven by hand gestures

Import the full set of libraries you have been working with (this means Simple-OpenNI, Serial, OpenGL, and KinectOrbit) and then declare the corresponding variables. You are using hand tracking, so you need NITE's Session Manager and Point Control. The Boolean variable `serial` can be set to false if you want to run the sketch without an Arduino board plugged to the computer.

```
import processing.opengl.*;
import processing.serial.*;
import SimpleOpenNI.*;
import kinectOrbit.KinectOrbit;

// Initialize Orbit and simple-openni Objects
KinectOrbit myOrbit;
SimpleOpenNI kinect;
Serial myPort;
boolean serial = true;

// NITE
XnVSessionManager sessionManager;
XnVPointControl pointControl;
```

```
// Font for text on screen
PFont font;

// Variables for Hand Detection
boolean handsTrackFlag;
PVector handOrigin = new PVector();
PVector handVec = new PVector();
ArrayList<PVector> handVecList = new ArrayList<PVector>();
int handVecListSize = 30;
PVector[] realWorldPoint;
```

You want to be able to control the position of the robot and the state of the gripper, so you need to declare the PVector motionVec and the two floats gripRot and gripWidth.

```
// Delta Robot
DeltaRobot dRobot;
PVector motionVec;
float gripRot;
float gripWidth;

private float[] serialMsg =  new float[5]; // Serial Values sent to Arduino
```

The setup() function initializes all the objects you have previously declared and enables all NITE capabilities you are going to be using. Note that you are only adding "wave" gesture recognition to the Session Manager because you want to have a better control of hand creation in runtime. If you include RaiseHand, as soon as your hand enters Kinect's field of view you trigger a hand creation event; by adding only waving, you can control when and where you start tracking your hand, which is useful in a later stage.

Of course, you also initialize the delta robot object to the dimensions (in mm) of the robot you have built and set the serial communication to the first port in your computer where your Arduino is connected.

```
public void setup() {
  size(800, 600, OPENGL);
  smooth();

  // Orbit
  myOrbit = new KinectOrbit(this, 0, "kinect");
  myOrbit.drawCS(true);
  myOrbit.drawGizmo(true);
  myOrbit.setCSScale(100);

  // Simple-openni object
  kinect = new SimpleOpenNI(this);
  kinect.setMirror(false);
  // enable depthMap generation, hands + gestures
  kinect.enableDepth();
  kinect.enableGesture();
  kinect.enableHands();

  // setup NITE
  sessionManager = kinect.createSessionManager("Wave", "Wave");
  // Setup NITE.s Hand Point Control
  pointControl = new XnVPointControl();
  pointControl.RegisterPointCreate(this);
```

```
  pointControl.RegisterPointDestroy(this);
  pointControl.RegisterPointUpdate(this);

  sessionManager.AddListener(pointControl);
  // Array to store the scanned points
  realWorldPoint = new PVector[kinect.depthHeight() * kinect.depthWidth()];
  for (int i = 0; i < realWorldPoint.length; i++) {
    realWorldPoint[i] = new PVector();
  }
  // Initialize Font
  font = loadFont("SansSerif-12.vlw");

  // Initialize the Delta Robot to the real dimensions
  dRobot = new DeltaRobot(250, 430, 90, 80);

  if (serial) {
    // Initialize Serial Communication
    String portName = Serial.list()[0]; // This gets the first port on your computer.
    myPort = new Serial(this, portName, 9600);
  }
}
```

Now add the XvN Point Control callback functions, which are pretty similar to the ones you used in previous examples for hand tracking. Set your hand tracking flag on or off and update your hand PVector and your hand position history ArrayList.

Add one more line to the onPointCreate function. Because you want to track your hand movements after you have waved to the Kinect, you need to set the handOrigin PVector to the point in space where the waving took place (Figure 13-34). This allows you to set your motion vector as the relative displacement vector of your hand from the hand-creation point.

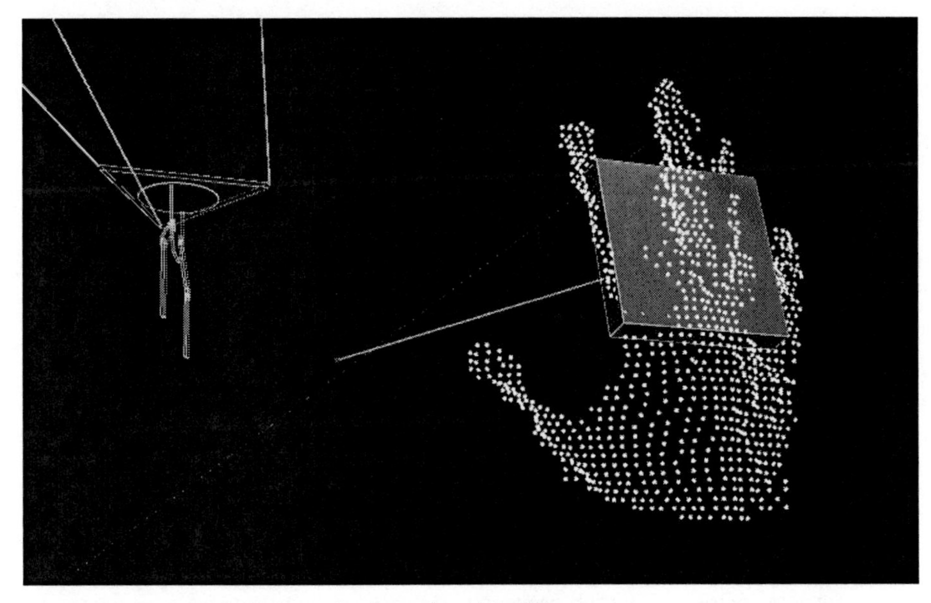

Figure 13-34. *Hand origin and current hand position*

```
public void onPointCreate(XnVHandPointContext pContext) {
  println("onPointCreate:");
  handsTrackFlag = true;
  handVec.set(pContext.getPtPosition().getX(), pContext.getPtPosition()
    .getY(), pContext.getPtPosition().getZ());
  handVecList.clear();
  handVecList.add(handVec.get());
  handOrigin = handVec.get();
}

public void onPointDestroy(int nID) {
  println("PointDestroy: " + nID);
  handsTrackFlag = false;
}

public void onPointUpdate(XnVHandPointContext pContext) {
  handVec.set(pContext.getPtPosition().getX(), pContext.getPtPosition()
    .getY(), pContext.getPtPosition().getZ());
  handVecList.add(0, handVec.get());
  if (handVecList.size() >= handVecListSize) { // remove the last point
    handVecList.remove(handVecList.size() - 1);
  }
}
```

The main draw() loop updates the Kinect object and sets a Kinect Orbit loop. Within this loop, you update the hand position and draw it, using two specific functions that you implement next.

```
public void draw() {
  background(0);
  // Update Kinect data
  kinect.update();
  // update NITE
  kinect.update(sessionManager);
  myOrbit.pushOrbit(this); // Start Orbiting

  if (handsTrackFlag) {
    updateHand();
    drawHand();
  }
```

Just to have an idea of the relative motion between the current position of the hand and the origin point (where the hand was created), draw a green line between the two.

```
  // Draw the origin point, and the line to the current position
  pushStyle();
  stroke(0, 0, 255);
  strokeWeight(5);
  point(handOrigin.x, handOrigin.y, handOrigin.z);
  popStyle();
  stroke(0, 255, 0);
  line(handOrigin.x, handOrigin.y, handOrigin.z, handVec.x, handVec.y,
  handVec.z);
```

And now you store that relative motion vector into the motionVec PVector and use it as a parameter to move the delta robot to its new position. You then proceed to draw the delta robot and send the serial data. Add a displayText function that prints on screen the data you are sending to the serial port for double-checking.

```
  motionVec = PVector.sub(handVec, handOrigin);// Set the relative motion vector
  dRobot.moveTo(motionVec); // Move the robot to the relative motion vector
  dRobot.draw();  // Draw the delta robot in the current view.
  kinect.drawCamFrustum(); // Draw the Kinect cam
  myOrbit.popOrbit(this); // Stop Orbiting
  if (serial) {
    sendSerialData();
  }
  displayText();  // Print the data on screen
}
```

Gripper Control

You want to implement a routine to control the delta robot's gripper with your hand movements. Here is the theory.

Using NITE, you have already worked out the position of your hand, that is the very center of your palm, and you are using it to set the robot's position in space. Wouldn't it be cool to open and close the gripper by opening and closing your hand? Well, there is a straightforward way of implementing this feature. You know your hand is defined by the set of points around the handVec position, so you can parse your point cloud and select all the points within a certain distance of your hand vector. After some testing, you set 100mm as a reasonable distance. If you display the points that fall within this rule, you get a "white glove" (Figure 13-35), which you use to extract the state of your hand.

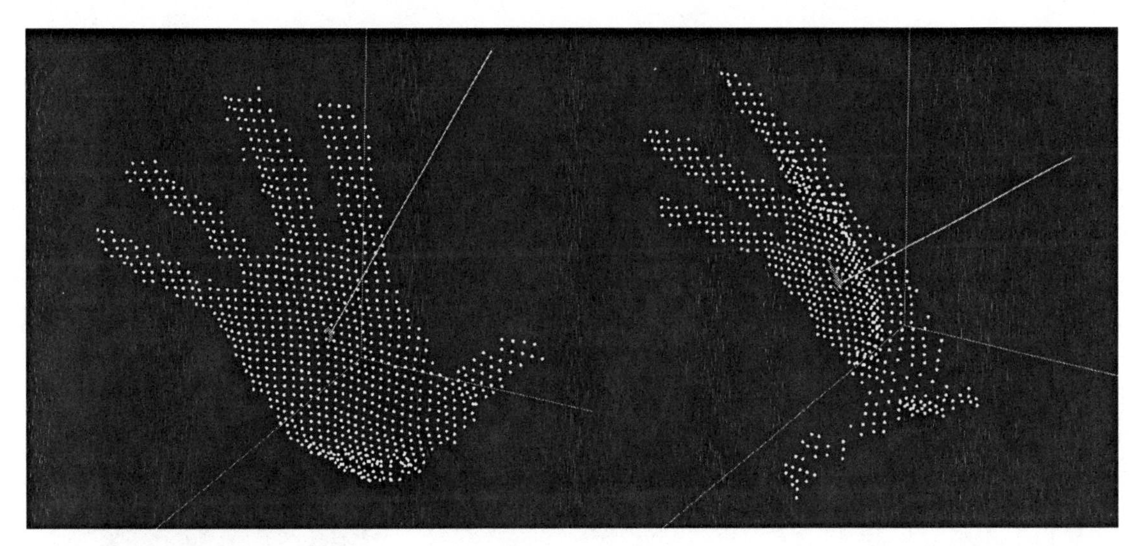

Figure 13-35. Hand point cloud

Once you know what points constitute your hand, you can think about what changes when you open and close your hand. No two ways around it: when you open your hand, the total width of your point cloud increases, which is to say that the horizontal distance between the rightmost and leftmost points (the tips of the thumb and little finger) increases (Figure 13-36).

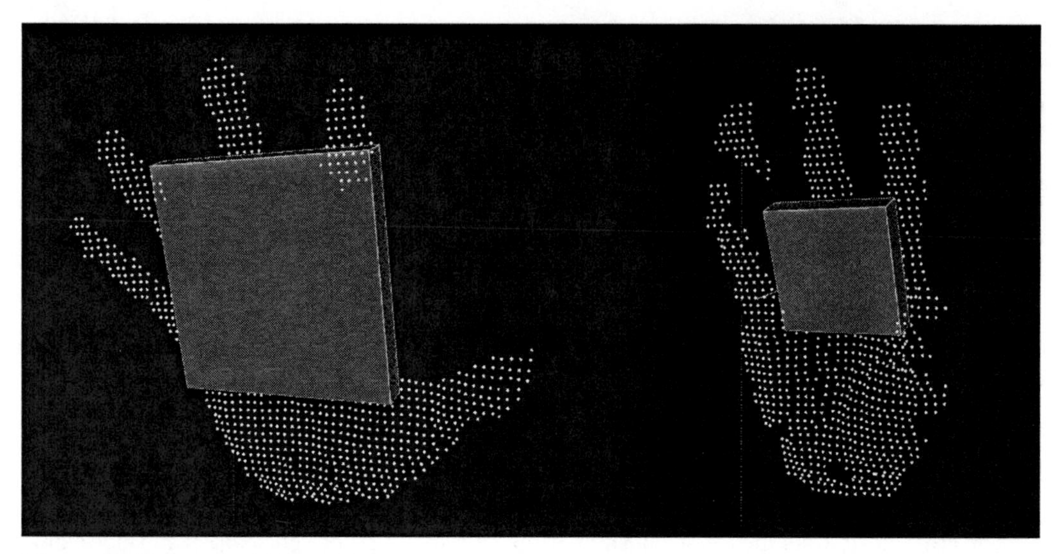

***Figure 13-36.** Hand width tracking*

In the same way, you can observe that when you tilt your hand forwards and backwards, the total height of the point cloud changes, which you can later use to define the rotation of your gripper (Figure 13-37).

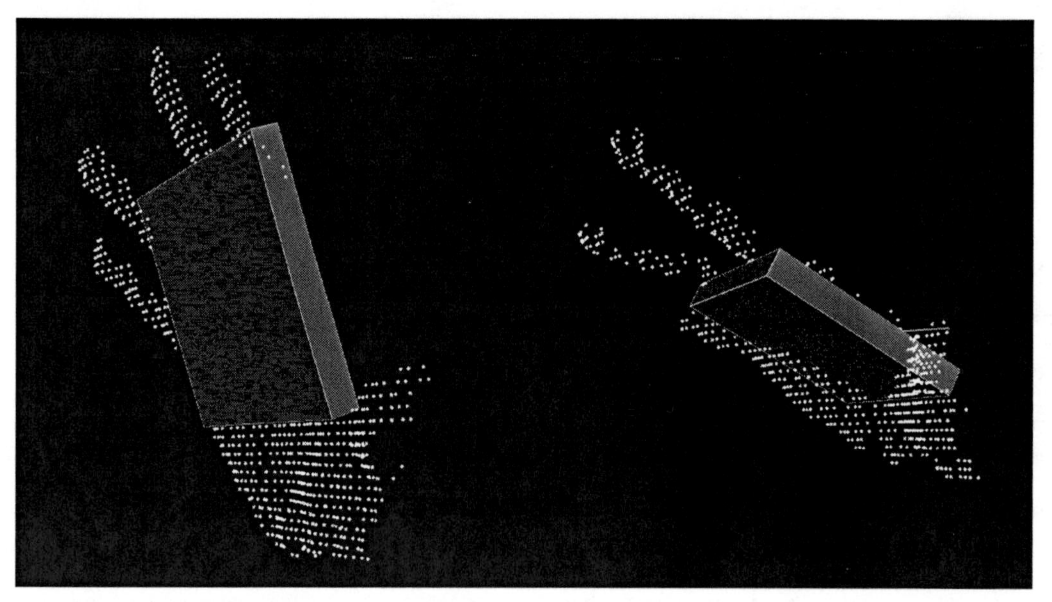

***Figure 13-37.** Hand tilt tracking*

You need to find these points and calculate their distances. Here's the routine.

```
void updateHand() {
  // draw the 3D point depth map
  int steps = 3; // You can speed up the calculation by using less points
  int index;
  stroke(255);
  // Initialize all the PVectors to the barycenter of the hand
  PVector handLeft = handVec.get();
  PVector handRight = handVec.get();
  PVector handTop = handVec.get();
  PVector handBottom = handVec.get();

  for (int y = 0; y < kinect.depthHeight(); y += steps) {
    for (int x = 0; x < kinect.depthWidth(); x += steps) {
      index = x + y * kinect.depthWidth();
      realWorldPoint[index] = kinect.depthMapRealWorld()[index].get();
      if (realWorldPoint[index].dist(handVec) < 100) {
        // Draw poin cloud defining the hand
        point(realWorldPoint[index].x, realWorldPoint[index].y, realWorldPoint[index].z);
        if (realWorldPoint[index].x > handRight.x) handRight = realWorldPoint[index].get();
        if (realWorldPoint[index].x < handLeft.x) handLeft = realWorldPoint[index].get();
        if (realWorldPoint[index].y > handTop.y) handTop = realWorldPoint[index].get();
        if (realWorldPoint[index].y < handBottom.y) handBottom = realWorldPoint[index].get();
      }
    }
  }
```

After running this loop, you have four PVectors storing the four points you need for your purposes. You are going to draw a control gizmo using these points. This gizmo is a cube, the size of which will change according to your hand's width. Likewise, the tilt will try to match the tilt of your hand, based on its height.

```
  // Draw Control Cube
  fill(100, 100, 200);
  pushMatrix();
  translate(handVec.x, handVec.y, handVec.z);
  rotateX(radians(handTop.y - handBottom.y));
  box((handRight.x - handLeft.x) / 2, (handRight.x - handLeft.x) / 2,
  10);
  popMatrix();
```

After running this code ad nauseam, we came up with some numbers that make a good range for the hand width and height values. A range of 65-200 works pretty nicely for both parameters. If you map this range to 0-255, you will have a pretty smooth value to be passed on to the Arduino board. You store the mapped values as `gripWidth` and `gripRot`.

```
  // Set the robot parameters
  gripWidth = lerp(gripWidth, map(handRight.x - handLeft.x, 65, 200, 0, 255), 0.2f);
  gripRot = lerp(gripRot, map(handTop.y - handBottom.y, 65, 200, 0, 255), 0.2f);
  dRobot.updateGrip(gripRot, gripWidth);
}
```

Include the `drawHand()` function from previous sketches to keep track of previous hand positions.

```
void drawHand() {
  stroke(255, 0, 0);
  pushStyle();
```

```
  strokeWeight(6);
  point(handVec.x, handVec.y, handVec.z);
  popStyle();

  noFill();
  Iterator itr = handVecList.iterator();
  beginShape();
  while (itr.hasNext ()) {
    PVector p = (PVector) itr.next();
    vertex(p.x, p.y, p.z);
  }
  endShape();
}
```

Sending the Data to Arduino

By now you have all the data you need to control your robot, so you can proceed to sending these values to the Arduino board. You will develop the Arduino code in the next section. Remember that you were calling a `sendSerialData()` function from your main `draw()` loop, so it's time to implement this function.

First, send the trigger character 'X' to indicate the beginning of the serial message. Then you have five values to send: three rotations for each arm's servo position, the gripper's rotation, and gripper's width.

Your servos accept a range from 500 to 2500, which gives you 2000 different possible positions. But Arduino reads the incoming serial data one byte at a time and, as one byte can only hold 256 values, that is the maximum resolution for your servo rotation that you can transmit in one single message. This is much lower than the resolution that your servos can use.

You want to match the resolution of your messages to the 2000 possible servo positions to get as smooth a movement of the robot as you can. The way you solve this problem is by splitting your integer into two single-byte numbers, sending the two bytes, and then recomposing the integer on the other end of the line (i.e. in the Arduino). What you are doing here is developing a protocol, as you learned in Chapter 4. To perform this splitting, you are going to make use of *bitwise operations*.

■ **Note** Bitwise operations handle binary numerals at the level of their individual bits. They are fast, primitive actions, directly supported by the processor. A detailed explanation of bitwise operations is beyond the scope of this book, but you can find more information on `www.cplusplus.com/doc/tutorial/operators`.

First, you write the trigger character 'X' to the serial buffer to indicate the start of the communication. For each leg, you work out your angle as an integer, mapped to the range 0-2000 in order to match its range to the range of the servos. You then store it in the `serialMessage[]` array to be displayed on screen later.

Now you know the integer `serialAngle` that you want to send via serial. You can proceed to decompose it into its more significant byte (MSB) and its less significant byte (LSB), which you send as two consecutive serial messages.

The bitwise operation `& 0xFF` masks all but the lowest eight bits, so you can use this to extract the LSB. The operation `>> 8` shifts all bits eight places to the right, discarding the lowest eight bits. You can combine this operation with `& 0xFF` to get the MSB.

You cast the two values to byte, because both operations return integers. After this, you are ready to write the two values to your serial port.

```
void sendSerialData() {
  myPort.write('X');
  for (int i=0;i<dRobot.numLegs;i++) {
    int serialAngle = (int)map(dRobot.servoAngles[i], radians(-90), radians(90), 0, 2000);
    serialMsg[i] = serialAngle;
    byte MSB = (byte)((serialAngle >> 8) & 0xFF);
    byte LSB = (byte)(serialAngle & 0xFF);

    myPort.write(MSB);
    myPort.write(LSB);
  }
```

For the grip rotation and width, 256 values are enough. If you are a perfectionist, you can take up the challenge of sending all the values as pairs of bytes.

```
  myPort.write((int)(gripRot));
  serialMsg[3] = (int)(gripRot);
  myPort.write((int)(gripWidth));
  serialMsg[4] = (int)(gripWidth);
}
```

Finally, you display on screen the values that you are passing to the Arduino. This is good practice for debugging purposes.

```
void displayText() {

  fill(255);
  textFont(font, 12);
  text("Position X: " + dRobot.posVec.x + "\nPosition Y: " + dRobot.posVec.y
        + "\nPosition Z: " + dRobot.posVec.z, 10, 20);
  text("Servo1: " + serialMsg[0] + "\nServo2: " + serialMsg[1]
        + "\nServo3: " + serialMsg[2] + "\nGripRot: " + serialMsg[3]
        + "\nGripWidth: " + serialMsg[4], 10, 80);
}
```

And that's it! The code you just wrote is ready to communicate with Arduino through serial. Now you need to write the Arduino code that will translate the information coming from Processing into physical movements of the physical robot.

Arduino Code

The code running inside your Arduino board has a simple role: it receives serial data that it remaps to the range of the servos and sends the appropriate message to each one of them.

You need variables to store the temporary values of the incoming data and then integers for the pin numbers and different pulses for the five servos. The longs `previousMillis` and `interval` deal with the spacing between messages sent to the servos.

```
unsigned int tempHandRot, tempGrip;
unsigned int servo1Pos, servo2Pos, servo3Pos;

int ledPin = 3;
```

```
int servo1Pin = 9;
int pulse1 = 1500;
int servo2Pin = 10;
int pulse2 = 1500;
int servo3Pin = 11;
int pulse3 = 1500;
int handRotPin = 5;
int handRotPulse = 1500;
int gripPin = 6;
int gripPulse = 1500;

long previousMillis = 0;
long interval = 20;

int speedServo1 = 0;
int speedServo2 = 0;
int speedServo3 = 0;

int handRotSpeed = 20;
int gripSpeed = 20;
```

Set all your pins as outputs and initialize the serial port.

```
void setup() {
  pinMode (ledPin, OUTPUT);
  pinMode (servo1Pin, OUTPUT);
  pinMode (servo2Pin, OUTPUT);
  pinMode (servo3Pin, OUTPUT);
  pinMode (handRotPin, OUTPUT);
  pinMode (gripPin, OUTPUT);

  Serial.begin(9600); // Start serial communication at 9600 bps

}
```

In the main loop, write a high pulse to the LEDs because you're having them on at all times, and then check the state of your serial buffer. If you have received more than eight values, which is what you are expecting, and the first of them is the trigger character 'X', you start reading the values as bytes. After every two values, you recompose the integer using the C function `word()`, which returns a word data type (16-bit unsigned number, the same as an unsigned integer) from two bytes.

```
void loop() {
  digitalWrite(ledPin, HIGH);
  if (Serial.available()>8) { // If data is available to read,
    char led=Serial.read();
    if (led=='X'){
      byte MSB1 = Serial.read();
      byte LSB1 = Serial.read();
      servo1Pos = word(MSB1, LSB1);

      byte MSB2 = Serial.read();
      byte LSB2 = Serial.read();
```

```
      servo2Pos = word(MSB2, LSB2);

      byte MSB3 = Serial.read();
      byte LSB3 = Serial.read();
      servo3Pos = word(MSB3, LSB3);

      tempHandRot = Serial.read();
      tempGrip = Serial.read();
    }
  }
```

And now you remap the pulses from their expected ranges to the servo range 500-2500.

```
  pulse1 = (int)map(servo1Pos,0,2000,500,2500);
  pulse2 =  (int)map(servo2Pos,0,2000,500,2500);
  pulse3 =  (int)map(servo3Pos,0,2000,500,2500);

  handRotPulse =  (int)map(tempHandRot,0,200,2500,500);
  gripPulse =  (int)map(tempGrip,0,220,500,2500);
```

And finally, if 20 milliseconds have elapsed, you send the pulses to your servo motors to update their angles. Remember that different servos have slightly different ranges, so you should test your servos and find the correct range that drives them from 0 to 180 degrees before remapping the values!

```
  unsigned long currentMillis = millis();
  if(currentMillis - previousMillis > interval) {
    previousMillis = currentMillis;
    updateServo(servo1Pin, pulse1);
    updateServo(servo2Pin, pulse2);
    updateServo(servo3Pin, pulse3);
    updateServo(handRotPin, handRotPulse);
    updateServo(gripPin, gripPulse);

  }
}

void updateServo (int pin, int pulse){
  digitalWrite(pin, HIGH);
  delayMicroseconds(pulse);
  digitalWrite(pin, LOW);
}
```

If all the steps have been implemented appropriately, you should have a working delta robot following your hand as you move it in space (Figure 13-38).

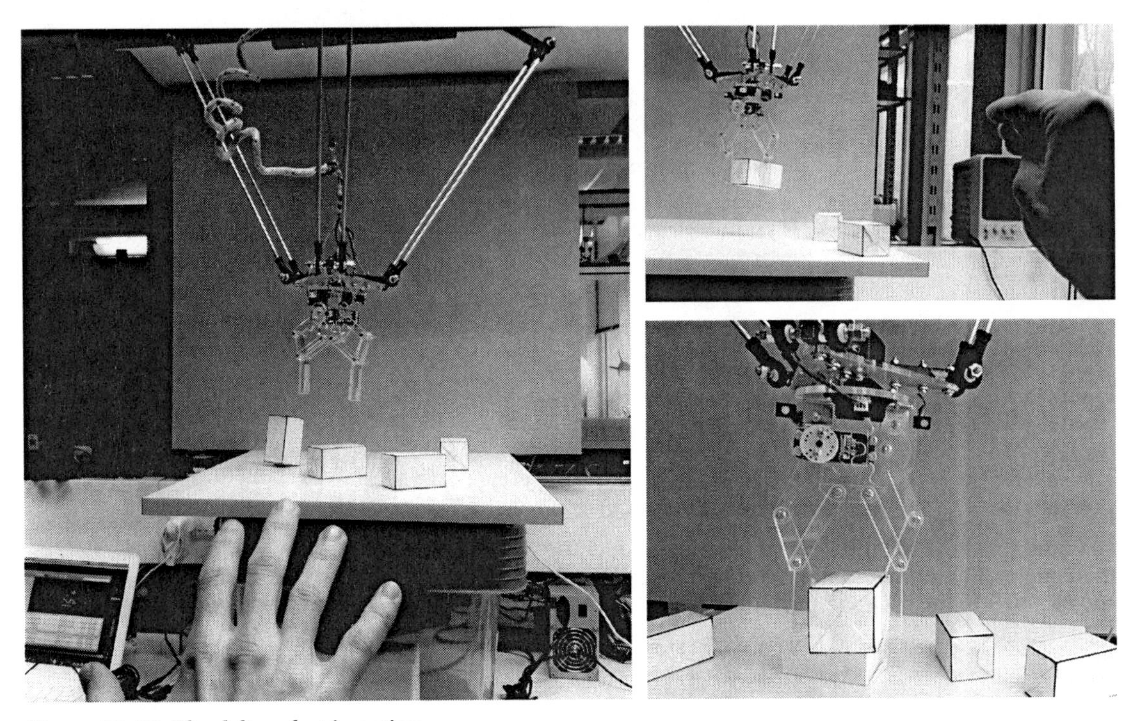

Figure 13-38. The delta robot in action

Summary

This has been quite a long project and its development has required you to master 3D transformations, inverse kinematics, hand recognition, point cloud analysis, serial communication, and a long list of programming techniques associated with the implementation of a physically accurate simulation of a robot. On the physical side, you worked with ball joints, laser cutting, servos, LEDs, and a series of other parts required by the making of a precise robotic platform.

The possibilities that natural interaction introduce in the world of robotics are yet to be explored. Not only are you delivered of using interfacing devices such as a mouse or keyboard, but you are also on the verge of a richer and more intuitive communication between humans and machines. Tools like Arduino and Kinect have allowed us to bridge the gap between our body kinetics and those of a machine within a pretty constrained budget and time scale. And this is only a start. You have now the tools to start tinkering with more complex applications for your robot, attaching different effectors, trying new forms of interaction. Your only limit is your imagination.

Index

A

E

F

G

H

I, J

K

O

P, Q

R

S

T

U

V

W

X, Y, Z

CPSIA information can be obtained at www.ICGtesting.com
Printed in the USA
LVOW131958220312

274350LV00002B/6/P

9 781430 241676